Heinz Brücher

# Useful Plants
# of Neotropical Origin
## and Their Wild Relatives

With 182 Figures

Springer-Verlag Berlin Heidelberg New York
London Paris Tokyo Hong Kong

Prof. Dr. HEINZ BRÜCHER
Instituto de Horticultura
CRICYT
Cas. de Correo 131
5500 Mendoza
Argentina

ISBN 3-540-18743-X Springer-Verlag Berlin Heidelberg New York
ISBN 0-387-18743-X Springer-Verlag New York Berlin Heidelberg

Library of Congress Cataloging-in-Publication Data. Brücher, Heinz 1915–  Useful plants of
neotropical origin : and their wild relatives / Heinz Brücher. p. cm. Includes index. 1. Plants,
Cultivated–Origin. 2. Plants, Useful–Origin. 3. Tropical plants–America. 4. Botany, Economic.
I. Title. SB73.B78 1989    633––dc20    89-11522

© Springer-Verlag Berlin Heidelberg 1989
Printed in Germany

Data conversion, printing and binding: Konrad Triltsch, Graphischer Betrieb, Würzburg.
2131/3145-543210 – Printed on acid-free paper

# Contents

Samples of "useful plants of neotropical origin" (Photo: Field Museum of Natural History, Chicago)

# Introduction

*"The greatest service which can be rendered to any country, is to add a new useful plant to its culture".*
*Thomas Jefferson, 1743–1826, Third President of the USA*

This book is the result of living and exploring over several decades in tropical latitudes on three continents, foremost in the neotropics. It presents various critical aspects of endangered species and natural resources of Latin America, where the biotic equilibrium is menaced as never before by the imprudent actions of one single species, Homo "sapiens". The destruction of natural habitats, especially the tropical forests between the rivers Parana, Bermejo, Paraguay, Mamoré, Madeira, Ucayali, Amazonas, Orinoco and Rio Magdalena progresses at the catastrophic pace of 100,000 km² per year (Hecht 1982, Jordan 1987, Raven and Axelrod 1975). By cutting down the woods, uncounted other plant species also disappear, still undescribed by science. Their potential benefits for mankind remain unknown, whilst some of them are only used by remote aboriginal people with intimate knowledge of their use (Herrera 1935, Perez-Arbalaez 1956, Pittier 1971, Lathrap 1975, Schultes 1979 and Prance and Kallunki 1984).

The genetic erosion of the ancestors of primitive cultivars and some world\-wide-diffused crops which originated in the mountain regions of Central and South America is tremendous, as we have observed in the course of our consecutive collecting expeditions since 1950. This refers especially to the gradual disappearance of the wild ancestors of pulses, pseudo-cereals, fruits, fibre and tuber plants, resins, spices, aromatics and drugs, from which the indigenous people have since time immemorial selected their food, some of them being of paramount importance as gene funds for modern plant breeding.

This book places special emphasis on the fact – often overlooked in Europe – that it was the Indians who domesticated and improved in quantity and quality the yields of numerous crop plants which today enrich the daily diet of highly developed industrial nations, who know little of their neotropical origin. Those contributions from Middle and South America are superior to other continents, for example from Africa (Brücher 1969, 1977).

If we compare Eurasia with the American continent, we notice a fundamental difference as far as the origin of our cultivated plants and domestic animals is concerned. Whilst almost all useful animals have been domesticated in the "Old World", a considerable amount of essential food plants and industrial crops originated in the "New World". This successful event would have been unthinkable without the active intervention of man. In this case, the reward for notable breeding abilities goes to Amerindian tribes in Meso- and South America (Hoehne 1937, Sauer 1959, MacNeish 1964, Parodi 1966, Schwanitz 1967, Ucko and Dimbleby 1969, Reed 1977 and Körber-Grohne 1988).

Before the arrival, not to say the pitiless invasion, of Spanish Conquistadores at the end of the 15th century, the civilized Old World was not aware of the hitherto tremendous hidden wealth of useful species on the American continent. Columbus and his crew were deeply amazed by the diversity of exotic fruits, vegetables and spices which the Indians presented to them on their first voyage (1492). Although the "expedition to India-Cipango" may be labelled as a failure in its aim to discover a short seaway to the "pepper islands" of Asia, Columbus was lucky to receive from the Amerindians other pepper plant (genus *Capsicum*), which after a few years of cultivation in Spain gained worldwide fame as the "Spanish pepper".

Besides this, a multitude of other useful plants soon arrived from America and spread to Africa and Asia. We mention only the most important. Tuber plants: batata, maniok, potato. Pulses: groundnut, swordbeans, limabeans, common beans, etc. Vegetables: tomato, squash, gourds, amarant, gautia spinach. Industrial crops: sisal, cotton, rubber. Fruits: ananas, avocado, chirimoya, passiflora, strawberry, guayaba. Aromatics and spices: vanilla, coca, cacao, chilli pepper and last but not least: maize. The diversity of the Indian patrimony of useful plants is, however, much greater. Recently –

as a result of my research work in Panama –
I have enumerated 150 species which crossed
the Isthmus of Dairén in one or the other
direction (Brücher 1988).

This book answers the question why the author is so interested in the search for, exper-
imentation with and improvement of van-
ishing or under-exploited useful plants and
trees of disappearing Indian cultures. Such
an endeavour may appear curious, because
at this moment the Northern Hemisphere
displays a saturation and even an overproduction of foodstuffs.

For more than two decades, I have been
ceaselessly pointing out (Brücher 1968,
1977, 1985 etc.) that in the tropics and subtropics plenty of "unused" vegetable food
reserves exist, especially in the Neotropics.
Some are known only to the natives, others
are already being gathered in thousands of
tons from the wild for industrial purposes
(like the kernels of the corozo palm) or for
food manufacture ("paranut" = Berthol-
letia). Whilst their manifoldness is tremendous, the number of species used by "developed nations" as food plants has declined
dramatically. The rich "palette" of a hundred autochthonous vegetables, fruits,
grains, pulses and roots of our forefathers
has shrunken to a mere dozen main crops,
some of them even of exotic or hybrid origin. A similar tendency exists also on a major
scale in the whole of the civilized world. It
seems incredible that at present only seven
main crops carry the burden of feeding five
billion humans (see Brücher 1982, Die
Sieben Säulen der Welternährung). The
yields of rice, maize, wheat, cassava, potato,
a group of pulses and sugar plants (beets and
cane) comprise three quarters of the plant
kingdom's contribution to world nutrition.
Unimaginable the catastrophe, if one of
these "monocultures" were struck by a
worldwide attack of a new disease or aggressive parasite! Such an event is by no means
a theoretical calculation; epyzooties on continental scale have already happened several
times, with devastating consequences. We
remember, for example, the potato blight
(*Phytophthora infestans*) which between
1846 and 1848 produced starvation and
death for a million Irish farmers, who had
chosen this exotic plant as their main mono-

culture. Only recently (1970) the major portion of the maize harvest in the southern
states of the USA was wiped out when a new
strain of corn blight (*Helminthosporium*) attacked hybrid corn monocultures. The devastating coffee rust (*Hemileia vastatrix*)
shook the economy of several Asiatic countries, when it spread from its African centre,
and condemned hundreds of thousands of
Hindu smallholders to poverty. A similar
drama is now beginning in tropical America,
where in Brazil, since 1970, infections with
*Hemileia* are increasing with menacing ve-
locity. Monocultures of several tree species
in North America and Europe are at the
present moment prey to voracious insects.

Rubber plantations in Brazil do not prosper
because they become heavily infested by the
ascomycete *Dothidiella*. Rubber production
of the world would break down if this fungus obtained a foothold in Malaysia. Nigeria's cacao plantations may soon be condemned, as in Ghana, by the swollen shoot
virus, where even the intentional elimination
of 162,000,000 cocoa trees could not stop
the rapid spread of this disastrous disease of
monocultivation. Strangely enough, this virus has not yet been observed in tropical
America, the homeland of several *Theobroma* species.

I think these few examples are sufficient to
underline the fragility of man-created biocoenosis with tropical crops and to emphasize the paramount importance of my points
of view expressed on the first page.

This book cannot resolve the paradoxical
dilemma of the actual food situation on this
our planet, which consists in the following:
According to F.A.O. statistics, 500,000,000
humans suffer from famine and approximately 40,000,000, most of them children,
die annually of hunger (cf. also Alwim and
Kozlowski 1977).

On the other hand, some of the highly advanced agro-industrial countries – mostly
situated in the temperate climate zone –
"suffer" from food surplus as a result of
efficient plant breeding, agrotechnique, animal husbandry and stimulation by Government subsidies. This overproduction of animal protein and fat, or vegetable carbohydrates and oil in the European Community and North America sometimes overshad-

ows the alarming food shortage of hundreds of millions of natives of the tropical belt. At present a race with time is being run between population growth without precedence in human history and feeding problems. Never before has the human species been threatened at the same time by overpopulation and undernourishment, as is now the case in the underdeveloped countries of tropical Africa, Asia and America. It is not rare to observe in certain regions of the tropical belt, exuberant in wild plant and animal life, that their human populations are starving through malnutrition, due to their very low food output and indolence. Is it not amazing, that whilst 80% of the people there are dedicated to agriculture and primitive animal raising, their governments implore food help from the Northern Hemisphere, where only 20% of the manpower is expended on the fields, under climatical stress and short growing periods?

In conclusion: from the global point of view it seems very disappointing to observe that the areas which produce maximal quantities of biomass are generally so poor in food delivery and fall more and more into depending for their food on help from overseas. Fortunately there are certain exceptions, as the reader may find in some of the following chapters.

December, 1988
Condorhuasi, my agricultural oasis in the desert of prov. Mendoza, Argentina
*Heinz Brücher*

## References

Alwim P, Kozlowski T (eds) (1977) Ecophysiology of tropical crops. Academic Press, New York

Brücher H (1968) Genetische Reserven Südamerikas für die Kulturpflanzenzüchtung. Theor Appl Genet 38:9–12

Brücher H (1969) Gibt es Gen-Zentren? Naturwissenschaften 56:77–84

Brücher H (1977) Tropische Nutzpflanzen. Ursprung, Evolution und Domestikation. Springer, Berlin Heidelberg New York, 528 pp

Brücher H (1982) Die sieben Säulen der Welternährung. Senckenberg Buch 59. Kramer, Frankfurt, 200 pp

Brücher H (1985) The South American gene pool of economical useful plants. Plant Res Dev 16:109–120

Brücher H (1988) Migration und Dispersion amerikanischer Nutzpflanzen über die Landenge von Darién (Panama). Naturwissenschaften 75:18

Hecht SB (ed) (1982) Amazonia: Agriculture and land use research. CIAT, Colombia, 430 pp

Herrera FL, Yacovleff E (1935) El mundo vegetal de los antiguous peruanos. Rev Mus Nacl 3:243–322; 4:31–102

Hoehne FC (1937) Botanica e agricultura no Brasil no seculo XVI. Edn Nacl Brasil 71, 410 pp

Jordan CF (ed) (1987) Amazonian rainforests. Ecosystem disturbance and recovery. Ecol Stud, vol 60. Springer, Berlin Heidelberg New York Tokyo, 133 pp

Körber-Grohne U (1988) Nutzpflanzen in Deutschland. Theiss, Stuttgart, 490 pp

Lathrap DW (1975) Our father the caiman, our mother the gourd. Emergence of agriculture in the New World. In: Reed C (ed) Origins of agriculture. Mouton, Paris Den Haag

MacNeish RS (1964) Ancient Mesoamerican civilization. Science 143:531–537

Parodi LR (1966) La agricultura aborigen argentina. Bibl de America. Cuadernos 4., Buenos Aires, 47 pp

Perez-Arbelaez E (1956) Plantas utiles de Colombia, 3rd edn., Madrid

Pittier H (1971) Manual de las plantas usuales de Venezuela, 2nd edn. Fund E Mendoza, Caracas, Venezuela

Prance GT, Kallunki JA (eds) (1984) Ethnobotany in the neotropics. Advances in economical botany, vol 1. New York Bot Garden, 156 pp

Raven P (1985) Botany and natural history of Panama. Miss Bot Gard St Louis USA

Raven PH, Axelrod DI (1975) History of flora and fauna of Latin America. Am Sci 63:420–440

Reed C (ed) (1977) Origins of agriculture. Mouton, Paris Den Haag

Sauer CO (1959) Age and area of American cultivated plants. Actas 33rd Int Congr Am 1:215–229

Schultes RE (1979) The Amazonia as a source of new economic plants. Econ Bot 33:258–266

Schwanitz F (1967) The origin of cultivated plants. Harvard Univ Press, Cambridge, Mass

Ucko PJ, Dimbleby GW (eds) (1969) The domestication and exploitation of plants and animals. Aldine, Chicago

# I. Carbohydrates from Roots and Tubers

For practical reasons – but not as a systematic classification – we divide this large group of tropical plants into two sections:

A. Major Tuberous Plants: batate, yam, cassava and potato. In the year 1985, world production of fresh roots and tubers of this group alone totaled 590,000,000 t, harvested from 47,000,000 ha. Nearly 70% of the typical tropical rootcrops (batate, cassava, yam) was produced by "developing countries" situated in tropical-subtropical latitudes, whilst nearly all the potatoes were produced in the temperate climate of the agriculturally more advanced regions of Eurasia, South Africa and North America.

B. Minor Tuberous or Rhizomatic Plants and Aroids. Most of these represent locally developed crops in Andine short-day regions such as *Oxalis tuberosa, Tropaeolum tuberosum, Ullucus tuberosus, Arracacia esculenta* and *Lepidium meyenii*, whilst others have wider adaptation, e.g. *Calathea* sp., *Canna edulis, Polymnia sonchifolia, Pachyrrhizus tuberosus, Maranta arundinacea, Xanthosoma sagittifolia* or *Helianthus tuberosus*. Some are of essential importance for the subsistence of indigenous peoples and should be protected from extermination in view of the sad fact that their areal is in continuous retrocession. Their share in the world production of starchy roots and tubers may be less than 5%.

## A. Major Tuberous Plants

CONVOLVULACEAE

### 1. *Ipomoea batatas* (L.)Poir

Sweet potato, batate-douce, batata-doce, camote, batata, cumara, apichu(kechua), camote (nahuatl), Süsskartoffel

$2n = 90$

NAME, ORIGIN AND HISTORY

This tuberous fruit is known in South America under several native names. The most discussed term is "cumara", which occurs both in Peru and in Polynesia. Several anthropologists and plant explorers have used this linguistic argument for far-reaching speculations and migration theories of human races in the southern hemisphere. Recent investigations, however, indicate that this name was not used in Peru until after the arrival of the Spaniards. Heyerdahl (1966) also referred to "cumara" and insisted that *Ipomoea* must have been brought in pre-Columbian times from the Peruvian coast to Polynesia and the Easter Islands. This intrepid explorer of the Pacific Islands has certain geographical arguments in favour of his theory. For example, it is astonishing that the Maoris have a surprisingly great number of local *Ipomoea* varieties, some of them extending in area as far as 45° lat S., in a cool climate, and far beyond the usual growing conditions of this neotropical crop. It is still an enigma how these oceanic races or cultivars of *I. batatas* originated. The assumption of Purseglove (1965) is that sweet potato capsules can float for a long time in seawater without the enclosed seeds losing their viability. *Ipomoea* capsules may have been carried by currents from the Pacific coast to Polynesia.

There are several archaeological proofs of *Ipomoea* tubers from coastal Peru (Paracas) dating back as far as 4000 years B.P. (Lanning 1966). Thus we may assume that "apichu" belonged to the basic food of early Indian agriculturists, long before the Inka time. Much older are fossil batatas from the cave Puna de Chilca (Peru), dated by Engels (1970) at 10,000 years. The postulation of a Peruvian centre of origin, claimed as early as 1925 by Safford, is, however, not supported by botanical arguments. There are no wild-growing *Ipomoea* in this region which fit in a phylogenetical scheme or could be considered as "ancestrals". Actually, any claim for a South American centre of origin is supported more by archaeological and historical evidence than by the criteria of botany and plant genetics (Stone 1984, Willey 1960). The hypothesis of a Central American origin has a weak basis: the word "batata" is of West Indian origin. The Nahua Indians called this tuber "camote". There is also the term "aje". Oviedo described it as an inferior sweet potato with large roots, but the use of this West Indian word was discontinued in favour of "batata" and "camote", which was generalized under Spanish colonization. The well-known North American botanist Merill added to the existing confusion by his assertation that Ipomoea *batatas* is of African origin, a hypothesis which found some supporters in Europe also, who speculated that sweet potatoes may have reached Polynesia from the African continent. There is, however, no proof of the pre-Portuguese existence of *I. batatas* in the Old World.

On the other hand, there is one irrefutable fact: Columbus found batatas on the 4th of November 1492, when he arrived at the coast of Cuba. Even if his description seems confused ("the fields there held an abundance of yams resembling carrots, possessing the taste of chestnuts"). Columbus employed the term "yam" because he had not seen the tubers before. Las Casas, who had access to the logbooks of Columbus (which later disappeared), confirms that the Spaniards found "batatas" in Cuba. The term batata appears for the first time in print in the year 1516 in a publication of Peter Martyr (who worked at the Council for the Indias, but never lived in America), where he stated that he had eaten these tubers in Spain and that they resembled turnips. As place of origin he cites the Isthmus of Darién. There must have been two distinct varieties, one of which was called "agi = aje" and was more starchy, whilst the true batata was described as more sugary.

Columbus took *Ipomoea* back to Spain on his voyage with the *Pinta*, which lasted 48 days and allowed the tubers to survive until their arrival at the harbour. From there the new food crop spread quickly on the Iberian peninsula. The botanist Clusius, who travelled to Spain in the year 1566, found three different cultivars of *Ipomoea* in Andalusia. All had white flesh, but different skin colour, red, rose and white.

The Portuguese seafarers soon recognized the value of batatas as ship's proviant and antidote to the feared scurvy, and spread the crop to Africa (Gambia, Sierre Leone, Nigeria) and even to Malaysia and Oceania (Yen 1974).

*Ipomoea* is a pantropical genus with more than 500 taxa described in botanical treaties. In the course of speciation polyploid series arose ranging from $2n = 30$, to $2n = 60$ and $2n = 90$ chromosomes.

Several species are used as ornamentals and herbs. Most notable is *Ipomoea aquatica* Forsk; selected in Asia as an important vegetable. Its young shoots are used as spinach and fish food by the Chinese. It is astonishing that this valuable herb is not sufficiently known and cultivated in tropical America. Several other *Ipomoea* species acquired similar importance as food and green fodder in Asia and Africa, for example, *I. pes-tigridis* L. *I. pes-caprae* (L.) Sweet. Others like *I. purpurea* have acquired a bad reputation. Morning glory seeds, even the flowers, are used as a narcotic in America.

It may be mentioned that the Guayami Indians of Northwest Panama domesticated the much-branched and leafy wild *I. acuminata* Vahl. This perennial species produces thickened roots with starchy pulp of agreeable sweet taste. Plants persist in the prov. Chiriqui and Bocas del Toro as weed crops in abandoned native gardens.

At a considerable distance away, in Northwest Argentina, natives use another wild batata, *I. calchaquina*, which forms considerable thick starchy roots. A similar use has been reported for *I. chacoense* O'Donnell and *I. platense*.

It is not certain whether natives of the tropical lowlands of Mexico use *I. triloba*.

## Morphology

Twining herbs with long trailing stems (80–300 cm), which make roots on contact with the soil. The so-called tubers of the batata plant are in a strictly anatomical sense thickened adventitious roots. Not all the cultivars of *Ipomoea* have a slender creeping habit. There are in Peru, for example, upright, compact-growing varieties. The thickened roots are globular or fusiform and may attain an astonishing size and weight (2–3 kg). They have a bright-coloured periderm, often red, purple, brown, yellow or white, and the flesh also has different colours, the most appreciated being the white and orange pulp.

The leaves are simple (5 × 15 cm) with long petioles, spirally arranged on the stem. The lamina shape varies from heart-formed to digitately lobed (see Fig. I.1). The differences are genetically controlled. In contrast to the common potato, where the leaves die back when the tubers ripen, the sweet potato plant continues to produce fresh leaves. This is important for the yield, because dry matter production in the batata depends on the leaf area and its assimilatory efficiency. During long day conditions, leaf production increases.

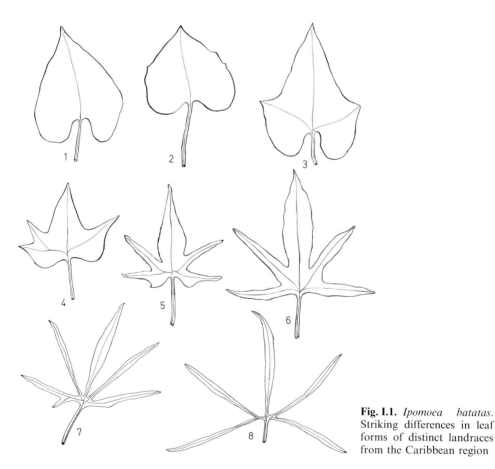

**Fig. I.1.** *Ipomoea batatas.* Striking differences in leaf forms of distinct landraces from the Caribbean region

Inflorescences: in certain varieties completely absent, in others cymose. Flowers are in general solitary, inserted in leaf axils. Corollas showy (4–7 cm long) with a purple or lavender limb and darker throat. Five stamens are attached near the base of the corolla, surrounding the filiform 2-cm-long style. Calyx 1–2 cm long, deeply five-lobed, acuminate. Fruits are dehiscent capsules (5–8 mm diam), with two to four seeds, but infrequently formed. Fruit setting is uncommon in temperate climate and long-day conditions. Auto-sterility and cross-incompatibility are common in commercial varieties (Montaldo 1972).

*I. batatas* is an excellent starch producer, but its roots are rather poor in protein (Table 1). For this reason we consider it a valuable

**Table 1.** Nutritive value of the enlarged roots of *I. batatas*. (After Martin 1984)

|  | Roots (%) | Leaves (%) |
|---|---|---|
| Water | 73–68 | 86 |
| Carbohydrate | 25–28 | 8 |
|  | (also 5% sugar) |  |
| Protein | 1–2 | 3 |
| Fat | 0.3 | 0.7 |
| *Mineral content* (mg) |  |  |
| Calcium | 21–33 | 86 |
| Phosphorus | 38–50 | 81 |
| Potassium | 210 | 562 |
| Sodium | 31 | 5 |
| Iron | 1–2 | 4 |
| Thiamin | 0.09–0.4 | 0.11 |
| Riboflavin | 0.04–0.05 | 0.22 |
| Niacin | 0.7 | 0.7 |
| Ascorbic acid | 21–37 | 17 |
| Carotene | 35,00–2400,00 | 2215 |

b

a

c

**Fig. I.2.** Three varieties of *Ipomoea batatas*. **a, b** from the tropics of Venezuela. **c** Peruvian w. Elyda selected by Del Carpo-Burga (1984)

improvement that a Peruvian plant geneticist, in two decades of breeding research, selected promising cultivars whose protein content exceeds 10% (Carpio Burga 1984) (Fig. I.2).

The high chromosome number of *I. batatas* indicates a polyploid origin of this neotropical tuber crop, but the circumstances surrounding the creation of its allo-hexaploid genome are still under discussion. Several wild species have been named as the putative parents. However, we do not know the place of such a hybridization, nor the possible interaction of man in the species synthesis. The result, similarly to other amphyploid crops of America, was a functional integration of chromosome sets from different species, which resulted in a high degree of biological potency, superior yield and wider colonizatory ability. The hexaploid status of *I. batatas* was discovered by King and Bamford. Further cytogenetic research by Martin (1974) revealed genetic affinities between the cultivated sweet potato and several wild species, such as *I. gracilis, I. lacunosa, I. tiliacea, I. trichocarpa* and *I. triloba* (Austin 1978). Nishiyama (1971) made the following experiments: he crossed *I. litoralis* (which is tetraploid) with *I. leucantha* (diploid) and obtained triploid progenies. Selected triploid individuals received artificial treatment for duplication of their genomes. Some hexaploid offspring were rather similar to cultivars of *I. batatas* and could be easily crossed with them.

A discussion is still going on between Japanese and North American authors about the discovery of a so-called wild "*Ipomoea* K.123" in Yucatan by Nishiyama, claimed as an ancestor of the cultivated sweet potato. In favour of his hypothesis is the fact that both have a common genetic system for incompatibility. Dr. F.W. Martin, however, informed me in a letter that "wild forms of sweet potato exist in many parts of Puerto Rico. These, I am sure, are escapes from cultivation. When the sweet potato is permitted to breed freely, many types of wild-appearing materials are observed". This is the opinion of an experienced batata researcher, who expressed the hope of resolving the question by electrophoretic techniques.

## PRODUCTION AND IMPROVEMENT

As with several tropical crops, the main production of batata has shifted from its original homeland to other continents, and has achieved its major importance as a human food in Asia, where two thirds of the global yield is at present produced.

Of the total world production of 146,000,000 t, China alone contributes 126,000,000, followed at a great distance by Indonesia (2.4 mio t) and Vietnam (1.7 mio t) (Villareal and Griggs 1982).

The planting technique in tropical countries is based exclusively on stem cuttings. These are bundled and taken to the prepared ridges and then hand-planted. The cuttings develop roots in 4–5 days. The first harvest can be gathered after 4–6 months.

In temperate zones, the batata crop is grown from root slips. Cool nights are favourable for tuber development.

For high yield, long growing periods are essential. It has been observed that shortening the growing season even by one day reduces the yield visibly. The advantage of late ripening can, of course, only be achieved in frost-free regions. Under such ecologically favourable growing conditions, yields as high as 35,000 kg/ha have been recorded, but in general the yields in native fields oscillate around 7000–9000 kg.

Based on high-yielding varieties from Puerto Rico, a pioneer in sweet potato improvement, Julian Miller from the Louisiana experimental station, created for example Unit I Puerto Rico, which for several decades was the genetic basis for sweet potato production in the southern states of the USA. The usual clonal multiplication of batata favours the manifestation of somatic mutations. Easily detectable are mutations in flesh colour (Table 2), whose appearance has been calculated by Miller at 3–6%.

Drought resistance is of great importance in tropical regions with shifting rainfall and drought spells. Screening of germplasm collections revealed that genotypes with smaller-sized stomata are less drought-affected. The selection TLB-5 in Japan has good drought tolerance due to its well-developed root system.

**Table 2.** New varieties of *I. batatas*. (Martin 1984)

| Variety | Geographical source | Characteristics |
|---------|---------------------|-----------------|
| Malesco 2 | Guadeloupe | Yellow-fleshed |
| Jabrum IV | Guadeloupe | Yellow-fleshed |
| Centennial | USA | Orange-fleshed |
| Nemagold | USA | Orange-fleshed; root-knot-resistant |
| Jewel | USA | Orange-fleshed; broad adaptability; widely grown in USA |
| Gem | USA | Orange-fleshed |
| Miguela | Puerto Rico | White-fleshed |
| Brondal | East Africa | White-fleshed |
| Eland | East Africa | Orange-fleshed, nematode-resistant |

Frost resistance cannot be expected in a tropical crop, but a certain hardiness against low temperatures has been reported from wild types growing in Colombia at 1500 m altitude.

The recently initiated breeding work of the international CIP organization (International Potato Centre in Lima, Peru) with a collection of 1000 new introductions for future genetical improvement of this tuber crop deserves special mention.

Very common in sweet potato plants is pollen sterility and poor seed setting. Martin (1984) investigated the self-incompatibility, which is controlled by various genes. Heterostyly inhibits self-fertilization. Considerable differences in the length of the pistil have been measured, in the range from 8 to 29 mm.

Recent breeding is interested in sweet potato, with its high vitamin and protein content. For example, cv. La Cobre in Puerto Rico has double carotene content. Others contain up to 8% protein (Bouwkamp 1985).

DISEASES AND PESTS

In comparison to the "white potato" (*Solanum tuberosum*), sweet potatoes suffer less from fungus diseases. We mention the following:

*Ceratocystis fimbriata* (black rot) attacks batata cultivation in the tropics.

*Fusarium oxysporum* is always present during the humid season in the southern states of the USA.

*Rhizopus* and *Monilochaetes infuscans* cause putrefaction of the roots.

Viruses

Vein mosaic is widely diffused in South America, causing considerable loss of yield. For example, in the Argentinian province Cordoba it was stated that the totality of all plantations of var. *colorada* and *blanca brasileira* are infected with vein mosaic virus. Locally such plants are called "batata crespa", after the curly deformation of the leaves. It has been reported that Folquer's Tucumana lisa selection possesses genetical resistance to this disease.

Other viruses, like virus internal cork and feathery mottle, are becoming more important when batata is planted in extended monocultures.

Pests

Damage by tropical insects is considerable. The mothborer *Megastes grandalis* lays its eggs in the axils of young batata plants. Its larvae burrow holes in stems and tubers and can ruin the greater part of a harvest.

Several weavels, like *Euscepes batatae, Prodenia* and *Cylas*, can cause total loss of leaves during the rainy season.

Nematodes are a severe problem worldwide in sweet potato plantations. Four different species of *Meloidogyne* attack *Ipomoea*, but also other nematodes, like *Rotylenchus reniformis*, are associated with sweet potatoes, and diminish their production. For this reason, breeding for resistance to nematodes is of great importance. As a result the tolerant Nemagold in the USA was released.

EUPHORBIACEAE

## 2. *Manihot esculenta* Crantz

Cassava, mandiu, mandioka, maniok, yuca, cazabe

$2n = 36$

## Name, Origin, Distribution

The correct Latin binome was created already in the year 1766 (emending Linné) by the German naturalist Crantz, and must be maintained for nomenclatorial reasons. There have been repeated attempts to replace it by other names such as *Manihot edulis, M. utilis, M. utilissima, M. dulcis, M. aipi, M. multifida*, etc., which are not valid. The vernacular names are derived from native terms. "Mandiu-mandioka" are Guarani words; "yuca" is an Arawak-Taino term, "cazabe" is an Indian word modified in African languages to "cassava", which is now the English denomination. This important root crop belongs to the botanical family of Euphorbiaceae, which has given several other essential species to mankind, like the rubber trees, *Hevea brasiliensis, Manihot glazovii, Ricinus comunis, Aleuritis fordii, Caryodendron orinocense* etc.

The neotropical genus *Manihot* very early aroused the attention of botanists. Pohl, who spent several years travelling in Brazil, published in 1827 the first monograph of the genus, describing 48 taxa, many of them still recognized. Mueller-Aargau described 72 species in Martius (*Flora Brasiliensis* 1874 XI). Pax (1910) established in his monograph 128 taxa, divided into 11 sections.

The most recent treatment is that of Rogers and Appan (1973). The authors re-examined the classical collections and performed extensive fieldwork, which culminated in 19 sections with 98 species, dispersed from North Argentina to the border of the USA. It is remarkable that all wild species which have been screened cytologically have the same chromosome number (2n = 36) as the cultivated cassava plant. *M. esculenta* hybridizes easily in artificial crossings with these wild species, forming regularly 18 bivalents (Magoon et al. 1970; Nassar 1978) (Fig. I.3).

Some of the wild-growing biotypes are of considerable importance for hybridization with cassava for improving its resistance to pests and diseases. The areal of these wild species extends from Paraguay to Mexico. It is therefore nearly impossible to establish a geographical centre of genetic diversity, where the wild forms surround the cultivars,

**Fig. I.3.** The author collecting wild manihot in a Brazilian forest, for future breeding work

as older writers thought. The two so-called genecentre No. 7 (highland of Mexico) and genecentre No. 8 (Andine highlands) are beyond discussion for the origin of *M. esculenta* for climatic reasons. In reality we do not know where and when manioc was first cultivated. As it is easily propagated by cuttings, migrating Indians probably carried stems of Manihot on their wanderings, exchanged them with neighbouring tribes as an ideal food for travellers, and planted them at their resting places, where they crossed with local forms.

The overwhelming quantity of wild-growing *Manihot* species belongs to South America. This entitles us to postulate the origin of the cassava culture as South of the Panama Isthmus, most probably in the eastern region of the Amazonas. We take as working hypothesis that a "radiation centre" for cassava domestication exists in the Amazonian basin, where different wild species of *Manihot* are indigenous. Due to the extremely humid climate of this region, however, we cannot expect to find archaeological re-

a         b         c

**Fig. I.4. a** Natives from the Upper Orinoco preparing cassava flour with a "tipiti" (*left*). **b** Cassava stalks stored under the roof of an Indian hut for future plantation. **c** A fresh harvested root of a primitive manihot variety

mains. What was found are several very interesting artefacts related to cassava processing: graters, griddles or "budares" (Fig. I.4).

According to Reichel-Dolmatoff (1965), pottery griddles similar to those used today for toasting manioc flour have been excavated in Northern Colombia and dated at 1120 years B.P., at Malambo. The "budares mas antiguas del Continente" were discovered at Rancho Peludo near Lago Maracaibo, dated at 2700 years B.P. (Rouse and Cruxent 1963).

Most astonishing is the quantity of archaeological material from Peru, where in the ruins of Paracas remains of *Manihot* roots were found. A ceramic has been recovered from graves in Chimbote depicting a frog (or a toad) together with mandioka. Also from Nasca, an artistic reproduction of *Manihot* plants has been reported.

Towle (1961) mentioned several phytomorphic finds of potteries from the Mocha and Chimu cultures.

Recently a unique sample of (sweet ?) cassava has been discovered in the Valle de Casma (Peru) about 18 km from the Pacific coast (Ugent and Pozorski 1985). This place was probably an important dwelling centre of the Chimu Indians. The collection consists of 197 pieces of root and bark fragments and four fruit capsules. Based on radiocarbon assay, the oldest finds have been dated at 1800 B.P. The authors believe that this was "sweet" mandioka, because this variety is still – and exclusively – cultivated on the eastern flanks of the Peruvian Andes.

Lathrap (1973) favours a South American origin for cultivated *Manihot*. He identified stylized plants of this crop on a very important archaeological document of the Chavin culture: the Tello Obelisk at Chavin de Huantar. It may be that the Chavin Indians crossing the Andes and settling on the Pacific Coast of Peru (in approximately 2900 B.P.) also introduced mandioka from the tropical forests of eastern Peru. For bioclimatical reasons we must exclude absolutely that the arid coasts of Peru could be claimed as origin of *M. esculenta*.

It seems strange that there are no archaeological references to mandioka in Brazil, whilst some have been reported in Venezuela and Colombia. The finds of the Momil I epoch on the Caribbean coast of Colombia contain a mandioka griddle, dated at 2000 B.P. (Schwerin 1970)

In this context, Rouse and Cruxent (1963) pointed out that a wide genetic variation of

*Manihot* species still exists in this region to-day.

For similar reasons, Sauer (1952) considered the savannas of Venezuela as a possible centre of origin of edible *Manihot*.

Looking to Meso-America, Rogers and Appan (1973) expressed the opinion that cassava arose there (including certain Caribbean Islands), based on the existing wild species *M. aesculifolia* and *M. pringlei*, and spontaneous hybrids between them. Useful clones resulting from these crosses would have been selected by Arawak Indians, perhaps as early as 10,000 years ago. Several objections have been raised to this hypothesis. Referring to Mexico, the archaeological evidence of mandioka cultivation in prehistoric times is rather scanty, if not non-existent.

The often-cited discovery by McNeish of seed and leaf remains in caves of Sierra de Tamaulipas in N.E. Mexico, 2100 years old, has been criticized. As Flannery (1973) wrote: "Over-eager archaeologists continue to cite it as an example of cultivated maniok"; but the botanist Earle Smith, who re-investigated the find, identified the material as seeds of one of the wild Mexican *Manihot* species, which have never been domesticated.

Pickersgill and Heiser recently expressed (1977) their doubts that Mexico/Guatemala was the original centre of cassava domestication. Even Rogers, who initially promoted the idea of a Central American cradle of mandioka cultivation, no longer favours this hypothesis.

In conclusion: in fact no reliable evidence exists of an independent domestication of *M. esculenta* from local wild species in the so-called Mexican gene centre.

Middle America accounts for a dozen wild-growing *Manihot* species, mainly *M. aesculifolia, M. gualanensis, M. isoloba, M. pringlei* and *M. oaxacana*. With respect to the wild status of *M. pringlei*, the objection exists that its HCN content is very low; in general this toxic substance is considered as good natural protection. Some of the taxa described from Meso-America seem too primitive to be considered as ancestors for the cultivars, others have proved to be synonyms. In conclusion, the numeric presence

of *Manihot* species in Mexico is only 10% of the total in America. Following a general rule in agro-botany, we may expect that domestication occurred "in the midst of the cradle of wild forms" and not at the periphery.

In any case, mandioka reached the humid tropics of Yucatan-Guatemala and the Maya and Aztec civilization long before the arrival of Columbus. The Mayas from Yucatan, also the Oltecs, had established a "trade network" with Southern coastal tribes (Kuna?). The expansion of these tribes with sea-going canoes along the Central American coast is an impressive and astonishing fact. We believe that even in very early times, Chibcha-speaking Indian tribes, who lived on both sides of the Isthmus of Darién, could have spread *Manihot esculenta* across this landbridge which was crucial for the early migration of useful plants (Brücher 1988). This is supported by the fortunate evidence of the presence of *Manihot* pollen in a deep-borehole at Gatun Lake (Panama), reported by Bartlett and Barghorn (1973) and dated at 4200 B.P. (see also Renvoize 1971).

*M. esculenta* does not exist in a true wild state. We consider it as hybrid swarms and do not exclude a *polyphyletic* origin in distant parts of America. Nassar (1978) established four different and geographically very distant "places of domestication" in South America. These are related with the migratory movements of Indian tribes, between them the Ge-speaking groups of South Brazil, the Tupi, the Guarani from Paraguay and the Arawaks from Central Brazil (their name means "people who eat tubers"). The Arawak tribe finally migrated in the 11th century to the West Indian Islands and entered Central America.

The highest concentration of wild species (about 38 taxa) has been found in the central part of Brazil (Western Minas Gerais and South Goia). There exists not only an impressive botanical diversity, but also many forms with primitive characters (dioecious inflorescences and non-lobed leaves, with short petioles) (Fig. I.5). With a certain degree of probability this area thus constitutes the "primary centre of diversity" of *Manihot* in South America. However, this does not

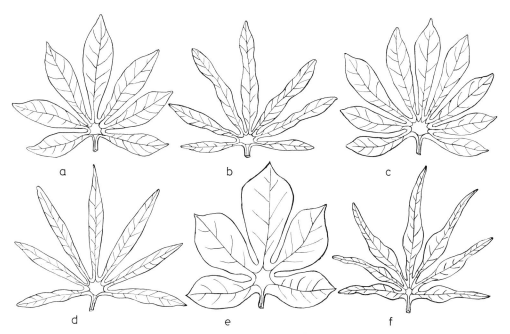

**Fig. I.5.** Great genetic variability among *Manihot esculenta* leaves

imply that the domestication of edible *Manihot* strains must also have occurred there. As we have stated on p. 12, it is not possible to determine, from among the almost 100 wild-growing manihots, which is its direct ancestor, for lack of cytogenetical and physiological indicators.

The next important area of species concentration is the northeast of Brazil, which includes the states of Bahia, Pernambuco and Ceara, mainly semi-arid biotopes. The list of names in Nassar's publication includes 18 taxa. Finally, more than 2500 km distant from Ceara, there is a third group of wild *Manihot* in the border region between Paraguay, Bolivia and Brazil. Here six to eight different species live, some with frost tolerance and high HCN content. It is noteworthy that *M. guaranitica* roots are gathered by nomadic Indians of the Chaco and still regularly used as food. This interesting fact is, however, not sufficient for claiming *M. guaranitica* as a direct ancestor of cassava.

Summing up, we believe that more than 100 species of *Manihot* live on the American continent, of which more than 150 different taxa have been described by botanists.

The following wild species of *Manihot* are related in some way to the cassava cultivars:

### *M. carthaginensis* (Jacq.) Müller Arg.

This has been described for more than 100 years from Cartagena in Colombia. It is apparently not used in South America, but natives of Costa Rica plant this "yuca del monte" sometimes for extracting starch from its roots. In Venezuela it grows wild in the states of Falcon, Zulia, Aragua, Sucre and Bolivar and the coastal islands (Isla de los Patos). The species also exists in Trinidad and Tobago.

### *M. aesculifolia* (H.B.K.) Pohl

Tall (up to 7 m) shrubs, with swollen roots, stout inflorescences, smooth stems which lack the typically swollen nodes of *M. esculentum*. Its distribution range extends from Panama to Costa Rica and Mexico.

As it crosses easily with cultivated cassava, Rogers considered it a close relative, even claiming it as the wild ancestor of *M. escu-*

*lenta*. As its main areal is in Mexico, it was even thought that Mexico may be the centre of origin of edible cassava, but recently Rogers has expressed doubts.

We have no indication that *M. aesculifolia* has ever been used in cultivation or selected by the Indians.

### M. guaranitica Chodat and Hassler

While living in Paraguay, where "mandiu" still represents the daily diet of the rural population, I had occasion to study the primitive use by Indian natives of wild and semi-cultivated *Manihot*. For a long time, Paraguayan botanists, like Hassler, Rojas and recently Pastor-Arenas (verbal communication), had reported the existence of true, wild cassava and certain Indian habits related to it. The plants live in sandy soils between the dense thicket of "mattorales". The main root is woody, but also contains starch. The natives call it "yam-si-po", which means that it is not cultivated. The Maskoy Indios of the Chaco search the strongly ramified shrubs and excavate the roots during their daily nomadic food gathering.

According to Pastor-Arenas, the Lengua Indios plant the variety Panko of *M. esculenta*, which is a very primitive one and highly toxic. As their fields are not protected against a number of predatory wild animals, the poison represents a certain protection. The Shamans of the Lenguas make use of the plants "para intoxicar los jovenes durante la ceremonia de iniciacion masculina". On this occasion they chant the "song of mandioka", which indicates that this crop, besides its nutritive value, also has a ritual importance for them.

### M. tripartita (Sprengel) Muell. Arg.

Known in Brazil under the vernacular name, "mandioquina do mato", the plants are similar to a real cassava, 30–80 cm tall, with tripartite leaves and extremely long petioles. The fruit capsules are hirsute. The whole plant has a high content of HCN.

During our *Manihot*-collecting trips in Brazil

and Paraguay, natives informed us that grazing cattle mistake this wild mandioka for the real cassava and eat it. At first the polygastric animals show no negative effects. Only when the ruminated, half-digested mass of leaves reaches the "fourth stomach" with its very acid reaction, do the animals suffer severe convulsions and even die from hydrocyanic acid intoxication.

### M. dichotoma Ule

These are tall shrubs, known in Brazil as "jequie manicoba" because their stems shed copious latex when cut or hurt. This wild *Manihot* has immunity against viruses and for this reason has been crossed repeatedly with cassava. However, the tuber production was low in the resulting offspring.

### M. glazovii Muell. Arg.

This tree is common in the Catinghas of Brazil. It does not develop any enlarged roots. As a rubber-producing species it is known worldwide as so-called Cera rubber. Due to its immunity to mosaic virus and brown streak, this species has been used in many hybrids with *Manihot esculenta*. In certain backcrosses of the third generation, surprisingly, clones appeared with high tuber yields.

### M. catingae Ule

This is also a tree-like species from Brazil. It has similar immunity against diseases, wood roots and the same vigorous growth as the above-mentioned species.

In Venezuela the following species have been collected, some of which we had the opportunity to compare in their native habitat with cultivated cassava, and to notice their similarity:

### M. orinocensis Croizat

This taxa grows at the rocks ("lajas") on the border of the Orinoco river, near Puerto Ayacucho.

## M. surinamensis Rog. and App.

This has been collected at the Isla de cazabe near Puerto Ayacucho and grows in the occasionally dry savannas of Venezuela.

## M. filamentosa Pittier

The natives call it yuca sibidigua. It has been found in the semi-arid province of Barquisimeto.

MORPHOLOGY

There is no anatomically defined "model" of *M. esculenta*. Cassava-mandioka is a short-lived, perennial shrub, with erect stems. Sometimes the stalks are very tall (3–5 m). Some varieties are without ramifications. Others have short stems with many side branches, giving them an "umbrella" habit. The angle of ramification and branch insertion is genetically controlled and is used as a varietal characteristic. Stems are round, with a diameter of 2–6 cm, covered with spots of different colour, which may characterize the cultivars. The stalks die with frost, but the underground parts survive during the winter.

The leaves have long petioles and are inserted on typical, protruding nodes. The lamina of leaves varies according to variety and ecological circumstances. It may be divided in five to ten narrow segments, which are linear, ovoid, rhomboid, or ovalate according to cultivar. The colour of the surface is always darker than the underface. It deserves special mention that the leaves are rich in protein (7 g raw protein in 100 g fresh leaves). Its use as vegetable should be recommended to starving people in the tropics. The leaf petioles are long (18–40 cm) and thin, curved or straight, depending upon the variety.

Inflorescenses: male and female flowers are separated, but borne together in the same panicle. There are always more staminate than pistillate flowers, which sit on the lower part. The pistillate flowers open earlier (they are one-week protogyne) and are ready for fecundation before the staminate flowers shed their pollen. For this reason, outcrossing is the rule in *M. esculenta* and the clones do not produce identical offspring from seeds. Fecundation is caused by wind and small insects, which are attracted by the abundant nectar. As an exception some varieties have hermaphrodite flowers and others are completely male sterile.

Fruits are tripartite capsules, which need several months for maturation. When ripe, the three seeds are expelled, in the typical manner of most Euphorbiaceae. The fleshy caruncles are appreciated by small animals.

The anatomy of cassava roots is comparatively simple. The inner xylem, which represents the edible part with a high starch content, is surrounded by the cortex, composed of the phloem and cortical layers. In the ripening roots the epidermis disappears, and some phelogene tissues cover the roots in thin layers.

The starch grains of cassava are round and have a diameter of 10–20 μ. Cassava is propagated exclusively by stem cuttings, and not from roots. For this purpose mature stems are cut in 30-cm-long pieces. At each site of a petiole abscission is an axillary bud with a growing point. Under good climate conditions and planted below soil, new cassava plants regenerate from these buds.

Since Graner (1935) determined for the first time the correct number $2n = 36$ for *M. esculenta*, very few cytologists have investigated the chromosomes of its wild relatives. We mention especially Magoon et al. (1970) and Nassar (1978). From a dozen species in total they reported the basic number $n = 18$. Chromosome associations were completely regular in metaphase I. No laggards or irregular separation of bivalents occurred in anaphase I of the pollen mother cells examined. Regular meiosis, even in interspecific hybrids, indicates that *Manihot* is phylogenetically a relatively young genus, in which cytological barriers have not yet been established.

*Manihot esculenta* is a complex hybrid and exists only in domesticated form. We may thus assume that spontaneous introgression and even (why not?) intentional crossings occurred when migrating Indians took cuttings of their basic staple crop to other places.

## Problems of Toxity

The fresh root contains toxic substances, mostly cyanogenetic glycosides (linamarin and lotaustralin) in varying amounts. Some varieties are considerably less toxic, those with less than 0,01 % of hydrocyanic acid being called sweet. The degree of toxity is not only genetically fixed, but it varies considerably with the potassium level and the moisture of the soil.

Hydrocyanic acid is a true protoplasmic poison and affects the human respiratory process. Cyanides cause oxygen to be unavailable to the cells and cause death by asphyxia, in lesser cases headache, nausea and goitre (Erdmans 1980).

Indians have observed the toxic effect of their main staple food since time immemorial and have tried to overcome it with homemade inventions.

1. The "cassava root grater" in its most primitive form is a rough stone. More highly developed are tools made from prickly palm roots and wood sticks, covered with rough, spiny shark skin, or grater boards equipped with sharp quartz splinters. With the help of these implements Indian women spend hours rubbing and grating the freshly harvested cassava roots after having peeled them.
2. The resulting cassava mash is then washed. For this purpose there are handwoven baskets of palm or liane fibre supported by a tripod. The resulting product is toasted or used as flour.
3. Certain Indian tribes very early in their history invented an elaborated instrument called aruyuba tipiti or sebucan. This is a long sleeve-like tube, woven of flexible fibres. When pulled, it constricts. The Indians fill the grated root mass into this sebucan, which is alternately stretched and squeezed. During this continuous moving of the tipiti, the poisonous juice runs down and the concentrated starchy mass remains in the sebucan. It is not easy to describe the technique with these 2-m-long tipiti but they work efficiently, as I have observed during my stay with the tribes of Warao, Makarikare and Waika in the Upper Ori-

**Fig. I.6.** Indian women working with a "tipiti" to produce cassava

noco-Rio Negro River system (Fig. I.6). The toxic sap is not thrown away, but carefully collected and cooked for preparation of fermented beverages, or as fish poison.

## Use and Marketing

It deserves special mention that *M. esculenta* grows well even on poor, acid, tropical soil known for its low phosphorus and zinc content (in acid soils Zn application doubles, even triples, the tuber yield; Nestel 1974).

Even if it does not appear in world food statistics, an uncounted number of rural populations live from cassava. We estimate that the small-scale farmers around the tropical belt of our globe are the main producers of *Manihot* roots. It is hazardous to give figures, but some food experts maintain that half of the world population in the tropics depends on cassava. We consider it as the second most important source of calories in underdeveloped tropical countries.

We have been told that cassava consumption decreases as income – especially in urban areas – increases. In fact, we found proof of this on the Caribbean Islands or in Panama, where there is widespread media propaganda for white bread consumption.

An unfortunate example is Jamaica, where an alarming decrease in cassava exists, which from 1977 with 40,247 t diminished to 20,851 in 1982 and even less (18,900) in 1983. The reason is the continuous importation of maize and wheat, mostly at below-cost prices from industrial countries or international help projects wishing to get rid of their grain surplus under the disguise of philanthropy.

Without doubt, cassava is the cheapest, most popular and also one of the most easily cultivated tropical crops. It gives ten times more starch than maize over the same agricultural area.

The objection has been made that cassava is mainly a starch producer with a low protein content. This is, in fact, an objection which can be made to nearly all tuber crops. It can, however, easily be balanced by mixture with pulses. Improvement of protein content in cassava is in fact an urgent need. Cultivated strains of *M. esculenta* have 2.2–2.7% crude protein in dry matter. In the literature several wild manihots have been reported with apparently relatively high amino acid content. However, such earlier estimations of total nitrogenous matter may have been influenced by the presence of cyanogenic glucosides. This cannot be excluded with *M. saxicola* and *M. melanobasis* with supposedly 8–10% protein.

The wild species *M. oligantha* subsp. *nesteli* with 7.10% and *M. tripartita* with 6.88% crude protein, both from Goyas, seem really promising for future hybridization with cultivars (Nassar 1978). The last-named species also has the advantage of a low HCN percentage. Furthermore, we should mention also the Central Brazilian wild species *M. gracilis* Pohl, which figures in literature with 8% crude protein.

Yields vary extraordinarily. Based on a global area of 13.7 mio ha, the mean yield is 9.4 t/ha. It can, however, be even less: 6 t/ha in exhausted soils of Paraguay. On the other hand, it is not unusual to obtain 30–40 t from well-managed plantations. Experimental plots in Venezuela gave as much as 100 t/ha with selected clones and adequate fertilization. Such root yields may be unparalleled in world agriculture (Montaldo 1972, 1979).

On a global scale, we observe a continuous increase in cassava output. In the year 1960 the annual world production was 70 mio t. In 1970 it went up to 92 mio t. In the year 1980 it exceeded 126 mio t. For 1985 the estimates reach 136 mio t (Cock 1982).

Since 1981 the major part of production has shifted from its original South American home (with 11 mio t annually) to the Old World tropics, where ten times more are produced.

During the last decade it was observed that cassava plants give higher yields in the presence of Mycorrhiza in tropical soils. They considerably improve the uptake of phosphorus, which is essential for root production. Various vesicular-arbuscular (VA)-mycorrhizal species act in this manner. Especial mention is deserved by *Glomus manihotis*. In new cassava plantations in recently cleared forest land, it is recommended to make artificial inoculation with selected mycorrhizal biotypes.

Indigenous fields in South America generally have native mycorrhizal fungi populations.

## DISEASES AND PESTS

Until 1930, *M. esculenta* was considered "a healthy crop", because it was nearly free of epizotics and epiphytotics, which are now so wide-spread in major monocultures. As long as cassava was planted on a small scale, even if there were millions of native backyards around the world, no serious phytopathological problems were reported.

However, with the initiation of large monocultural plantations in Africa and Asia, the problems began. On the African continent appeared two new virus diseases that were immediately transmitted by infected clones to India and Java.

### Fungus

Different fungus diseases now have worldwide expansion in cassava fields:

a) *Cercospora henningsii* and *C. caribae*.
b) *Colletotrichum* causes stem anthracnosis.

c) *Fomes lignosus* acts as a root fungus.
d) *Sphaceloma manihoticola* (sexual phase *Elsinoe*), is also called the "superelongation disease". It occurs in different physiological races, which makes it difficult to select resistant cultivars. This fungus provokes typical elongations of the internodes. It is suspected of acting similarly to a growth regulator of the gibberelin group, which also originated in a fungus.
e) *Diplodia manihotis* provokes stem rot and also destroys the roots. The presence of picnidios on the epidermic tissues is a symptom of infection. Propagated by stem cuttings, it is at present extending in Colombia and Brazil, and has recently also been reported from Africa and India, where *Diplodia* is causing serious damage in cassava monocultures.

## Bacteria

*Xanthomonas manihotii, X. campestris*:
Infected plants begin to wilt early. Field resistance has been found in the Brazilian cv. Manjari.

*Cassava bacterial wilt* = CBB = *añublo bacterial*
This bacterial disease is the most serious problem in *Manihot*. Its agent is still under investigation and has not yet been determined. Using serological methods, it has been shown that CBB is not caused by *Erwinia, Pseudomonas* or *Xanthomonas* bacteria. The pathogen is still an enigma. The infection is transmitted mechanically. From Brazil, where it was observed 70 years ago, it has spread to all cassava-producing countries. In view of the fact that prophylactic measures did not give reliable results, it seems that only the future use of genetical resistance can resolve this serious problem. Lozano and Booth (1976) reported resistant clones in the CIAT germplasm collection.

## Viruses

The cassava culture has recently been invaded by various virus diseases, some of them so severe that they can destroy the whole yield (Jennings 1963).

a) Virus "frogskin disease" FSD = *"cuero de sapo"*

Infected cassava plants suffer from a blockage of the vascular system, which produces disorders of the epidermis. Damage can be high, e.g., the valley of Cauca in Colombia lost 20, 50 and even 80% of the cassava yield through *cuero de sapo*. Recently it has been discovered that certain South American *Manihot* cultivars, e.g. M. Col. 1468, are resistant to this disease, which is easily transmitted by "white flies" (*Bemisia* spp.).

b) Brown streak virus = CBS
This disease makes the root unfit for human consumption. Up to now observed only in East Africa, in lower altitudes.

c) Common Mosaic Virus = CCM
Its origin is probably Brazil/Colombia. This virus has a wide host range in South America, mainly in Malvaceae and Euphorbiaceae, but no natural vector is known so far. The disease is mostly transmitted by knives and machetes and produces chlorosis of the leaf blade.

d) Mosaic Virus = CMD
Occurs in Africa and India and causes heavy losses (20–90%). Its vector is *Bemisia* and the host plants are escaped *Manihot* plants. The symptoms are distortion of young leaves, with bright yellow areas. This virus is considered to be the most dangerous in Asia and it should by all means be prevented from entering South America.

## Pests

Several insects, *Acaros, Lagochirus* spp. and last but not least, *Phenacoccus* spp. destroy the *Manihot* plants. Especially dangerous is *Phenacoccus* because this "piojo harinoso" multiplies parthenogenetically. It lives underneath the leaves and produces rosettelike deformation of the apical leaves and heavy distortions of the whole metabolism. The origin of different *Phenacoccus* species is tropical America. In North Brazil cassava fields lose up to 80% of the root yield. Similar losses are occurring now in Africa, where this insect has been inadvertently introduced.
The impact of the parasites *Phenacoccus manihotii* ("el piojo harinoso") and *Phenacoccus herreni* was dramatic. Both are of

neotropical origin, and caused no damage as long as they had to live and survive on small farms. Now these two insects cause 70–80% damage in the big plantations of Africa.

Fortunately a series of natural enemies have been discovered which live as parasite on *Phenacoccus*, and also a fungus of the genera *Cladosporium* acts in a similar way.

---

DIOSCOREACEAE

### 3. *Dioscorea trifida*

Synonym: *D. brasiliensis* Willd.
Mapuey, cush-cush, yampi
2 n = 18 also 2 n = 54, 72, 82

This is the only commercial yam of neotropical origin, therefore also called the edible American yam. Several other *Dioscorea* taxa also exist which have long been used as an emergency food by natives of South and Central America. We may mention: *D. hastata* (Brazil), *D. panamensis* Knuth, *D. standleyi* Morton, a weed plant and *D. racemosa* Klotsch and Uline, already described from Central America in the last century and found in Indian gardens. The tubers of these wild yams are used when other root crops fail. Due to its sporadic and spontaneous appearance, the natives of Panama believe that this food plant is sent from heaven in response to prayer. I have seen these wild yams in the Darién rainforest sprouting in forest clearings, and I noticed the considerable size of their root tubers. Their vines are genetically determined as dextrorse-climbing, and can thus be easily distinguished from cultivated *D. trifida* which has left-climbing stems.

We mention also the many wild Middle American *Dioscorea* species which are used in pharmaceutics and folk medicine. They contain small amounts of sapogenin (diosgenin) and also some poisonous alkaloids (dioscorine) and are therefore collected in the wild by the natives. In Central America, mostly in Mexico, *D. composita, D. floribunda* and *D. mexicana* are found. In view of the excessive exploitation in their wild stands, they have now been cultivated.

It is not necessary to stress the fact that world consumption of yams is based on African and East Asiatic species, like *D. esculenta, D. alata, D. hispida* and *D. japonica* from Asia, and *D. cayensis, D. bulbifera, D. dumetorum* and *D. rotundata* of African origin. In comparison to the only species of neotropical descendency, the Old World yams represent more than 90% of the trade (for more details see Martin 1973, Brücher

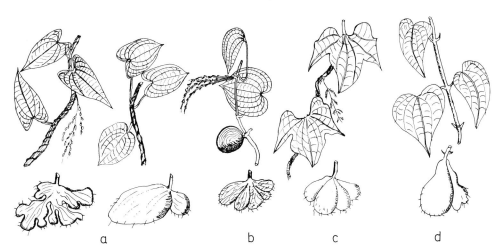

**Fig. I.7.** Comparison of different yam species **a** *Dioscorea alata* (Asia). **b** *D. bulbifera* (Asia). **c** *D. trifida* (Central America), **d** *D. cayennensis* (Africa)

1977, Onwueme 1978, Miege and Lyonga 1982) (Fig. I.7).

## Morphology

*D. trifida* plants are in general dioecious; male plants have a different habitus from females. Their stems are several metres long and cannot carry themselves, needing support from surrounding vegetation. Its twining direction is to the left, in contrast to most other yam cultivars, which climb in clockwise direction, (e.g. *D. alata*), to the right. In cross-section the stem of *D. trifida* is rectangular, often with extended wings. Stems rise from a corm, which on its opposite side produces roots. They emerge close to the surface of the soil and penetrate scarcely deeper than 1 m, at which depth the tubers also develop. According to Onwueme (1978), the yam tuber is neither a stem structure nor a root enlargement. Whilst Burkhill (1960) and others maintain the opinion that yam tubers are derived from stem structures, Martin and Ortiz (1963) concluded from their morphogenetic studies that it is a lateral outgrowth from the hypocotyl region of the plant axis. Mature tuber roots are surrounded by several layers of cork, protecting the interior portion, which is filled with starch. The size of tubers is irregular, often cylindrical. Raphid crystals may cause irritation if the raw tuber is eaten. Tubers can be used as propagating material.

Leaves are cordate, with a three-lobed lamina. Their petioles are winged. The leaf construction resembles a dicotyle plant, nevertheless some authors consider the whole genus as a monocotyl (Ayensu 1972; Coursey 1968). For this reason *Dioscorea* occupies botanically a unique place in the plant kingdom.

Inflorescences: rather inconspicuous; male and female flowers are produced on different plants. Male flowers have three sepals and three small petals. The female flower is somewhat larger and produced in the leaf axils. Its ovary has three locules. In spite of the small size they are pollinated by insects. The fruit is a deshiscent capsule and contains winged seeds, which are dispersed by wind.

We studied *D. trifida* plantations on the Caribbean Islands. Before producing ripe tubers, each plant forms a large canopy of intertwined vines and leaves. Natives are accustomed to cut the vines at this stage, which I consider detrimental to good yield, as it is the vines and leaves which produce the photosynthetic reserves to be sent down to the roots and tubers. The custom is to plant *D. trifida* in recently cleared forest land, supported by a few remaining bushes. This form of cultivation of course excludes any mechanical harvesting. Due to its inferior yield the neotropical *I. trifida* is being replaced by other *Dioscorea* species, introduced from Asia and Africa (Coursey 1967).

## Production

Most of the yams produced in the "yam belt" of our globe are consumed locally by the natives within the countries of production. This is especially the case with the neotropic *D. trifida*, which is not entered in international crop statistics like the African and Asiatic yams. Its utility lies in its local use as source for carbohydrate and some protein with the following values: Starch: 25%, crude protein: 2%, fat: 0.3%. The vitamin content is low (vitamin C: 6–10 mg per 100 g of tuber pulp). Calcium, phosphorus and iron are also present.

Without doubt, the different yam cultivars belong to the most tasty tropical tubers, but the yields cannot compete with other tuber crops, and their planting is very laborious, due to the staking. The development period is slow; tubers reach maturity only after one year. Yam needs soil of good fertility, in contrast to mandioka and batata, which prosper even on poor, acid soils. The ultimate solution to boost global yam production would be mechanical harvesting and breeding of bushy types, instead of the present climbing ones.

Storage temperatures between 20° and 25° C are dangerous; on the other hand, storing at 15° C may cause irreversible damage. Compared to other tropical tuber crops, however, yam has longer and better keeping quality. Propagation is comparatively easy, and can be done by stem cuttings, tuber pieces, small entire tubers, or by sowing true seeds.

**Fig. I.8.** *Dioscorea composita,* a Central American wild yam famous for its high content of diosgenin and therefore cultivated (Photo Montaldo)

The most economically important world culture of yams is concentrated in West Africa, mainly in Cameroun and the Ivory Coast, where *D. cayenensis, D. dumetorum* and *D. rotundata* originated from their respective wild forms. On a world scale, approximately 1,800,000 ha are planted, giving a yield of 17,000,000 t. In comparison with this amount, the neotropical yam production with *D. trifida* is only 1%, although admittedly of better quality.

Several Central American *Dioscorea* species appear in world trade, not for human nutrition, but as a supply of naturally derived diosgenin. *D. floribunda* and *D. composita* are perennial climbing vines, which produce 4% sapogenines in their rhizomes (Fig. I.8). World demand for diosgenin, from which a variety of sex hormones and contraceptives ("the pill") are prepared, is continuously increasing. In the mid-1970's, the world demand was 180 t, in the mid-1980's 500 t; but the pharmaceutical industry anticipates for 1990 the need for 3000 t for the production of contraceptive pills.

## DISEASES AND PESTS

In general, cryptogamical diseases are of less importance in yam than in other tropical tuber crops (Kranz et al. 1979).

*Cercospora* causes leafspot. The infection begins as dark brown spots on the leaves and spreads during the growing season to such an extent that the leaf assimilation may be damaged.

*Virus* infection consists mainly of mottle mosaic and so-called shoestring virus, the latter with typical dwarfing and lanceolate deformation of the leaves. The incidence of the two viruses in the fields is not very serious, mostly few scattered plants are attacked. The mode of transmission is still under investigation.

More serious are different insects which damage the tubers, especially scale and meal bugs; they cover the yam roots with thick whitish layers and suck the sap, thus reducing the turgency.

Nematodes, especially *Scutellonema bradys*, live on yam tubers and destroy the meristem, thus inhibiting further sprouting. The damaged tubers easily succumb to fungal or bacterial infections.

SOLANACEAE

## 4. *Solanum tuberosum* L.

*Synonym: S. andigenum* Juz. and Buk.
White potato, papa, patata, pomme de terre,
Kartoffel,
aksu (kechua), choke (aymara), yomsa (chibcha),
poñi (araukana)
2 n = 48

### NAME, ORIGIN AND HISTORY

The potato has developed within less than
500 years from a neglected Indian food plant
of the subtropical highlands to the most important tuber crop of European and North
American agriculture. Even if its origin is
still surrounded by unresolved enigmas, one
thing is very clear: *S. tuberosum* has an exclusive South American origin. There is no
scientific record that potatoes existed before
the Spanish Conquest in North or Middle
America, or even in South Africa, as Correll
(1962) once suspected.
Due to its wide cultivation in prehistoric
times, which embraced nearly 50° latitude
along the Andean mountain range, from
Colombia to Patagonia, uncounted indigenous names existed for this tuber fruit.
The most frequently used we have noted
above; but in fact the Indians have had
many more different denominations for
their local selections, i.e. hundreds of kechua
and aymara words (see Bertonio 1612;
Febres 1765; Vargas 1954; Cardenas 1969
etc.)
The first authentic description of Indian
potato fields by Spanish explorers was given
by Juan Castellanos in his *Historia del
Nuevo Reino de Granada* (1537). He was a
member of a Spanish expedition which penetrated through the valley of Rio Magdalena to the Altiplano. In the vicinity of Soro-
cota (coastal Cordillere), he observed huge
plantations of "trufas" (potatoes) together
with maize and beans. Similar reports were
given by Cieza de Léon (1538) from Po-
payan (Colombia) and Kollao (Peru). The
latter author stated in his *Cronica del Peru*
(1553) that the potato was the staple food of
the "kollas" in Peru. ... "Lo llaman papas,
que es a manera de turmas de tierra, el cual
despues de cocido, queda tan tierno por den-

**Fig. I.9.** This archaic "agritechnology" with a
wooden footplough (chakitakla) still exists in the
Altiplano. It is used for planting frost-resistant
potatoes in altitudes above 3800 m, like "luki" or
"shiri" which are later dried and stored as
"chuño" (Poma de Ayala 1520)

tro como castaña cocida". There is, however, no early report on potato cultivation
from Chile (Fig. I.9).
In contrast to other Indian cultigens (maize,
beans, peanut) we have only scanty ar-
chaeobotanical records of *S. tuberosum*.
Plant remains do not exist. Relics of tubers
which may be potatoes have been recovered
in the Peruvian highland at Chiripa (dated
2400 years B.P.) and on the coast (1000 B.P.
Towle 1961). Vessels and indigenous ceramics, where potatoes are depicted, exist in big
quantities, but they are not very old. Some
Mochica pottery with potato tubers has
been dated between 1900 and 1300 years
B.P. According to Tello, the oldest potato
depiction on ceramics dates back to the Pro-
to-Chimu epoque (second horizon of
Muchik).
Of course, South American Indians had
worked potato fields thousands of years before. Indications for this are stone hoes
which have been excavated near Ayacucho
(southeast of Lima, Peru) and dated as early
as 5000 years B.P. (Chihua period) and later,

from the Cachi period (4800–3700 B.P.). Conservatively, we may consider that for 6000 years Indians have eaten potatoes. I do not, however, believe that the cradle of *S. tuberosum* and its many tetraploid varieties was in the region of Lake Titikaka (in 3900 m alt.); from my personal observations, there is a rather marginal high-mountain climate with short frost-free periods reigns. Much more indicated are Andine valleys, where many small rivers and creeks shed their water down the Cordillere, in eastern as well as in western direction, creating the luxurious vegetation of the Ceja de Montaña.

Discussing Indian potatoes, many agrobiologists placed their main emphasis on the impressive quantity of *S. tuberosum* clones (= andigenum forms) in Bolivia and Peru, but underestimated the local varieties of Ecuador, Colombia and Venezuela. Bukasov (1966) referred already in 1930 to native Colombian potatoes. Brücher (1969, 1970) for the first time described the autochtonous cultivars of Venezuela, which had been overlooked so far. Due to the heavy land erosion and other civilizatory impacts, these interesting varieties at the northernmost end (10°N latitude) of the South American distribution arc of *S. tuberosum* are condemned to early and regrettable extermination. Questioning the small mountain farmers in the Merida, Zulia and Tachira states confirmed the decrease and even extermination of century-old local varieties. The surviving clones have an extremely long growing time (8–9 months) combined with high resistance against late blight (*Phytophtora infestans*), an advantage in the humid mountain climate where they are cultivated, between 2600 and 4000 m altitude. We tested them at our Department of Genetics in Caracas and found, for example, that "arbolona negra" is frost-tolerant ($-2°C$) and "Cucuba" has resistance against *Meloidogyne* eelworms. The first-named cultivar exceeds all known Indian potatoes in corolla diameter (7 cm!). We put especial emphasis on these autochthonous Venezuela mountain potatoes, because the British slave trader John Hawkins took potatoes from the small Venezuelan harbour Santa Fé (hidden in a bay of prov. Sucre) to England. This

historically important fact has been overlooked in the discussion about the origin of the European potato, due to a confusion of names. There are two places called Santa Fé in the then Virreinato de Nueva Granada: Santa Fé de Bogota in the high plateau of the present Republic of Colombia and the small harbour Santa Fé on the north coast of Venezuela.

Hawkins reported (1565) his adventures as slave trader with his ship *Jesus of Lübeck* (!) and that he was highly delighted to barter with the natives of the hidden harbour of Santa Fé fruits and vegetables (essential at this time against scurvy on the long ocean cruise) when he wrote: "... these potatoes be the most delicate roots, that by me eaten, and do far exceed our parseneps and carrots...".

Most probably Hawkins delivered some tubers to the English botanist Gerard, who gave an illustrated description of an extremely large-flowered potato plant in his *Herball* (1597). Such large corollas do not occur in the andigena group of Bolivia and Peru, but they exist in the Cordillere of Venezuela. This coincidence indicates the mountains and the harbour Santa Fé in Venezuela as the source of the first English potatoes.

We have published this theory already in an extensive contribution, in German, on *Migration and Domestication of Solanum tuberosum* (1975), but it must have been overlooked due to language barriers. Our disclosure may change the usual opinion about the introduction and history of the early potato in Europe, which still cites the names of the pirates Drake and Raleigh as the "bringer of the potato". Decades ago Safford (1925) and Salaman (1949) were also in doubt about it. The exact geographical origin of the cultivated potato (*S. tuberosum*), i.e. the place where Indian peasants for the first time transformed wild-growing tuberous *Solanum* into a useful field crop, has been the subject of speculations and heated controversies for decades; it seems that we still do not have the correct answer. The main reason for this uncertainty is the presence of over 150 tuber-forming *Solanum* species, distributed in almost every country of the American continent, from about 40°

northern latitude to about 45° South. Wild potatoes exist in various states of the USA from Colorado to the border with Mexico. They are represented in all Central American countries. Southward from the mountains of Venezuela-Colombia, their quantity increases towards Peru, Bolivia and North Argentina. Several wild potatoes are found even in the hot plains of Paraguay, Brazil and Uruguay. Finally, some grow on the Chilean Pacific coast and into Patagonia.

Bukasov (1980), the Russian authority on *Solanum* systematics, recently established 34 series of wild potatoes: 14 from the Andean Altiplano, 12 from Central and North America, 5 from the Atlantic and 3 from the Pacific coast. North American (Correll 1962; D'Arcy 1972) and British (Hawkes 1979) solanologos claim fewer series, but recognize also that in America more than 120 tuber-forming species exist.

No agreement has been reached about the ancestral species which gave rise to the tetraploid cultivated potato and its geographical zone of origin (Anderson 1979).

Nearly all South American countries (with the exception of Argentina and Brazil) have claimed on one or another occasion to have been "the homeland" of the potato.

During the last century, Mexico counted some outstanding naturalists in its favour. A. von Humboldt in 1801 considered it as the possible region of origin, but was astonished to find no historical data about it from the Aztecs. In 1848 the British botanist Lindley received some Mexican tuber-bearing *Solanum* and expressed the opinion that they might be related to *S. tuberosum* and play a part in its origin. The German solanologos Wittmack and Bitter, after having planted *S. stoloniferum*, which has rather large tubers, maintained at the beginning of this century that this species was the "Stammpflanze der Kartoffel". Correll (1952) also did not a priori exclude *S. demisssum* and *S. stoloniferum* from the phylogeny of *S. tuberosum*. Ugent (1968), after having performed exhaustive fieldwork, definitively rejected any relation between the Mexican *Solanum* species and the cultivated potato.

The whole of Central America must be rejected as a cradle of *S. tuberosum*, notwithstanding the fact that 33 wild potato species exist, some of them with large tubers, which could have stimulated prehistoric Indians to try some selections. Correll (1952) reported, for example, that *S. stoloniferum* has 4-cm-long tubers, *S. verrucosum* develops tubers 6 cm in diameter, similar to *S. demissum*, whose tubers are sometimes collected in the mountain forests by natives as an emergency food; they have, however, never tried to domesticate them.

*Solanum tuberosum* is a relative newcomer to Mexico and seems to have been brought there by Spaniards only in the 16th century. From the Aztec civilization there exists neither any phytomorphic representation of potatoes, such as are common in Peruvian archaeology, nor ethnobotanical or historical evidence from pre-Columbian times.

The first historian of Mexico, Lopez de Gomara (1606), did not include the potato in his list of local food plants.

Having excluded Central America, the most eager contestants for the geographical origin are now *Chile* and *Peru*.

Chile's claim, in fact, is based more on myth than on reality. There are some old writings (Bomari 1774, Molina 1782), which in rather vague form declare that "the potato grows wild in wide extensions of South Chile". This phrase is based on a very superficial observation about an abundant, non-tuber-bearing species (*S. brevidens*), with flowers similar to the common potato. The strongest support for the erroneous assumption is drawn, however, from casual remarks of foreign visitors, Darwin (1835) and Hooker (1844), which influenced A. De Candolle (1855–1884) to declare in an authoritative manner: "The potato grows in Chile spontaneously, and that in a form which persists also in our cultivated potato plants".

Several decades later, Vavilov (1926) strengthened the Chilean nationalistic feelings about "their potatoes", going so far as to declare the Island of Chiloé as gene-centre and cradle of *S. tuberosum* (supported by Bukasov, Juzepzuk, Kameraz etc.).

This was a rather hypothetical and unsubstantiated claim, because the Russians had done no original research on this – at that time inhospitable and nearly inaccessible –

island and never saw wild-growing potatoes there. They had to rely on third-hand information and collections. In 1955, we accompanied Academician Pjotre Zhukowsky personally to South Chile, but for climatical reasons (heavy rains in summer, which exceed 3000 mm) he also could not land on the Island Chiloé.

Decades ago Russian botanists and agrobiologists developed the hypothesis – that can still claim adepts and a considerable number of followers – that *S. tuberosum* and the so-called *"S. andigenum"* arose independently in two distant gene-centres (= Island Chiloé – High Andes of Peru). As wild progenitors for the Southern centre, they proclaimed the taxa *S. fonckii* Phil. *S. leptostigma* Juzep. and *S. molinae* Juzep. as putative real wild-growing species. These are, however, mere tetraploid primitive cultivars, growing in the midst of dozens of well-yielding island potatoes. Some of them have been traced back to original Altiplano varieties. As there was, after the Spanish Conquest, lively commercial contact between the Peruvian harbour Callao and the Island of Chiloé (both belonging at that time to the Virreinato del Peru), not much imagination is needed to realize that the sailors and workers, who were sent down from Callao to cut the big Fitzroya trees for ship-masts, took the native Peruvian potatoes with them. Occasionally they may have even planted them in the beneficial climate of the Chilean Archipelago, where some have survived until our days.

I investigated the Island of Chiloé in the course of two exploration trips. The first transect of the dense forest and bamboo thickets of the west coast was especially difficult (Brücher 1960, 1979). In spite of my disclosure that Chiloé has no wild potato which could be claimed as the wild progenitor of the cultivated potato, and that it is not at all a gene-centre of other cultivated plants (*Fragaria, Madia, Bromus* etc.) as the Russians asserted, such erroneous statements persist in the international literature. "Et voilà, justement comme on écrit histoire" (Voltaire).

*Peru* has many arguments in its favour: the impressive quantity of wild-growing tuberous *Solanum* (Ochoa 1989), the agricultural abilities of its aboriginal human races, the agro-ecological diversity of this country and the neighbouring Bolivia, and finally the pressing need for food of a huge population in the high Andine valleys and the Altiplano. With respect to potatoes, Peru still maintains an impressive clonal heterogeneity of different *Solanum* species, selected in ancient times for their culinary characteristics and tolerance to adverse field conditions. As we explained, Peru-Bolivia leads in the possession of edible potatoes of different genomic structure, i.e. diploids, triploids, tetraploids and pentaploids. For this reason Peruvians consider the potato as their "planta nacional". Strangely enough, however, we could find no decisive proof that the early European potato came from Peru. Nevertheless, we believe that "pappas peruanum" as they were denominated by Clusius (1588), whose hand-coloured drawing of them is kept in a private Museum in Anvers, arrived early in Europe. In 1970 I had the chance to examine this valuable picture (Brücher 1975) and came to the conclusion that details of its morphology coincide with the usual characteristics of an Andine highland potato.

At this time, the yields were low; according to Clusius, the plants were considered in Central Europe as a botanical rarity and not at all as a food crop. The exsiccate of a well-preserved potato plant in the *Herbarium* of C. Bauhin (1596) is very enlightening. This again is a typical Andine biotype with rather "open leaf" position and many interstitial secondary leaflets. Such Andine highland potatoes were introduced with Spanish ship cargoes to the Iberian Peninsula, but reliable dates are rare and inexact. A potato transport occurred in 1565, when a case with tubers was sent from Cuzco to King Philip of Spain. In the year 1570 potato plantations existed on a reduced scale around Seville. Fortunately the "inventory of purchases" of the Hospital of Seville from 1573 has not been lost; this registers the aquisition of 25 kg potatoes = patatas (?) as food for the sick. The opinion that potatoes have a healing effect must have been widespread at this time, because the same King Philip II who received the Peruvian potatoes from Cuzco sent some tubers to the then sick pope

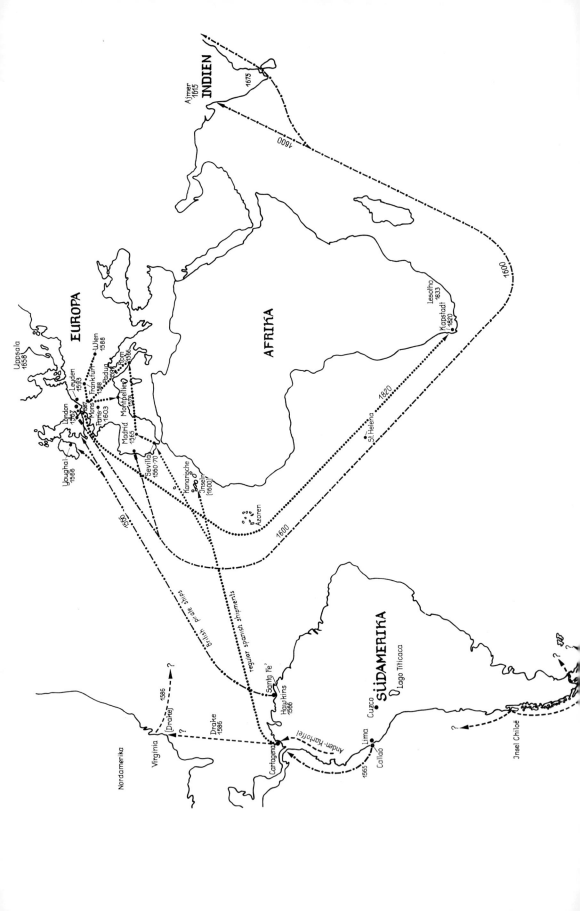

INDIEN

Ajmer
1615

1675

1600

EUROPA

Uppsala
1658

London
1567
Leiden
1593
Mons
Frankfurt
1588
Padua
Wien
1588
Paris
1603
Dorn
1566
Youghal
1566
Montpellier
1474
Madrid
1565
Sevilla
1560-70
Kanarische
Inseln
(1600)

Azoren

AFRIKA

St. Helena

Kapstadt
1820
Lesotho
1833

1820

1600

1600

1450

British pirate ships

regular spanish shipments

Nordamerika

Virginia
1586
?

(Drake)
?

Drake
1586
?

Cartagena

Santa Fe'

Hawkins
1566

Anden-Kartoffel

SÜDAMERIKA

Lago Titicaca

Cuzco

Lima
1565
Callao

Insel Chiloé

Pius IV, as was reported by Mellado in his encyclopedia.

Once the Europeans had discovered the nutritional value of *Solanum tuberosum* and selected clones with fair yields under their own photoperiodical and climatical conditions, this New World crop diffused astonishingly quickly. By the beginning of the 17th century, its cultivation had spread in Spain, Italy, and the different kingdoms of Germany. The population of France and Great Britain was still reluctant, but Ireland accepted the "spud" eagerly as a nourishing fieldcrop, and thanks to potatoes, the Irish people survived a severe famine in 1662. In 1697 Russia received its first introduction by order of Tsar Peter I, and developed its own varieties. Three hundred years later the USSR were planting 8,000,000 ha potatoes. India must have received its first potato shipments very early in the 17th century, because potatoes appeared in the year 1615 at a banquet given by the Prince of Ajmer (Fig. I.10).

The United States probably had its first shipment in 1621 from the Bermuda Islands. Irish immigrants took tubers with them when they settled in Virginia, in 1720.

MORPHOLOGY

From the tetraploid *Solanum tuberosum* so many (more than 2000 !) distinct ancient landraces and modern varieties of highly selected clones and cultivars exist that it is nearly impossible to give a general valid morphological description. It is customary that slight variations in habit, branching type, leaf dissection, flower and tuber colour are used by potato breeders to differentiate their varieties.

A tetraploid potato plant is a herbaceous annual, with stems 30–120 cm high, which may be upright or spreading on the ground; its internodes are hollow. The leaves are imparipinnate, with a varying quantity of leaflets and secondary "interjected leaflets". Very important for diagnosis is the angle of leaf insertion on the stem. Leaves may be dropping or rigid, and have either a short distance between the leaflets or large intervals. The texture of the leaves may be smooth or hard to the touch, but can be glossy or hairy.

The inflorescences in most of the ancient landraces are many-flowered, often beautifully coloured with large showy flowers. Modern cultivars blossom rarely, some even suffer from an early abscission of flower buds. The difference between compound flowers, as in many primitive clones, and simple inflorescences depends on the subdivision of the peduncle, which may be dichotomously branched or forked.

The corolla is pentagonal or rotate. Its colours have very wide variation: from white, pink, red, violet-red, dark violet to intensive blue-purple. Its diameter varies from 2 cm to 7 cm (in the Venezuelan landrace "Arbolona negra").

The androecium consists of five stamens, their filaments being joined at the bases. The anthers are often malformed and without fertile pollen in some modern cultivars. Viable pollen is usually produced in deeply orange-coloured anthers, whilst pale, whitish-yellow anthers indicate male sterility. The style of the gynoecium is of varying length, and bears a bi-lobed stigma. In some varieties the style may protrude several mm beyond the top of the anther cone.

Fruits are profusely developed in Indian varieties and in some European cultivars (e.g. Capella, Kerpondy, Industrie). The tubers are anatomically enlarged stems – not roots – and develop as a sign of maturity in the plant at the end of stolons. Primitive landraces have long-running stolons with dispersed and deep-eyed tubers of varying shape and size. In contrast, modern selection work is directed to short stolons, medium-sized round tubers with tough skin, suitable for mechanical harvesting.

The Indians exhibited astonishing versatility to select quite different biotypes. They created hundreds of cultivars with the most fantastic skin-colour patterns, shapes and pulp characters, some of a nutritional value

---

◄ **Fig. I.10.** Paths of introduction of the potato (*Solanum tuberosum* L.)

**Fig. I.11.** Tuber from a selected commercial variety (Kennebec) in comparison with an Argentinian wild potato *(Solanum ruiz-lealii* Brücher)

which considerably exceeds the European varieties (Bavyko 1982) (Fig. I.11).

The potato is the unique case among all the other American crop plants, where a whole series of polyploid genomes is of use for man. Since time immemorial, the South American Indians have selected (and still use today) diploid, triploid, tetraploid and pentaploid tuber-bearing *Solanums*. Chibcha, Kechua, Aymara, and Diagita tribes that have inhabited the Andine valleys for more than 6000 years, testified by the domestication of this polyploid *Solanum* to their creative ability, without – of course – knowing the cytogenetical background of this series, which has only recently been elucidated.

The phenomenon of polyploidy in Andean potatoes is extraordinarily striking when one visits indigenous market places in Bolivia and Peru. The piles of potato of the vendors, mostly women who bring their products in several days' travel from the Altiplano, contain – often neatly separated – diploid, triploid, tetraploid (the majority) and sometimes even pentaploid clones. As I know the indigenous words, in kechua or aymara, for landraces with different ploidy (of course previously determined cytologically), my visiting guests, during potato-collecting journeys, observed with the most astonishment that the vendors handed them in each case a separate series of polyploids, when I asked for diploid, triploid, and tetraploid tubers. To be sure, this trick was not so easy with plants in potato fields, with their immense variability and admixture of local strains and weed potatoes. A selected example for this genetical variability of Indian potatoes is presented in Fig. I.14.

**Fig. I.12.** *Solanum vernei* Bitter & Wittmack from Argentina and Bolivia. A possible precursor of primitive cultivated diploid Indian potato

### Diploid Potatoes

It seems logical that potato domestication began with the initial search for wild-growing diploids which excelled the others in tuber size, high yield and acceptable taste (low solanin alkaloid content, e.g. in *S. vernei*). As examples have been named by Hawkes (1958) *S. sparsipilum* Bitt. and *S. leptophyes* Bitt., whilst Brücher proposed another typical wild species *S. vernei* Bitt. and Wittm., all native wildpotatoes in the border region of Bolivia – Northargentina (Brücher 1975). No final decision has been reached so far.

Once taken over in cultivation, the offspring of these wild species naturally showed significant differentiation, according to different local environments. Some may have developed to high-Andine cultivars, others were selected for subtropical humid-warm growing conditions, with tubers lacking dormancy.

### S. *ajanhuiri* Juz. and Buk.

Its aymara names are "yari" and "ajahuiri". This species produces medium-sized, elongated tubers with good palatability, and is still an important food plant for inhabitants at high altitudes, especially between 3800 and 4200 m around Lake Titicaca. It tolerates night frosts of minus 5°. Some varieties have a strong anthocyan pigmentation, both in flesh and skin, of the tubers, some possess completely purple-coloured pulp, which has a bitter taste, often used for the preparation of "chuño" (Ross and Rowe 1969).

### S. *goniocalyx* Juz. and Buk.

This diploid cultivar displays astonishing adaptation to different agro-ecological environments. It spreads from its original Andine dispersion centre (Peru-Bolivia) as far as the Island of Chiloé, where it thrives well in humid-temperate maritime climate. We discovered it near the settlement Chacao (Collection Nr. 1173) during our International Expedition to the South American gene centres 1958 and described the find 5 years later after comparative cultivation together with Peruvian clones of "papa amarilla" (Brücher 1963). As the local name "papa mantequilla" indicates, the pulp has an intense orange-yellow colour. The flowers are white with a ribbed calyx base.

### S. *phureja* Juz. and Buk.

The most striking feature of this edible diploid potato is that its tubers lack dormancy. They begin sprouting a short time after harvest. The ecological range of "phureja" (a local name of Chibcha origin) are the humid-temperate valleys, below 2700 m in Colombia/Venezuela. In Bolivia the "yungas" and in Peru, the "Cejas de Montana".

Its quick growing cycle (ripe in 3 months) and ability to sprout immediately after harvesting allows three yields in one year under adequate environment conditions. S. *phureja* exists in many different local clones, some of which produce rather large tubers. Leaves are glossy green with rare pubescence. The calyx has long irregular lobes. Tubers are often fusiform.

### S. *stenotomum* Juz. and Buk.

This is perhaps the oldest of all cultivated diploid potatoes. It has been selected for the long dormancy and good yield of its tubers and also high tolerance to low temperatures. Tubers are often elongated, even snake-like, twisted, with many "eyes". It seems that the Amerindians of prehistoric times were motivated in their breeding and selecting work not merely for food, but with pleasure in colour variation and imaginative, striking forms and shapes. According to Vargas (1954) native farmers of the Altiplano developed several hundred local clones.

## Triploid Potatoes

### S. *chaucha* Juz. and Buk.

These triploid hybrid clones exist in many genetic combinations in the Andine valleys from Ecuador and Peru to Bolivia. They seem to have originated from crossings between S. *tuberosum* and S. *stenotomum* at different places and from distinct ancestors. Natives often plant the tubers (probably without knowing the difference) together with tetraploid potatoes of the *andigenum* group. They like "chaucha" for its quick-growing, smooth skin, often red, and shallow eyes. Tubers are often long-ovoid with good cooking quality and taste.

### S. *juzepzukii* Buk.

The plants have a rosette habit and small rotate corollas (similar to the tetraploid wild species S. *acaule*, which always grows at very high 3800–4500 m altitudes), and produce small bitter-tasting tubers with white flesh. Hawkes (1962) showed that artificial crossings between S. *acaule* and the diploid S. *stenotonum* produce hybrids rather similar to the cultivated S. *juzepzukii*, which inherited its high frost resistance from the wild potato S. *acaule*.

This triploid cultivar does not produce fruits. Its vernacular name is "rukki, choque-pitu, ayo", sometimes also "paya", or "luki", similar to the pentaploid species *S. curtilobum*. We observed some decades ago that natives of North Argentina (prov. Jujuy and Salta) planted this frost-resistant potato in the mountains, but its area is diminishing, whilst in Bolivia and Peru it persists in wider extension (Ochoa 1982). The natives of the Altiplano use the tubers for preparing dehydrated conserves, like "tunta, chuñu, moraya". The tubers are in general flattened, often with anthozyan-sprinkled pulp, deep eyes and violet skin.

### Tetraploid Potatoes

*Solanum tuberosum* (*tuberosum* group, *andigena* group)
To this group belong many hundred local "varieties" or Indian clones. They have been selected from auto-tetraploid ancestors. More details are discussed on p. 24–27.

### Pentaploid Potatoes

*S. curtilobum* Juz. and Buk.
This is a complex hybrid, where the short lobes (= *curtilobum*) indicate the heredity of *S. acaule*. The plants do not form rosettes, instead they have an elevated habit. The conclusion of Hawkes (1962) that *S. tuberosum* crossed with an unreduced (2 n = 36) gamete of *S. juzepzukii* produced this species, seems right. The hybrids are frost-tolerant and sometimes produce berries with aborted seeds. Its range reaches from Central Peru, Bolivia to North Argentina. The natives of the Altiplano (Peru-Bolivia) call it "luki" or "shiri" and use it mainly for production of dehydrated tubers ("tunta"). The tubers generally have a violet skin with slightly coloured eyes. Flowers and stem contain much anthocyan.

At several places in North Argentina we collected this pentaploid potato in small potato fields above 3000 m altitude, called by the natives there "monde luki" or "sipancachi".

Due to the enormous geographical extension of potato cultivation by South American natives (from 10°N to 45°S), one should not be astonished to encounter a wide amplitude of genetic segregation or morphological and physiological diversification. Russian authors, following their (now obsolete) gene-centre hypothesis, claiming a sep-

arate centre of origin at the Island of Chiloé and another in the Peruvian Altiplano, at the same time created two opposite potato groups. Juzepzuk and Bukasov suggested that the "Chilean group" was specifically distinguishable from the "Andean group". They claimed that South Chile (the Island of Chiloé especially) is the place of origin of the European potato and that *S. tuberosum* evolved there from genuine wild ancestors. The Russian agrobiologists did not, however, realize that the supposed ancestral potato species from Chiloé in fact had an Andine origin (Hawkes 1958, Brücher 1960, Ochoa 1989). In the South American gene pool of *Solanum* many biotypes exist which have a neutral reaction to daylength. They flower and tuberize under both short and long day conditions. Obviously the genes which govern photoperiodical reactions have never been decisively selected, nor have species barriers been created by daylength differences.

In opposition to this, we have for decades defended a different phylogenetical concept. We assert that all edible tetraploid potatoes belong taxonomically to the collective species *S. tuberosum* L. and have phylogenetically a common ancestry. There is no reason to establish another potato taxon, as Russian botanists did some decades ago, when they introduced the term "*S. andigenum*". As they could not present valid morphological or cytogenetical arguments in its favour, they claimed historical, photoperiodical and geographical reasons for a systematic separation of "*S. andigenum* Juz. and Buk." from *S. tuberosum* Linné.

Russian authors have placed much emphasis on the daylength factor in differentiating cultivars of *Solanum*, along the South American Andes. There is no doubt that the light factor has an influence on tuberization and even on tuber production but this cannot be used for taxonomical purposes. As support for their hypothesis that the European potato originated on the Islands of the Chilotéan Archipelago, situated between 41° and 43° lat.S, they indicated that this is the same daylength (15–17 h in the summer) as in Central Europe, whilst in the Peruvian Andes a rather short photoperiod (12–13 h) reigns.

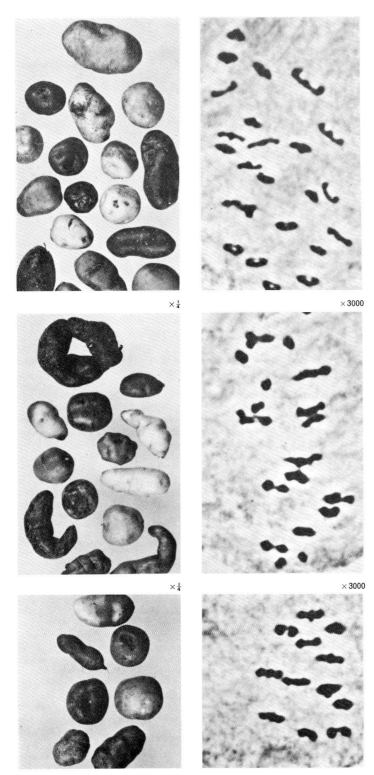

**Fig. I.13.** Chromosome differences in Indian potatoes (Dodds 1966). *Above* tetraploids; *middle* triploids; *below* diploids

Decades ago it was shown from many physiological experiments (e.g. Salaman 1949, Simmonds 1969, Plaisted 1972) that daylength reactions are an absolutely unsuitable tool for species separation, because it is rather easy to select from progenies of typical shortday plants after a few generations of inbreeding true longday potatoes and vice versa. Simmonds (1969), showed with a mass selection experiment that it is possible to re-create the "*tuberosum* group" from longday forms of the "*andigena* group".

Simmonds interpreted the results of his experiments as convincing proof for the identity of the taxa *S. tuberosum* and *S. andigenum*: "The experiment offers powerful support for Salaman's contention that the *tuberosum* potatoes evolved from *andigena* by selections for local adaptation in Europe."

In addition, there are no biochemical reasons to separate Andine cultivars from the "*tuberosum* group", as Zwartz demonstrated with electrophoretic experiments. No physiological difference in their protein components exists. "Although their appearance was indeed quite different, the protein distribution was not ... they showed the same migration velocities as the Dutch varieties" (Zwartz 1967).

Quite correctly, Dodds already in 1962 declared that the epithet "*andigenum*" or "*andigena* group" is superfluous and all tetraploid potatoes should be included in the historically well-defined term *S. tuberosum* Linné. It is perhaps possible to differentiate them as geographical "Rassenkreise" and designate the cultivars from the Andes as "formas andigenas" and those of Chiloé as "formas chileanum", a term quite acceptable and used by Hawkes formerly.

Although Hawkes strongly opposed the Russian hypothesis at the beginning, declaring quite rightly that the cultivars present in Chiloé farmlands were introduced from the Andes and that there is no wild ancestor on the Island of Chiloé, he apparently changed his opinion later when he created the terms "subspecies *S. andigena* (Juz. and Buk.) Hawkes" and "subspecies *tuberosum*", thus dissolving a well-established botanical unit. Mansfeld, Danert and Rothacker never accepted the arbitrary change from "*andi-*

*genum*" to "*andigena*" (see Mansfeld 1986) (Fig. I.13).

CONSUMPTION

**Table 3.** World production of potatoes between 1970 and 1980 (in 1000 ha and 1000 t) F.A.O.

| World area | | | | |
|---|---|---|---|---|
| 1970 | 1980 | 1983 | 1984 | 1985 |
| 21,000 | 21,006 | 20,523 | 20,376 | 20,170 |
| World production | | | | |
| 1970 | 1980 | 1983 | 1984 | 1985 |
| 275,758 | 291,356 | 287,565 | 311,293 | 299,132 |
| Production in South America | | | | |
| 1970 | 1980 | 1983 | 1984 | 1985 |
| 14,000 | 14,923 | 15,146 | 16,448 | 18,331 |

The actual world production (Table 3) has been estimated at 300 mio t, which represents an annual value of 106 billion US, a sum which considerably outstrips the total of silver and gold treasures extracted during the Spanish Conquest of South America, the homeland of the potato. When the conquerers and soldiers of Pizarro reached the Andean plateau in 1525 and met this Indian tuber crop for the first time they called it disrespectfully "turmas de tierra" and did not grasp the high nutritive value of a crop which 200 years later was to save several European countries from famine and starvation, and has now won a leading place in World nutrition, ranking fourth after wheat, maize and rice (Fig. I.14).

One hundred and thirty different countries of the world are already potato producers and cover a total of 20 mio ha with this tuber crop. Particular mention is deserved of two "newcomers" in this trade: India and China.

China alone plants approximately 4 mio ha with *Solanum tuberosum*, which has radically changed the eating habits of the world's most populous nation. The Asiatic potato crop has now reached an all-time record. In comparison with this, Western Europe, the former traditional potato region, shows a continuous decrease in acreage. Germany's 720,000 ha per year appear rather modest, but the yields per unit reached with 35 t/ha

**Fig. I.14.** Primitive potatoes from a Peruvian marketplace; some with different polyploidy

an impressive figure. Countries of the social-istic sphere produce in general only 10–15 t/ ha.

Potatoes have acquired an important position in the human diet, due to their rapid expansion in subtropical and tropical regions. The International Potato Centre (CIP) in Peru, according to its director Dr. Sawyer, considers that hundreds of millions of people in developing countries of Africa and Asia live in areas, where "tropical potato varieties" (still to be created on a major scale!) may grow well. However, several impediments exist so far; for example, potatoes are considered there as an expensive luxury vegetable, they are not yet accepted as a source of daily food, they have not been sufficiently popular, and many trials with tubers imported from Europe (selected there under completely different environments) have failed. With the intention of changing this situation, the CIP distributes "true seeds", sexually produced from flowers of hybrids, which are adapted to tropical conditions. The task is not easy, because *S.*

*tuberosum* has a genetical constitution which was selected in temperate climate with low humidity.

With respect to the nutritive value of *S. tuberosum* we would like to rectify some erroneous concepts (Leung and Flores 1961, Woolfe 1987).

Undisputedly, its main value is based in the calories produced by its carbohydrates, but clinical experiments have shown that also the protein content of European potato cultivars is an appreciable source of amino acids. Potato protein is nutritionally superior to cereal proteins. In addition, the potato is a valuable source of several important vitamins and minerals. We make special mention of certain Andine potatoes which contain more than 5% protein and some even having 15% protein on a dry weight basis, according to analysis from the International Potato Centre (CIP) in Peru. The nitrogenous constituents of 2 kg of boiled potatoes can cover the daily minimum requirements of humans, calculated in 6 g of nitrogen, and supply sufficient methionine and cystine.

The "biological value" of potato protein (i.e. its real absorption by the human body) surpasses that of beans ($= 50$) and is similar to that of soybean (with $72-73$), exceeded only by egg protein ($= 100$).

Feeding experiments with human volunteers over several weeks disclosed that 100 g of potato dry matter provide the human body with 300 calories. In general one assumes that 3000 calories are needed daily, so $2-3$ kg cooked potatoes would be sufficient to sustain an adult person. The potato is, however, also a source of vitamins (especially ascorbic acid), as was shown in former times by the many sailors living on potatoes during ocean cruises. According to Woolfe (1987), the average nutritional values of raw and cooked samples are given in Table 4.

**Table 4.** Potato: average nutritional values (Woolfe 1987)

| g/100 g | | | mg/100 g | | | | |
|---|---|---|---|---|---|---|---|
| | Water | Carbo-hydr. | Total N | K | Ca | Fe | P |
| Boiled in skin | 78 | 16 | 0.33 | 500 | 15 | 1 | 50 |

| mg/100 g Vitamins | | | | | |
|---|---|---|---|---|---|
| A | B$_1$ | Ribo-flavin | Nict. Acid | | C$^3$ |
| 0.01 | 0.1 | 0.02 | 0.5 | | 22 |

DISEASES, PESTS AND RESISTANCE BREEDING

Of all our root and tuber crops, *S. tuberosum* is the species most infected by fungus, bacteria, virus and various pernicious parasites. The gradual increase of these calamities can be traced back several hundred years and began shortly after the arrival of this New World crop in Western Europe.

Fungus

Blight (*Phytophthora infestans*)
This is "enemy No. 1" of all fungus diseases which attack the potato. The losses of yield

caused by this fungus can be catastrophic. As there are no varieties with all-round genetic resistance, spraying with suitable fungicides is still the best way to protect potato plantations against *Phytophthora*. The fungus has several reproductive mechanisms. It can produce sporangia asexually, but not uncommonly sexual mating also occurs between different biotypes, resulting in the formation of oospores, which on germination also make sporangia. Such sexually as well as asexually produced sporangia produce motile zoospores after germination. These different reproductive mechanisms are the cause of new pathogenic recombinants, called races. Such strains exist in dozens, which makes a defense strategy against late blight very difficult. The frequency with which new races of *P. infestans* have overcome certain promising potato clones has discouraged whole breeding programmes (see Schick and Klinkowski 1961).

Although breeding for resistance to blight began early in Europe, no long-lasting immunity has been achieved.

Planned immunity breeding by crossing and selection of tolerant potato seedlings was already performed in Scotland by Patterson (1856), who created his successful variety Patterson's Victoria. Nichol (1876), with his var. Champion, obtained a certain tolerance to the devastating *Phytophthora* of that period. Also Sutton (1879) produced hybrids between cultivars and South American wild potatoes with the aim of combining disease resistance with good yield. These early experiments were stimulated by the "blight pandemic" in Ireland of the years 1847/48 with its well-known consequences of millions of people starving, dying and emigrating (Burton 1966).

Wart, Cancer (*Synchitrium endobioticum*)
This infection is seldom visible on the green parts of the plant, but the tubers are often heavily attacked, and produce tumour-like excrecencies, sometimes the size of a fist. The sporangia can survive for more than a decade in the soil. All European countries have developed legislation with respect to wart disease and the planting of susceptible varieties is forbidden. Resistance is inherited as dominant by (probably) various genes.

As most of the present cultivated varieties have wart immunity, the existence of resistance genes in wild potatoes is of only secondary interest. Various South American species could, however, gain in importance in the case of the appearance of new virulent wart biotypes. *S. acaule, S. chacoense, S. commersonii, S. maglia, S. microdontum* and *S. vernei* have been mentioned in the literature as bearers of some degree of resistance.

## Stem Canker, Black Scurf (*Rhizoctonia solani*)

This fungus disease has become a serious menace over the last few years in many potato-producing subtropical countries. The symptoms are visible in tubers, stems and leaves. The infected plants develop a "leafroll" very similar to a virus infection. In both cases the free flow of sap in the vessels is interrupted, which provokes deficiencies in the normal leaf form and finally premature death. The affected shoots and stolons develop brown lesions, which on extension dramatically reduce tuber formation. Small greenish tubers often appear close to the surface of the soil or at the stem basis, which become covered with a whitish felt-like mycel. The tubers are covered with numerous black sclerotia. These bodies consist of a mass of fungus threads, which, after planting, seriously infect the soils and the plants of the next generation. The "black scurf" is difficult to combat because none of the industrial chemical products promises effective control, apart from their high costs. Neither has the American gene pool of *Solanum* contributed genetical immunity to *R. solani*, which unfortunately has a wide host range.

## Powdery Scab (*Spongospora subterranea*)

This tuber disease can be traced back to the cool, humid mountain regions of the South American Andes, where it is not uncommon in indigenous potato fields. It seems therefore quite natural that local *Solanum* species possess some resistance against powdery scab, as has been reported by authors of socialistic countries (see Rothacker 1961).

## Early Blight (*Alternaria solani*)

Early blight reacts quite differently on potato plants than the late blight disease. It needs a combination of high temperatures and high humidity to develop. The disease causes a brown-black necrosis on stems, and on leaves typical round, dark necrotic spots, which are surrounded by concentric rings. When tubers are infected, 1–2 cm sunken spots appear, which develop brown corky dry rot. In many screening experiments Argentinian biotypes of *S. chacoense* showed high resistance.

## Verticillium Wilt (*Verticillium albo-atrum*)

This disease is soil-borne and prospers during warm weather. The symptoms are dying and yellowing of the leaves from the base upward, until a cluster of green leaves remains at the top. Then the whole plant succumbs. Tubers from affected plants develop a gradual discoloration of the xylem. Ochoa in Peruvian cultivated potatoes, discovered various resistant varieties. Also the Argentinian wild potatoes *S. vernei, maglia, kurtzianum* and *S. chacoense* have been recorded as resistant to *Verticillium* wilt.

## Scab, (*Streptomyces scabies*), (Synonym: *Actinomyces scabies*)

Infected plants show no symptoms but the tubers are covered with rough warts which sometimes penetrate several mm into the periderm. The agent is an actinomycete which uses the lenticels as medium of infections. Light alkaline soils are favourable for infections.

Promising resistance was found in Argentinian tuber-bearing *Solanum* such as *S. chacoense, S. commersonii, S. microdontum* and *S. vernei* (Ross 1960).

## Bacteria

### Bacterial Wilt (*Pseudomonas solanacearum*)

Decade-long testing of *S. tuberosum* clones (nearly 9000 clones since 1947) gave meagre results and only moderate levels of tolerance were found. An improvement came when Rowe and Sequeira introduced a collection of diploid *S. phureja* from Colombia. The authors observed that resistance in hybrids between *phureja* clones with diploid and tetraploid *tuberosum* forms was rather simply inherited. Actually advanced crossings exist, which are resistant to bacterial wilt under field conditions, according to information from the CIP. These are promising results for the future extension of potatoes to humid tropical lowlands. Unfortunately, in such tropical soils the bacterial infection is enhanced by (*Meloidogyne* spp.) root-knot nematodes, which are nearly omnipresent there.

### Ring Rot (*Corynebacterium sepedonicum*) (Synonym: *Phytomonas sepedonicum*)

This is a highly infectious tuber disease which is easily spread by tools and mechanical equipment. Plants are often infected without showing symptoms. Usually in the advanced stage the infection consists of wilting stems and leaves. Heavily affected tubers are not only worthless for marketing, but act as vectors for further dissemination of the ring rot disease. Promising genetical resistance has been found in several South American wild potatoes, for example, *S. acaule* and *S. megistacrolobum* from the Argentinian mountains, and *S. vernei* from the humid forest regions.

### Soft Rot (*Erwinia atroseptica*)

It has long been known that treatment with antiobiotic substances (Streptomycin sulfate or Terramycin) give good control of bacterial decay in harvested tubers; genetic resistance would, naturally, be preferable.

### Black Leg (*Pectobacterium carotovorum*)

The stems of affected plants become black, combined with a continuous discoloration down to the tubers. In storage such tubers suffer wet rot, which spreads to healthy ones, causing severe losses. Infected plants have a stunted appearance and die after yellowing. The disease occurs especially under cool wet climate conditions.

## Viruses

### Virus A (PVA)

Transmission occurs by aphids and mechanical infection. The symptoms are difficult to detect and consist of a slight mosaic of the lamina, but in combination with other viruses the leaves can suffer severe damage.

The Mexican wild potatoes *S. demissum* and *S. stoloniferum* possess genes for immunity, also the Argentinian species *S. brevidens, S. chacoense, S. kurtzianum* and *S. simplicifoli-*

*um*(= *microdontum*) have valuable resistance (Ross and Rowe 1965).

### Virus Leafroll (PLRV)

This is probably the most widespread and serious potato disease. Yields may be reduced to one-half of normal in affected plants. Transmission occurs by aphids, mainly by the peach aphid (*Myzus persicae*). The symptoms, as the name implies, are a strong leafrolling, combined with rigidity of the lamina and dwarfed stems. The tubers in certain varieties suffer from phloem necrosis, which gives them a bad appearance if they are sold.

Resistance to leafroll virus is inherited in a polygenic way. Certain South American wild potatoes, when crossed with *S. tuberosum*, gave offspring with tolerance to PLRV virus, but it seems that no new field-immune potato variety with genetic resistance has been released so far.

### Virus M (PVM)

The symptoms are similar to a slight interveinal mosaic with some leafrolling. In combination with another virus, PVM may cause considerable losses. The disease was mainly known from Europe and the USA, but recently it spread with Polish potato shipments to Argentina, where in 1986 the seed production region of Balcarce suffered heavy impact, so Virus M has been declared by the Government "plaga nacional". Transmission occurs by aphids and infected sap.

### Virus S (PVS)

This virus can only be transmitted mechanically. The symptoms are a faint mottle and the formation of many small tubers.

Many North American high-yielding varieties, such as Katahdin, Pontiac and Sebago, are 100% latent-infected with Virus S. Genetical resistance in wild potatoes is not known so far.

### Virus Y (PVY)

Also called "rugose mosaic" for its severe disturbance of leaf growth. Plants which emerge from infected tubers are dwarfed and early leaf-drop occurs with premature death of the plant (Herold 1967).

Transmission occurs by aphids and also mechanically by infectious sap. Typical symptoms: leaves are mottled, with discoloured areas. The veins on the underside show necrotic areas, resembling black lines, and heavy distortion of the surrounding lamina. A valuable resistance for improvement of cultivars against virus Y was discovered decades ago in the North Argentine mountain species *S. vernei* and *S. microdontum* (forma *gigantophyllum*). Later it was shown that also *S. chacoense* of the Argentinian plains possess genes for immunity against virus Y. Several races of PVY have been established.

The very pernicious tobacco necrotic strain of Virus Y ( $= Y^n$ ) is endemic in North Argentina (Valle Humahuaca, Maimara), as has been described by Brücher (1969); this virus was apparently introduced from there with tubers of a potato-collecting expedition to Great Britain in the 1940's (Silberschmidt 1961). The consequences of the inadvertent spread of this new, aggressive virus in Europe were disastrous. Bode (1958) calculated the damage in tobacco plantations alone at many millions and called "veinal necrosis" the most important yield-depressing factor in Europe for tobacco and potato. Most surprisingly, in the 1970's, a German-Argentinian potato gene pool was established in this same highly virus-infected valley of Maimara, and potatoes have been distributed from there to many places.

Potato Virus X (PVX)

This virus is not insect-transmitted, only mechanically by contact or sap. Infected plants are often "symptomless", or the infection is masked under high light intensity. Older varieties are 100% infected (e.g. White Rose, Majestic, Triumph, Arran Consul). The foliage symptoms are a mild mosaic and sometimes necrotic spots on the lamina. Even if this virus is generally considered as rather harmless, in combination with other viruses it can reduce the yield considerably.

Argentinian wild potatoes proved to be a valuable source of genetic resistance against virus X: *S. acaule*, also *S. simplicifolium* = *S. microdontum*, *S. kurtzianum*, *S. chacoense* and *S. maglia* (often in a mixture of susceptible and resistant clones). By crossing with *S. acaule*, Ross produced a new potato variety with good virus-X resistance. Immunity of *S. acaule* ($2 n = 48$) has disomic inheritance.

Virus Effect or "Degeneration"?

Much confusion existed not long ago in Latin America about the so-called "degeneracion de la papa". This term was ascribed to senescence of clones and wrongly considered by certain plant physiologists as a curable phenomenon in *S. tuberosum*, as was claimed vehemently by the then president of the Soviet Academy of Agriculture, Lyssenko, who discarded genes and viruses.

The reality is, however, quite different. Without entering into the heated debate, we repeat only what leading plant pathologists had disclosed at the time, namely that the so-called degeneration of clones was in fact caused by increasing virus infection of the tissues. By rejecting this, the Soviet Union lost a considerable number of valuable *Solanum tuberosum* selections and old, high-yielding varieties. Even when he was removed from power, Lyssenko's writings induced a group of Argentinian agrobiologists to propagate the "cure of degeneration" by different temperatures and to reject the virotic cause. Unnecessary to say, they could not "cure" the degeneration with their widely propagated methods, which are detrimental to the scientific efforts to produce virus-free seed potatoes.

A virus-free *Solanum* clone which is maintained in natural isolation and protected from possible sources (vectors and virus) of infection should not "degenerate". Fortunately, there exist many reports which confirm this. Davidson (1936) showed this with old Irish varieties, introduced before 1860 and grown in isolated places in Ireland, which are still vigorous and good yielders. Herold (1967) performed virus tests with vigorous "Lesotho potatoes" which were introduced to South Africa in the year 1820 by Casalis from Holland. Several clones, which we collected in the mountain fields of the Basuto negroes, and which had been raised there under rather primitive multiplication maintain an astonishing vigour and health, with very low virus infection after more than 150 generations (Brücher 1975; Harris 1978).

PARASITES

Golden Eelworm (*Globodera rostochiensis*)
(Synonym: *Heterodera rostochiensis* Wollen-
weber)
Root eelworms cause very serious losses in
Europe and Asia, which can exceed 50% of
the yield. Dry years and sandy soils provide
optimal living conditions. The United States
is still free from the invasion of this originally South American parasite, thanks to a very
strict quarantine in all harbours and air-
ports.
A soil once infected may contain viable cysts
of *Globodera* (*Heterodera*) for several years,
up to a decade, even in the absence of further potato crops.
The existence of eelworms is indicated by
dwarfed plants and yellowing patches in affected potato fields (Rothacker 1961).
Genetical resistance has been provided from
our Argentinian wild potato collections
EBS 250 and EBS 510 of *S. leptophyes*
(= then *S. famatinae*), *S. vernei* (polygenic
inherited) and *S. kurtzianum* (with genetic
resistance to several *Heterodera* biotypes,
Hujsman and Lamberts 1972).

Rootknot Eelworm (*Meloidogyne* spp.)
Several species of rootknot-causing eelworms parasitize *Solanum tuberosum*. The
infection begins with small galls, followed
by excessive branching of the attacked roots.
The larvae pierce the cells with their stylets
and feed on them, provoking giant cells
which disrupt the xylem, and the plants may
die. Tubers develop an ugly appearance with
a rough wart-like surface.
Breeding for resistance is difficult due to the
presence of various species and pathogen
races of *Meloidogyne*. In the Republic of
South Africa Brücher (1967) performed extensive test series with a representative
*Solanum* collection (wild and Andine potatoes) artificially infected with four *Meloidogyne* species in greenhouse containers. Considerable genetic resistance was found in Indian potato landraces of North Argentina.

Colorado Beetle (*Leptinotarsa decemlineata*)
Genetical resistance to this devastating insect, which came from the United States in
1921 to SW France, and has now spread
throughout Europe, has been found in wild

potatoes from Mexico and Argentina. The
chemical background of immunity are glu-
co-alkaloids (demissine, leptine etc.), which
are toxic to the larvae and the beetle. Whilst
hybrids of *S. tuberosum* with the hexaploid
*C. demissum* lost their resistance in following generations and backcrosses, the *S.
tuberosum* x *S. chacoense* (diploid) crosses
fared much better. As *S. chacoense* exists in
Argentine in numerous local populations
and clones, one cannot expect that this is a
genetically identical gene pool. Inheritance
of the immuntiy factors is polygenic and
some *S. chacoense* strains may even lack
such factors. For further information,
we refer to Buhr and Schreiber (1962) in
the Potato Handbook of Schick and
Klinkowski, which contains 2000 contributions on *Leptinotarsa* and resistance breeding.

# B. Minor Tuberous, Rhizomatic and Aroid Plants

## 5. *Arracacia esculenta* DC.

Synonym: A. xanthorrhiza Bancr.
Peruvian carrot, arracha, arrakacha, apio
peruviano

Like other horticultural crops of the Umbel-
liferae family, for example carrot, parsnip
and celery, arrakacha forms tasty aromatic
roots, 20 cm long and 8 cm thick, which
contain some volatile oils ("carvone").
We presume that its early domestication occurred in the cool mountain region of the
South American tropics, where we have observed wild forms of this vegetable in temperate valleys of Peru. Its present actual
main cultivation is situated in the Andine
valleys of the cloud forest region of Colombia and Venezuela, with 150,000 t yearly.
Thanks to its high resistance against cryptogamical diseases, it could replace potato

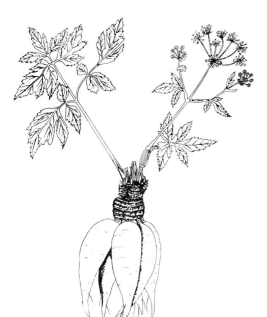

**Fig. I.15.** *Arracacia esculenta*, an attractive South American root vegetable

cultivation, which is heavily affected in this region by several diseases. Arrakacha is photoperiodically sensitive and needs short-day conditions for tuberization. For these reasons, repeated experiments to introduce this promising "Indian carrot" to Europe have failed. Its early use as a root vegetable in pre-Columbian times has been testified by archaeological finds in Peruvian tombs.

The fleshy roots, golden yellow or red in colour, are rich in vitamin A, contain 25% starch and have an exquisite flavour, but have also the disadvantage of high enzymatic activity in their respiratory system, twice that of other edible roots. Therefore arrakacha roots have short storing ability.

Future breeding work of this attractive root crop should be directed towards creating biotypes physiologically indifferent to daylength and reduced enzyme activity. In this way, the recommendation of the famous French-Swiss botanist A. Decandolle may become realized, when, more than a century ago, he indicated *A. esculenta* as a genuine enrichment of the French cuisine (Fig. I.15).

## 6. *Calathea allouia* (Aubl.) Lindl.

Lairén, leren, macu, chufle, bijao (kuna), topi-tambo, cocurito

### NAME AND ORIGIN

Several *Calathea* taxa have been described as growing wild in Mesoamerica and Colombia/Venezuela. Therefore we assume that the origin of the only cultivated species, *C. allouia*, may come from this region. We mention the following related taxa: *C. insignis* Peterson, *C. lutea* (Aubl.) Meyer, *C. villosa* (Lindl.) *C. casupito* Meyer, *C. discolor* Meyer.

These wild species have many applications among the natives. The most important seems the luxuriantly growing *C. lutea*. The underside of its big leaves is coated with a valuable wax (similar to carnauba wax) in rather high percentage. Natives of Amazonia collect it with stiff brushes for sale. For other tropical tribes the leaves are valuable for thatching huts, and finally, as I was told in Panama, natives of Dairén used the water-repellent, waxy leaves to cover the bodies of their dead and protect them against tropical rains.

At certain times the use of *C. allouia* must have been so universal that the tribal mythology of certain Indians designated the plant as a "gift from heaven". For this reason "bijao" still deserves high esteem among the Kuna Indians.

"Lairén" cultivation spread from Amazonia through the Isthmus of Panama to Honduras, but also to the West Indian Islands (Martin and Cabanillas 1976).

### MORPHOLOGY

Plants reach 100–150 cm, sometimes even 200 cm in height, with several stems rising from the corm. Leaves are large (40–60 cm), oblong-elliptic, glabrous, with long (4–20 cm) petioles.

Inflorescenses emerge from a long sheath. Spikes terminate the leafy stem. Flowers are covered with violet or yellow bracts.

The tuberous roots are produced around the

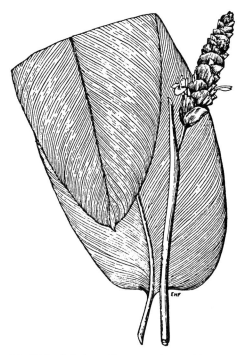

**Fig. I.16.** *Calathea allouia,* a plant which produces the starch rich "lairen" roots (from Flora of Panama)

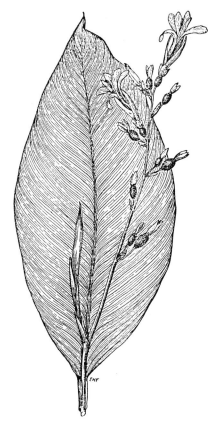

**Fig. I.17.** *Canna edulis,* a plant with flowers, producing large rhizomes (from Flora of Panama)

plant basis, and need one year for growth. They are globular (5 cm diam.) with thin yellow skin, and excellent keeping qualities once harvested and after the foliage has died. They lack fibres and sprouting points. Their starch has a special quality because it is "levulose" (which polarizes the light towards the left) and is considered of high nutritional value. Experiments in Puerto Rico gave good yields with 10 t/ha (Fig. I.16).

CANNACEAE

## 7. *Canna edulis* Ker.

Purple arrowroot, achira, capacho, marant, imocona, toloman

2 n = 18

NAME AND ORIGIN

More than 50 species belong to the exclusive neotropical genus *Canna*. Mostly they are erect-growing herbs with large leaves, showy inflorescenses and strong rhizomes.

*C. edulis* appears very early (about 4300 years B.P.) in archaeological sites of the Peruvian coast (Ugent and Pozorski, 1984), but no finding exists in Central America. Therefore Gade (1966) excluded this region as origin when he wrote: ... "the total lack of archaeological *Canna* material from Mesoamerica indicates a South American origin for Achira." We presume that domestication occurred in the subtropical eastern valleys of the Cordillera, where in North Argentina and Bolivia we often observed wild-growing *Canna* plants. It may be that *C. glauca* L. is the progenitor, but this question is not easy to decide in view of the abundant *Canna* selections and cultivars, mostly for ornamental and floricultural purposes, which have been distributed over the whole world. Indian natives have loved this plant since time immemorial and have used its flowers and leaves as ornaments on vessels (e.g. in the Nazca ceramics). Even today they celebrate the "Raimi" festival at Cuzco, where thousands of achira roots are offered to the ancient fertility gods. The small seeds (which are astonishingly regular and equal in weight) are still used by gold diggers and native goldsmiths as a weight unit for small gold quantities (Fig. I.17).

## MORPHOLOGY

Plants reach 150–250 cm in height and reproduce each year from perennial corms. Leaves are oval-elliptic, large (50 × 15 cm) with coriaceous texture. Flowers zygomorph, with three sepals and petals and a prominent style with three locular ovaries. The most visible parts of the flower are the transformed stamens, two of which are sterile and more or less connate, forming coloured labellum, emulating corollas, which attract colibris (humming birds). One stamen is free and bears a one-locular, fertile anther.

*Canna* has been known since 1570 in Europe as an ornamental flower, but not as a starch plant. Only recently has the fecula been appreciated as food for babies and sick people. The starch grains of *Canna* are the largest of all almidon-producing plants and may be distinguished with the naked eye. The rhizomes contain 24% starch, 1% protein and considerable quantities of phosphorus and calcium. *Canna* is planted in Australia and India as a promising starch crop, with yields of 25 t/ha; in its native countries of South America only on a small scale, mostly as garden plants.

Due to its high biomass production and natural freedom from diseases, *Canna edulis* deserves more attention from plant geneticists.

COMPOSITAE

## 8. *Helianthus tuberosus* L.

Topinambur, topinambour, Erdbirne
$2n = 102$

## NAME AND ORIGIN

Sometimes called Jerusalem artichoke in English, a completely misleading name, that should be eliminated. The plant has neither any similarity with the artichoke (*Cynara scolymus*), a well-known vegetable, whose flower capitula can be eaten, nor any connection with Jerusalem. Since pre-Columbian times, North American Indians have excavated the tubers and cooked them as an emergency food. From there European explorers brought topinambur in the 17th century to France, where it has been cultivated on a reduced scale until today.

According to Heiser (1976) the genetical origin of *H. tuberosus* is still obscure. Probably a 68-chromosome wild species (*H. decapetalus* or *H. hirsutus*) is one progenitor, whilst the other may be the common sunflower ($2n = 34$). A triploid form then arose, which after chromosome doubling produced hexaploid offspring (Fig. I.18).

*H. tuberosus* is not a tropical crop by origin, but some selections do exist which grow well in hot climate and shortday conditions. As *H. tuberosus* exists in its North American homeland only in the wild state, France may really be considered as its secondary gene centre. There 150,000 ha are planted with different clones, between Poitiers, Limoges and the Atlantic coast; mostly as forage for cattle, but also for alcohol production in the cognac industry (Delhay 1979).

## MORPHOLOGY

The perennial plants reach 2 m in height. Stems are stout, covered with coarse hairs. Leaves oblong-obovate (12–18 cm long) simple, hirsute, on long petioles inserted on the stem, in opposite position.

Inflorescences are terminal on the main stem and many lateral branches. Flowers have yellow capitula (5–10 cm diam.). Disk florets are hermaphrodite. Scarce seed production occurs, in spite of the high polyploidy, which in most cases provokes sterility. Topinambur is propagated in the field in a clonal way.

Tubers are produced on long stolons, spindle-shaped, (10–18 cm), fleshy, without a protective epidermis. They are easily bruised

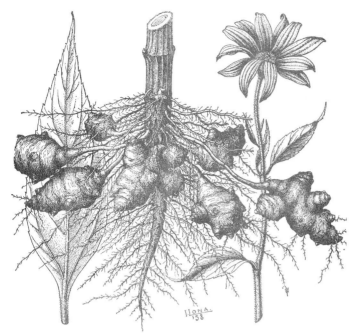

**Fig. I.18.** *Helianthus tuberosus* with tubers (topinambur)

and rapidly lose moisture once harvested. Tubers have white flesh, but a different epidermis colour, yellow, red or brown. They contain 76% water, 2.4% protein and 16% carbohydrates (incl. 8% inulin, a polysaccharide composed of fructose). The existence of inulin in topinambur tubers is important for diabetics. Tubers can be processed to fructose syrup, also to alcohol.

FUTURE BREEDING

Until now this tuber crop has received little attention from professional phytogeneticists and has been mainly handled by hobby breeders. The breeding trend should be reversed. Instead of producing clones for game feed and pigs under European longday conditions, emphasis should be given to high yield under shortday environment and tropical climate. Topinambur has only few diseases and suffers from few pests and could therefore compete well against its main competitor, the potato.

MARANTACEAE

## 9. *Maranta arundinacea* L.

Arrowroot, marante, aru-aru, caualla, chuchute, yuquillo, ara-ruta, Pfeilwurz

$2n = 48$

NAME AND ORIGIN

The strange English name is claimed to have been derived from the native use of its roots as a remedy against poisonous arrows, but we believe it comes from a mis-spelling of the Caribbean word "ara-ruta" which means mealy root.

Species of Marantaceae have contributed worldwide in beautifying inumerable gardens and homes with ornamental plants which come from the Old and New World. In the American tropics two starchy root crops originated, which are highly appreciated by Indian natives: *M. arundinacea* and *M. allouia*, the last named now separated in an extra genus as *Calathea allouia*. Both live in wild state in Brazil and Venezuela, where Indians of rainforests domesticated them (Coursey 1968, Erdman and Erdman 1984).

**Fig. I.19.** *Maranta arundinacea* (right) growing with Xanthosoma sagittifolium. (Photo Dr. J. de Greus)

Paraguayan natives used a lesser-known endemic species of the Parana region, *M. divaricata*, as an emergency food. In times of famine the Guarani collected the large starchy rhizomes and cooked them. Oviedo and also Cobo knew of the use of this starchy root, but it is not clear if it was Lairén or *M. arundinacea*.

MORPHOLOGY

Plants are erect, 100–180 cm high, of ornamental aspect. Stems arise from perennial rhizomes, (30 cm long, 5 cm thick) of fleshy consistence, covered with scales. The rhizomes are somewhat fibrous and consist of 19–21% starch, 1–2% protein and some sugar. The starch grains are oval (15–17 $\mu$m in length) and very fine-grained. They differ from other tuber starch by their high viscosity.

Leaves are dark green obovate-oblong and sit on long petioles which are sheathed at the base and have a typical joint (= pulvinus) at the junction with the lamina. Inflorescences: terminal with few zygomorphic flowers (2 cm long) and white tubular corollas. Stamina reduced in their function, with only one fertile stamen. Style welded to the interior of the flower, which inhibits self-pollination. Therefore fruits are seldom produced and there is no seed propagation of the cultivars, which must be multiplied vegetatively.

The rhizomes mature after 10 months and produce good yields with an average of 7–13 t/ha, which may rise with fertilizer to 37 t/ha. The same plantation gives repeated yields for 5–6 years, according to reports from the Caribbean islands of St. Vincent and Antigua, where arrowroot starch has been extracted on an industrial basis. In view of its exceptional quality and high digestibility as baby food, besides its social value as a crop for many smallholders, *Maranta* cultivation should receive a new stimulation. The island of St. Vincent alone once had a maximum production of 45,000,000 kg per annum. In the meantime the main production has been transferred to S.E. Asiatic countries (Fig. I.19).

OXALIDACEAE

## 10. *Oxalis tuberosa* Mol.

Oka, aipilla, ibias, cuiba
$2n = 66$

NAME AND ORIGIN

From the several hundred taxa of the genus *Oxalis* which are extended worldwide, only very few are of economical importance. It is therefore all the more interesting that Mountain Indians of South America were able to select high polyploid cultivars from one of the many Andine species described by Brücher (1968). Actually *O. tuberosa* is still the second most important root crop of the Andes, next to the potato. In prehistoric times, oka cultivation spread from the Colombian mountains southward to the Is-

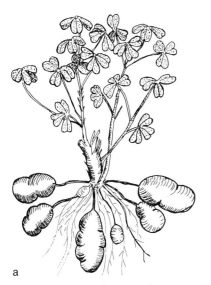

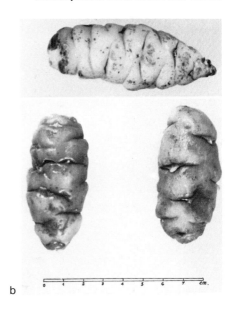

a                                              b

**Fig. I.20.** *Oxalis tuberosus,* plant and tubers

land of Chiloé, covering more than 50° latitude in South America. However, it never
crossed the Isthmus of Dairén, nor was it
introduced later in the mountain regions of
Mesoamerica.
We believe that it was the Inka Empire
which spread the oka so far south to Chile.
We were astonished to find *O. tuberosa* on
an expedition to the Island of Chiloé in the
year 1955, growing there in the moist-cool
climate of the Pacific coast.

MORPHOLOGY

Plants are erect-bushy with succulent stems.
Leaves are typically trifoliate, as in all other
Oxalidaceae. Flowers are yellow and show
heterostyly in their fecundation system. The
fruits are five-locule capsules, which eject
their seeds.
The tubers are many-coloured (white, red,
yellow etc.) and have a smoth skin, bearing
many scales which cover the deep eyes. They
contain a fair percentage of carbohydrates
(13–15% average) and ascorbic acid, but
the cv. huila zapallo has 20%, together with
1% protein.
Oka is notably frost-resistant and may be
planted even above the 4000 m limit.
Around the 3600-m-high Lake Titicaca it is

very common to see huge plantations of
oka, calculated at 15,000 ha. This is greater
than the potato area. The actual production
of the Altiplano states Bolivia, Ecuador and
Peru may exceed 100,000 t yearly. The
breeding work of the Experimental Station
Patacamaya at 3800 m alt. in the Bolivian
Cordillere deserves international support
(Fig. I.20).

|  | LEGUMINOSAE |
|---|---|

## 11. *Pachyrrhizus tuberosus* (Lam.) Spreng.

Potato bean, pois-patate, cacara, jicama
$2n = 22$

NAME AND ORIGIN

Its origin has been traced back to the humid
tropical valleys of the Upper Amazonas. Its
areal extended in pre-Columbian times
through Brazil and Venezuela to the Carribbean Islands northwards, but also reached
the nearly tropical region of East Paraguay.
The Guarani Indios planted it as a tuber
crop and medicinal plant, according to
Montenegro (1710).

He recorded that the "mbacucu" was sown each year, that its roots were used as a preserve prepared with salt. The Paraguayan botanist Bertoni saw plantations of *P. tuberosus* at the margin of the Parana River and mentioned "that the Indians consumed the tuberous roots like potatoes" and that their weight was 6–8 pounds.

During our explorations of the flora of Paraguay (1970–1973) we tried in vain to recover this interesting species, which most probably has been exterminated in Eastern Paraguay by the violent advance of modern agriculture.

In Mexico, roots often are eaten raw, so that obviously they are free of the toxic sustances which complicate the consumption of the green part of jicama plants; the tubers can be harvested after 6–7 months. They have low carbohydrate content (10%), but a fair part of sugar (2% reduced, 3% non-reduced), 1.5% protein, and several minerals, especially calcium and phosphorus.

MORPHOLOGY

Stems long (5–7 m) climbing on bushes and trees, at the base lignified.

Leaves tripartite, with leaflets of different size and form. The terminal leaflets are rhomboid, whilst the laterals are asymmetric-triangular. Inflorescences many-flowered with violet-coloured flowers (seldom white). Fruits large (15–30 cm long) with rather large seeds (11–14 mm diam.). Tubers large (50 cm diam.), watery, with edible fleshy pulp, 20–30 kg in weight.

It is noteworthy that the plants do not show insect damage in an environment which is heavily infested by all sorts of pests and parasites. This may be the result of the slight natural content of rotenone, which is well-known for its "insecticide effect".

## 12. *Pachyrrhizus ahipa* (Wedd.) Parodi

Aymara: huitoto, kechua: ajipa

2 n = 22

NAME AND ORIGIN

While studying ancient Peruvian plant descriptions and painted ceramics of the Mo-chica culture, we came across stylized depictions of *Pachyrrhizus*. This indicated that this species must have played an important role in the daily life of the Andine Indians. Probably they selected it from regional wild forms in the Ceja de Montaña, which are now exterminated. *P. ahipa* has a double utility, as both its tubers and seeds are consumed. Unfortunately, its use is rapidly decreasing. Occasionally one can find some samples in the house gardens at 1000–2500 m altitude in Peru and Bolivia. The sites in the North Argentine Prov. Jujuy,

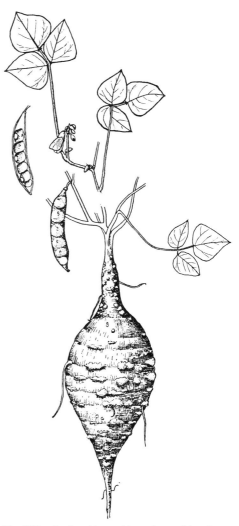

**Fig. I.21.** *Pachyrrhizus ahipa*, plant with tuber

which Parodi reported in his taxonomic description, may have been exterminated.

MORPHOLOGY

The plants are of compact growth, with short (30–90 cm) stems.

Leaves have extremely asymetric lateral leaflets. The short terminal leaflets are broader than long.

Inflorescences: with many flowers mostly violet, some white. Fruits are short (8–10 cm long) with 4–5 seeds. Seed colour is brown-black, their diameter is 10 mm. Their raw protein content is approximately 16%.

The roots produce 10–15 cm tubers (500–800 g in weight) with sweet white pulp and yellow skin. They may be consumed raw or boiled. Burkart (1952) considers *P. ahipa* as the most indicated for future genetic improvement and rational cultivation, due to its compact habit, and comparatively quick growth. Sown in December, it can be harvested – both seeds and tubers – already in April. In spite of its origin in tropical short-day regions, it adapts well to other environmental conditions (Fig. I.21).

Once improved genetically, *P. ahipa* could make a considerable contribution to the diversification of protein supply in tropical countries. In view of the near extinction of ancient landraces in the high Andine valleys, the existing remains of a formerly rich gene pool should be collected as soon as possible and made available to advanced plant breeding institutions for further improvement.

---

TROPAEOLACEAE

## 13. *Tropaeolum tuberosum* R.P.

Añu, Isaño, Mashua

$2n = 42$

The American genus *Tropaeolum* is known worldwide for its garden flowers. Only one species from among the 50 which are found from Patagonia to the Colombian mountains has aquired practical use. This is the tuber-bearing *T. tuberosum* (Fig. I.22), which has been selected in the Peruvian high

valleys of the Andes, probably from the wild-growing *T. tricolor*.

Isaña exists in various cultivars with attractive coloured tubers, mostly lemon yellow, red or marble-veined. The tuber form is oblong (10–18 cm), with deep eyes, from which many sprouts grow, once planted. The pulp is watery, with only 12% carbohydrates, 1–2% protein and a high content of ascorbic acid (67 mg/100 g).

The plants have a creeping and twining habit. The leaves are shiny, glabrous, five-lobed with a pronounced "nasturtium" smell. Flowers are red and only seldom produce seeds. Plants have pharmaceutical use, with a sedative effect due to the myrosin. At high altitude (3000–4000 m) they resist light frosts.

---

BASELLACEAE

## 14. *Ullucus tuberosus* Lozano

Ruba, melloca, ulluque, papa lisa, ulluko, olloko, ulluma, papa verde

$2n = 24$ and $36$

NAME AND ORIGIN

Besides the potato, *U. tuberosus* is one of the Andine Indians' oldest useful plants, as is confirmed by prehistoric finds and an exceptionally large number of local varieties – over 70 – whose areal extends from Northern Argentina to Venezuela. Stylized representations of ulluku plants and tubers are to be found on clay vases from the Tihuanako culture and on numerous other pre-Columbian depictions.

Spanish chroniclers of the time of the Conquest tell how – around Quito, for example – they came upon wide fields of *Ullucus* and were astonished at the bright colours of their tubers. In Cordillerian countries many local varieties with white, yellow and brilliant green tubers ("papa verde") still exist, whilst others have carmine-red or violet colours and are also rather different in shape (Fig. I.23).

In the year 1958 I discovered the wild form on one of my expeditions in the South American Andes; and, after cultivating it

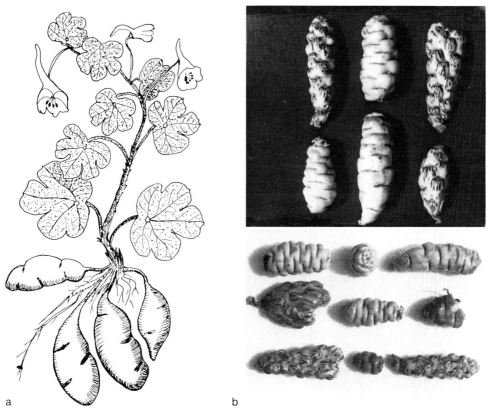

**Fig. I.22.** *Tropaeolum tuberosum,* distinct land races with tubers

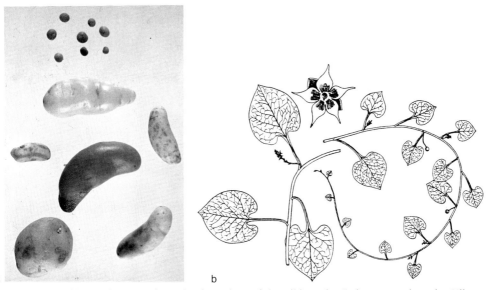

**Fig. I.23. a** *Ullucus tuberosus,* above the tiny tubers of the wild species, **b** the ancestral species *Ullucus aborigineus* Brücher

**Fig. I.24.** Tubers of *Ullucus* in the center, surrounded by *Oxalis* tubers

for several years, described it as *U. aborigineus* (Brücher 1967). The existence of this wild-growing species was confirmed later from Bolivia and Peru. In contrast to the cultivated species, the wild *Ullucus* is a creeping plant with 2-m-long branches, which entwine themselves with other herbs. It has tiny tubers, only 5–10 mm in length, with a bitter taste. Their chromosome number has been determined as $2n = 24$. Using this wild form Indians accomplished step by step domestication, selecting for better tuber size and bushy compact habit. Today intermediate primitive forms still exist called by the Indians quite-lisa, as witness of their long breeding efforts. These semi-wild biotypes grow under the protection of human artefacts, in the walls of cemeteries, in "pirkas", or on trails between fields. Their tubers are larger than those of *U. aborigineus*, but bitter, like the wild species. The natives have no use for them and do not gather this "quite lisa" either.

MORPHOLOGY

Plants bushy, compact, 20–40 cm tall, with fleshy watery stems.
Leaves cordate, 5–14 cm long, rather succulent, shiny green, producing thick foliage. Inflorescences axilliar, with inconspicuous yellowish flowers and five perianths. Fruits are triangular capsules with one seed, but seldom observed. Propagation is practically asexual.

At the basis, the plants develop plagiotrope estolones which at their tips produce tubers, often more than 20 per plant, with 5–8 cm diameter. There are reports of high yields (10–11 t/ha) in good mountain soils at 3000 m altitude. Tubers have 1% of protein and 12–14% starch, a high content of vitamin C (23 mg/100 g), and a fair quantity of phosphorus and other minerals. Natives sometimes eat the tuber raw, but mostly boiled or dried. The pulp is somewhat mucilaginous. Ulluko is still an appreciated tuber crop in the high Andine valleys, where it often replaces the potato, due to its better frost resistance and good immunity against diseases. Travellers can find it from North Argentina, Bolivia, Peru, Ecuador, Colombia to Venezuela (state of Merida) on the native markets. In Peru alone the yearly yield is calculated at 30,000 t.

Only few diseases and pests attack *U. tuberosus*; the most damage is caused by the coleopterous *Premnotrypes solani* ("el gorgojo de los Andes"). Its good performance against many diseases and parasites was the reason that in the past century the British Administration introduced *Ullucus* to South India. The potato plantations in the Nilgiris mountains suffered at that time heavy losses, as we know now, due to *Phytophthora* and nematodes (*Heterodera rostochiensis*). When I worked there in 1969 as adviser for the German Government, however, I could find no single *U. tuberosus* plant left.

---

ARACEAE

## 15. *Xanthosoma sagittifolium* L. (Schott)

Tannia, yautia, calaku, cocoyam, belembe
$2n = 26$

NAME AND ORIGIN

*Xanthosoma* is a neotropical genus, but due to its morphological similarity with the Old-World group of *Alocasia* and *Colocasia*, it is frequently confused with these tuber crops (Plucknett 1976).

A quantity of indigenous names exist throughout tropical America, for example, caraku, malanga, kiskamote, mafafa, otoé, tannia, yautia, cocoyam, okumo (Morton 1968).

*Xanthosoma* cultivars were selected from several American wild species many thousand years ago. There are, however, practically no archaeological remains, because its main use was always in the rainforests with their quick decomposition of organic material. Putative wild species from which Indians selected garden forms are: Central America *C. robustum* and in South America *C. jaquinii*.

## Morphology

Perennial herbs of strong habit, 120–300 cm tall, with large leaves on long petioles. Stems are formed on the ground and arise from large corms.

Lamina sagittate, with strong intramarginal veins (in contrast to the similar *Colocasia* sp.), often with two basal lobes. Contain needle-shaped crystals of calcium-oxalate, which provokes irritation on the human skin.

Inflorescences are seldom present. They consist of flag leaves and spadix, and after pollination produce red berries. Growth is slow, with corms ripening after 1 year to be harvested. In plantations we observed high yields of 18–20 t/ha fresh rhizomes (Fig. I.25). It is chiefly a home garden plant for casual use of leaves and rhizomes. *Xanthosoma* needs strong continuous rainfall and high temperatures and prospers best in heavy wet soils, even water-submerged, with much organic matter. Its propagation is by corm offshots.

There are different *Xanthosoma* varieties (a) with yellow-orange corm (b) with white flesh and large corms. (c) Small corms with purple flesh, which have been separated as: *atrovirens, caracu* and *violaceum*.

Plants are highly resistant to leaf diseases and grow even under ecologically adverse conditions.

*Xanthosoma* entered the African west coast during the last century, and spread from

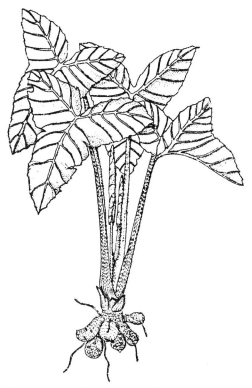

**Fig. I.25.** *Xanthosoma sagittifolium,* plants and cormus

there to the whole continent as a much appreciated vegetable. Finally it extended also to Indonesia and Hawaii, being generally called Cocoyam. The nutritional value of the tannia corm is given in Table 5.

**Table 5.** Nutritional value of the tannia corm

|              | % Fresh wt.              |
| ------------ | ------------------------ |
| Moisture     | 70–77                    |
| Carbohydrate | 17–26                    |
| Protein      | 1.3–3.7                  |
| Fat          | 0.2–0.4                  |
| Crude fibre  | 0.6–1.9                  |
| Ash          | 0.6  1.3                 |
| Carotene     | 0.002 (2 mg per 100 g)   |

In spite of its high yield possibilities, *Xanthosoma* lacks scientific breeding work. Selection should be directed to eliminate oxalate and to diminish mucilaginous substances in the root corms. Due to poor flowering, hybridization between selected strains is difficult. For this reason one should induce artificial mutations.

For reasons of completeness we mention here finally the following rhizome-producing species:

*Mirabilis expansa* 2 n = 58 (Nyctaginaceae) = Mauka, in Bolivia, Peru
*Lepidium meyenii* (Cruciferae) = Maca, in the Altiplano of Bolivia
*Jacaratia hassleriana* (Caricaceae) = Ybia, cipoy, in Paraguay
*Polymnia sonchifolia* 2 n = 60 (Compositae) = Yacon, aricuma, jiquima, from North Argentina, all Andine States to Columbia.

These are plants which produce "emergency food" for people abandoned in the Altiplano or remote Andine valleys. In such cases natives may have survived thanks to rhizomes of these species, and taken seeds or tubers to their home garden for further production and improvement. However, after studying them, we do not consider them worth future phytogenetic improvement. For more details see: Brücher (1977), Leon (1968), Montaldo (1972); these books contain descriptions and pictures of these crops.

# References

Anderson GJ (1979) Systematic and evolutionary consideration of species of *Solanum*, sect. *Basarthrum*. In: The biology and taxonomy of Solanaceae. Linn Soc Sympos 7:445–454
Austin DF (1978) The *Ipomoea batatas* complex. Taxon Bull Torrey Bot 105:114–129
Ayensu ES (1972) Anatomy of the monocotyledons VI. Dioscoreales. Oxford University Press
Bartlett AS, Barghorn ES (1973) Phytogeographic history of the isthmus of Panama. I. Graham A (ed.) Vegetational history of Northern Latin America. Chap 7:203–299. Elsevier, Amsterdam
Bauhin G (1596) Phytopinax. Basel
Bavyko NF (1982) Cultivated species of South American potatoes, their areal and value for breeding (in Russian) Tr Prikl Bot Genet 73:109–113

Bertonio L (1612) Vocabulario de la lengua aymara. Impresa Fco. del Canto. Prov Chuquito, Puno, Peru
Bode O (1958) Research on the virus of the tobacco brown rib disease. Proceed II Intern Sci Tobacco Congr Brussels:93–96
Bolhuis GG (1953) A survey of some attempts to breed cassava varieties with high content of protein in roots. Euphytica 2:107–112
Bomari C (1774) Diccionario de Historia Natural. Santiago de Chile
Bouwkamp JC (1985) Sweet potato products, a natural resource for the tropics. CRC Press, Boca Raton, Florida, p 280
Brücher H (1956) Critical observations on the taxonomy of Argentine wild potatoes. *Solanum vernei* and its synonym *S. ballsii* Hawkes. Anal Dept Invest Cient Mendoza 2:1–12
Brücher H (1960) Problematisches zum Ursprung der Kulturkartoffel aus Chiloé. Z Pflanzenzücht 43:241–365
Brücher H (1963) Das südlichste Vorkommen diploider Kulturkartoffeln in Südamerika auf der Insel Chiloé. Qual Plant Mater Veg 9:187–202
Brücher H (1967) Rootknot-eelworm resistance in some South American tuber-forming *Solanum* species. Am Potato J 44:370–375
Brücher H (1968) Poliploidia en especies sudamericanas de Oxalis. Bolet Soc Venez Cienc Nat 115:145–178
Brücher H (1969) Observations on origin and expansion of $Y^n$-Virus in South America. Angew Bot 43:241–249
Brücher H (1970) Chromosomenzahlen argentinischer, chilenischer und venezolanischer Wildkartoffeln (*Solanum Tuberarium*) Cytologia (Tokyo) 35:153–170
Brücher H (1975) Domestikation und Migration von *Solanum tuberosum*. Kulturpflanze 23:12–70
Brücher H (1977) Tropische Nutzpflanzen, Ursprung, Evolution und Domestikation. Springer Berlin Heidelberg New York
Brücher H (1979) Das angebliche "Genzentrum Chiloé" 50 Jahre nach Vavilov. Z f Pflanzenzüchtg 83:133–147
Brücher H (1988) Migration und Dispersion amerikanischer Nutzpflanzen über die Landenge von Darién (Panama) Naturwissenschaften 75:18–26
Brücher H (1988) Las especies tuberiferas del genero Solanum de Argentina, Brasil, Chile, Paraguay and Uruguay. Buenos Aires (manuscript)
Buhr H (1962) Leptinotarsa. In: Schick and Klinkowski (eds) Die Kartoffel. VEB Landw Vlg, Berlin

Bukasov SM (1966) Die Kulturarten der Kartoffel und ihre wildwachsenden Vorfahren. Z Pflz 55:139–164

Bukasov SM (1980) Obzor taxonomii vidov kartofelja sekcii *Tuberarium*. Bull inst rast im N Vavilova Leningrad 105:3–6

Burkhill IH (1960) Organography and evolution of Dioscoreaceae, the family of yams. J Linn Soc Bot 56:319–412

Burton WG (1966) The potato, 2nd edn. Veenman & Z Wageningen

Burkart A (1952) Leguminosas argentinas, Acme Buenos Aires II edit 520 p

Candolle A de (1884) Origine des plantes cultivées. Paris

Cardenas M (1969) Manual de plantas economicas de Bolivia. Impr Icthus, Cochabamba

Carpio-Burga R (1984) Nuevos logros en el valor nutricional del Camote. Fondo Fomento, Agrop Chincha. Peru

Castellanos J de (1536) Historia del Nuevo Reino de Granada Madrid (not published until 1886)

Cieza de Leon P (1538) published by Markham (1864) Hakluyt Soc Pub Nr 33

Cock JH (1982) Cassava, a basic energy source in the tropics. Science 218:755–762

Correll D (1952) Section *Tuberarium* of the genus *Solanum* in North America and Central America. Agric Monogr 11, USDA Washington

Correll D (1962) The potato and its wild relatives. Texas Res Found, Renner pp 605

Coursey DG (1967) Yams. Longman, London pp 230

Coursey DG (1968) The edible aroids. World Crops 20:3–8

Davidson WD (1936) The history of the potato and its progress in Ireland. J Dep Agric (Repub Irel) 34:286–307

D'Arcy WG (1972) Typification of subdivision of *Solanum*. Solanaceae studies. II Ann MO Bot Gard 59:262–278

Delhay R (1979) *Helianthus tuberosus* L, a potential root crop for the tropics. Int Symp Trop Crops, Manila, Philippines

Dodds KS (1962) Classification of the cultivated potatoes. In: Correll (ed) The potato and its wild relatives, Renner, Texas

Engels F (1970) Exploration of the Chilca canyon, Peru. Curr Anthr 11:55–58

Erdman MD, Erdman BA (1984) Arrowroot (*Maranta arundinacea*) food, feed, fuel and fibre resource. Econ Bot 38:332–341

Erdmans AM et al (1980) Role of cassava in the etiology of endemic goitre and cretinism. Int Dev Res Centre, Ottawa, Canada

Flannery T (1973) Mesoamerican food plants archeology. In: Reed (ed) The origins of agriculture

Gade D (1970) Ethnobotany of *Chenopodium pallidicaule*, a rustic seed crop of the Altiplano. Econ Botany 24:55–61

Gerard J (1633) The Herball, or General History of Plants. London Harris PM (1978) The potato crop. Sci basis for improvement. London p 730

Graner EA (1935) Contribuicão para o estudo citológico da mandioca. Esc Sup de Agricultura, USP, Piracicaba, Brazil p 28

Harris PM (1978) The potato crop. Scientific basis for improvement, London p 730

Hawkes JG (1962) The origin of *Solanum juzepczukii* and *S curtilobum*. Z Pflanzenzücht 47:1–14

Hawkes JG (1979) Evolution and polyploidy in potato species. The biology and taxonomy of *Solanum*. Symp Linn Soc Serie 7, London

Heiser CB (1976) Sunflowers-*Helianthus*. In: Simmonds (ed) Evolution of crop plants. Longman, New York

Herold F (1967) Virus investigations of potatoes from South Africa. Amer Potato J 44:22–33

Heyerdahl T (1966) Prehistoric voyages as agencies for Mesoamerican and South American plant dispersal to Polynesia. Bishop Museum Press, Honolulu

Hujisman CA, Lamberts H (1972) Breeding for resistance to the potato cyst nematode in the Netherlands. CIP-Lima, Peru 161–171

I.I.T. A (1986) Yams. Annu Rep Int Inst Trop Agric Ibadan, Nigeria

Jennings D (1963) Observations on virus diseases of cassava. II Brown streak disease. Emp J Ex Agric 28:261–270

Kranz JH, Schmutterer H, Koch W (1979) Krankheiten, Schädlinge und Unkräuter im tropischen Pflanzenbau. P. Parey, Berlin, pp 723

Lanning E (1966) American aboriginal high cultures. Peru. 36 Congr Int Americanistas, Sevilla I:187–191

Lathrap D (1973) The antiquity and importance of long distance relationships in the moist tropics of pre-columbian South America. World Archeol 5:170–186

Leon J (1968) Fundamentos botanicos de los cultivos tropicales. Inst Interam Cienc Agric San Jose, Costa Rica

Leung WW, Flores M (1961) Food composition table for use in Latin America. Nat Inst of Health, Bethesda, M.D.

Lozano IC, Booth RH (1976) Diseases of *Cassava*. CIAT, Cali-Colombia, p 46

Mac Neish RS (1971) Early man in the Andes. Sci Amer 224:36–46

Magoon ML et al (1970) Cytogenetics of the F-1 hybrids between cassava and ceara rubber and its backcross. Genetica 14:425–436

Mansfeld R (1986) Verzeichnis landwirtschaftlicher und gärtnerischer Kulturpflanzen. Springer Berlin Heidelberg New York, p 1998

Martin FW (1973) Protein content and amino acid balance in yams. J Agric Univ Puerto Rico 57:78–83

Martin FW et al (1974) A wild *Ipomoea* species closely related to the sweet potato. Econ Bot 28:287–292

Martin FW, Ortiz (1963) Chromosome numbers and behaviour in some species of *Dioscorea*. Cytologia, Tokyo 28:96–101

Martin FW, Cabanillas J (1976) *Calatheae allouia*, a little-known tuberous crop of the Caribbean Marantaceae. Econ Bot 30:249–256

Martin FW (ed) (1984) Handbook of tropical food crops. CRC, FL, Boca Raton, p 296

Miège J, Lyonga SN (eds) (1982) Yams, Ignames. Clarendon, Oxford, p 411

Molina I (1776) Compendio della storia geografica naturale e civile del Regno del Chile. Stamperia di S. Tomas d Aquino, Bologna

Montaldo A (1972) Cultivos de raizes y tuberculos tropicales. Lima, ICA, p 284

Montaldo A (1979) La Yuca, cultivo, industrializacion, aspectos economicos y mejoramiento. ICA, Costa Rica, p 386

Morton JF (1968) Cocoyams (*X. caracus*), an ancient root and leaf vegetable growing in economic importance. Proc Fl State Hortic Soc 81:318–328

Nassar NMA (1978) Conservation of the genetic resources of *Cassava*. Econ Bot 32:311–320

Nassar NMA (1978) Wild manihot species of Central Brazil for cassava breeding. Can J Plant Sci 58:257–261

Nassar NMA, Costa CP (1977) Tuber formation and protein content in some *Manihot* species native to Central Brazil. Experiencia (Brazil) 33:1304–1305

Nestel BL (1974) Current trend in cassava research. IDRC-36 Int Dev Canada, Ottawa, p 32

Nishiyama (1971) Evolution and domestication of the sweet potato. Bot Mag Tokyo 84:377–387

Ochoa C (1962) Los *Solanum* tuberiferos silvestres del Peru. Lima, p 279

Ochoa C (1989) The potatoes of South America, vol I. Bolivia, Impr Univ Cambridge, London

Onwueme IC (1978) The tropical tuber crops. J Wiley, Chichester, p 234

Pickersgill B, Heiser CB (1977) Origins and distribution of plants domesticated in the New World tropics. In: Reed (ed) Origins of agriculture. Mouton Paris

Plaisted H (1972) Utilisation of germ plasm in breeding programs. Use of cultivated tetraploids. CIP-Lima, pp 90–99

Plucknett DL (1976) Edible aroids Alocasia, Colocasia, Cytrosperma, Xanthosoma. In: Simmons (ed) Evolution of crop plants. Longman, London

Purseglove J (1965) The spread of tropical crops. In: Genetics of colonizing species. Academ Press, New York

Purseglove JW (1968) Tropical crops, Longman, London

Reichel-Dolmátoff G (1965) Colombia, ancient peoples and places. New York

Renvoize BS (1972) The area of origin of *Manihot esculenta* as a crop plant. Econ Bot 26352–360

Rogers DJ, Appan S (1973) Manihot and Manihotoides. Flore Neotropics. Monogr 13, Hafner, New York, p 272

Ross H (1960) Über die Zugehörigkeit der knollentragenden *Solanum*-Arten zu den pflanzengeographischen Formationen Südamerikas und damit verbundene Resistenzfragen. Pflanzenzücht 43:217–240

Ross RW, Rowe PR (1969) Utilizing the frost resistance of diploid *Solanum* species. Am Potato J 46:5–12

Rothacker D (1961) Die wilden und kultivierten mittel- und südamerikanischen Kartoffelspecies. In: Schick-Klinkowski (ed) Die Kartoffel, pp 353–558

Rouse IB (1963) Archeology of Venezuela. Rep 13 Annu Caribbean Conf Curtis Wilgus, Gainesville

Rouse IB Cruxent M (1963) Some recent radiocarbon dates for Western Venezuela. Am Antiquity 28:537–540

Safford W (1925) The potato of romance and reality. J Heredity 16:4,5,6

Salaman RN (1949) Social history of the potato. Cambridge University Press, p 684

Sauer CO (1952) Agricultural origins and dispersals. Amer. Geogr Soc, New York

Schick R, Klinkowski M (1961/62) Die Kartoffel, ein Handbuch. VEB Deutscher Landwirt-Verlag, Berlin, p 2112

Schwerin K (1970) Apuntes sobre la Yuca y sus origines. Bol Inf Antropol Venezuela 7:23–27

Silberschmidt K (1961) The spontaneous occurrence of strains of potato virus X and Y in South America. Phytopathol 42:175–192

Simmonds NW (1969) Prospects of potato improvement. Scott Plant Breed Rep, pp 18–38

Stone D (ed) (1984) Pre-Columbian plant migration. Peabodoy Mus Arch and Ethnol Harv Univ Press, vol 76

Towle M (1961) The ethnobotany of pre-Columbian Peru. Aldine Chicago

Ugent D (1968) The potato in Mexico: geography and primitive culture. Econ Bot 22:109–123

Ugent D, Pozorski T (1985) Archeological *Manihot* from Coastal Peru. Econ Bot 40:78–102

Ugent D, Pozorski T (1984) New evidence for ancient cultivation of *Canna edulis* in Peru. Econ Bot 38:417–432

Vargas C (1954) Las papas sudperuanos. II parte. Cuzco, Peru

Vavilov I (1926) Studies on the origin of the cultivated plants (in Russ). Bull Appl Bot Plant Breed 16:1–245, Leningrad

Villareal RL, Griggs TD (eds) (1982) Sweet potato. Asian Veget Research Center Tainan, Taiwan, p 481

Willey GR (1960) New World prehistory. Science 131:73–86

Woolfe JA (1987) The potato in human diet. Cambridge Univ Press, 256 pp

Yen DE (1974) The sweet potato and Oceania. Bishop Museum Press, Honolulu, Hawaii

Zwartz J (1967) Characterization of potato varieties by electrophoretic separation of the tuber proteins. Medd Landb Hog 9:1–127 Wageningen

# II. Farinaceous Plants

## 1. *Amaranthus* spp.

The genus *Amaranthus* has a worldwide distribution; at least 50 species have been identified. In general they are sturdy, fast-growing plants with a high phytomass production. Several are aggressive weeds (= "pigweeds"), very difficult to eradicate when they invade cultivated fields. The genus is cytogenetically dibasic with species of $x = 16$ and $x = 17$ chromosomes, which do not cross in free nature. Taxonomically the genus is a very difficult object. Aellen (1967) and Sauer (1967) tried to order its systematics. They recognized two sections: *Blitopsis* and *Amaranthotypus*.

The first category has mainly $n = 17$ chromosomes, which are separated from the group of $n = 16$ chromosomes. Most are wild-growing cosmopolites, useful as emergency vegetables. Such potherbs are: *A. tricolor, A. blitum, A. lividus, A. spinosus, A. gracilis, A. oleraceus, A. gangeticus, A. albus*. Their origin is Asia and Africa, where some of them are used in local dishes ("calalu") or as animal feed. Under tropical conditions they have high yields (= 15–20 t/ha of fresh green leaves in one month), allowing repeated cuttings if the initiation of flowering is prevented. Such vegetable amarants have a fair content of protein and are rich in Vitamins A and C, as well as in minerals; but they contain also slight amounts of anti-nutritional factors, especially oxalates and nitrates.

These leaf-producing amarants are adapted to many different ecological environments. The physiological reason is their very efficient type of photosynthesis which uses $C_4$ carbon fixation, superior to and different from the usual $C_3$ carbon fixation pathway. They convert a higher ratio of carbon-dioxide from the atmosphere to plant sugars.

The section *Amaranthotypus* is of superior importance as a food crop. Its cytology has been studied by Khoshoo and Pal (1972). Cytogenetically, the investigated taxa are not strongly differentiated and have bivalent pairing. In the case of artificially produced interspecific hybrids, one can observe an interchange of long segments of distinct chromosomes. The species separation seems due to minor genic and cryptic structural differences. Pal et al. (1982) crossed a white-seeded cultivated *A. hypochondriacus* ($n = 16$) with black-seeded African *A. hybridus* ($n = 17$). The resulting offspring had $2n = 32, 33$ and $34$ chromosomes, in relation $1:2:1$. The authors concluded that the species with $n = 17$ chromosomes originated in $n = 16$ species by primary trisomy.

The genealogy of the neotropical crop *Amaranthus* is still under study. Allozyme analysis proved that the species *A. caudatus, A. cruentus,* and *A. hypochondriacus* are more closely related inter se than they are to a putative weedy progenitor (Kulakov et al. 1985). (One exception is the weed-crop pair: *A. quitensis, A.caudatus)* (Jain et al. 1984).

### MORPHOLOGY

Annual herbs of coarse, erect or bushy growth, (100–200 cm tall), with monoecic habit. Leaves simple, rhomboid-cordate, in alternate position on the stems and side branches. Inflorescenses terminal with red, yellow, green flower stalks, with many glomerules. Each glomerule contains 200 female flowers with one male flower; mostly self-pollinating. They produce huge quantities of seeds, 200,000 to 500,000 each. These are covered by short bracts. In the cultivars the bracts are of soft texture, whilst the weedy species have prickly bracts, which make seed collection by hand difficult.

The seeds have different colours and are lense-shaped (1 – 2 mm diam). One gram of seed contains approximately 1000 – 1500 seeds. They have a relatively large embryo which surrounds the centrally located perisperm in a circular form. The seed coat is thin and contains different pigments, black, brown, red, beige and white. The selected cultivars in general have clear colours. Most of the protein is located in the embryo, in general 12 – 13 % of the whole seed.

*Amaranthus* species have been cultivated since time immemorial in both Middle and South America, mainly for their edible seed grains; but also as an essential potherb in the Altiplano, which is devoid of vegetables. Especially in the Aztec Empire, amarant was an integral part of the daily diet of the Mexican lower classes. Its paramount importance is demonstrated by the huge annual tribute (5,000 t) which the peasants had to contribute to the Aztec emperor at Tenochtitlan. This was equal to the amount of maize and beans. These yearly contributions are confirmed by the Codex Mendoza (1541).

Some cruel rituals were involved with amarant grains. Blood of sacrificed humans was mixed with amarant flour to form replicas of Aztek godnesses, which were ceremonially distributed by the Aztec priests among the people. The new priests from Spain suspected some sacrilege of their Christian sacrament of blood and banned this pagan ceremonies. Anyone growing amarant was condemned to death by the conquistadores. Even its normal use as food was punished. So "gautli" disappeared in Mexico. Only casual small plots survived in remote valleys, which now represent an invaluable genepool for future breeding and improvement of this once essential staple food.

Thanks to North American entrepreneurs and breeders, *Amaranthus* research has recently gained unexpected momentum. Amarant grains are now considered a health food in the USA, with an estimated consumption in 1984 of 70,000 kg. Mexico is producing 50,000 kg for export. The National Academy of Sciences in Washington declared it in 1975 "a promising, under-utilized food source, which should receive immediate study". In this way a forgotten and even prohibited food plant of the Indians has now re-surfaced.

Essentially the different useful species of genus *Amaranthus* are still not more than semi-domesticated cultivars, similar to the landraces used by the natives 500 years ago (Wildenow 1790; Hauptly and Jain 1984).

In this context the amarant breeding and divulgation work by a private North American enterprise deserves special mention: Rodale, Inc., besides that of INCAP in Guatemala. Although the use of amarant grains as human food has millenarian antecedents in both Americas, it is not easy to adapt this grain to modern use.

The main obstacle for remunerative farming is the lack of well-developed mechanical harvesters. Until now, many farmers have to cut and shake out the seeds by hand, which is uneconomic (Weber et al. 1988).

The yield, even with continuous improvement, is still critically low in comparison with high-yielding cereal varieties in North America, and the actual price costs to produce amarant grains are higher than for wheat.

However, there are also several factors in its favour (Senft et al. 1982, Vietmeyer 1982): Amarant starch is composed of very small granules, with amylopectin as its major component.

The seeds have approximately 6% fat, about 40% of which is linoleic acid. With this the amarant has a higher oil content than other cereal grains. The total protein content has been determined as 10 – 17%. Its lysine percentage is astonishingly good (Becker et al. 1981, Downton 1973).

Only such processing and milling methods should be used which do not destroy the quality of amarant protein. Popping or flaking are the best methods to make the essential amino acids available.

Grain amarants are easy to produce. For one hectare only one pound of seeds is necessary. Dense sowing may result in 360,000 plants/ha. Its grain yield can vary between 1000 and 2400 kg/ha. Breeding should be directed to medium-sized plants, because the usual big ones (150 – 200 cm tall) have an excess of stems and leaves and consequently have a tendency for lodging. The plants tolerate rather high ambient temperatures (40 °C) and drought in the soil, but they have no genetic resistance to frost. Even at temperatures less than 4 °C young plants may be injured. Only *A. caudatus* of the Andes is more resistant to chilling, and grows well above an altitude of 2500 m (Anonymous 1984, Grubben and Van Slooten 1981).

Due to their ages-long selection under relatively shortday conditions (30° lat. north and 30° south of the Equator), various *Amaranthus* species are sensitive to longday in their flowering habit. Thus, the grain amarants from Peru or Mexico will not set flowers in the European or North American summer, but remain there for a long time in a vegetative phase. When experimentally the daylight was cut to less than 8 h, the flowering process was initiated.

Few diseases and pests are known: *Pythium* attacks young seedling plants. *Alternaria* may damage adult plants. In North America some sucking insects, like *Lygus lineolaris*, inflict damage on immature grain stalks.

In the following we give a short description of several species, considered by ethnobotanists as Indian domesticates. These are the South American *A. edulis, A. caudatus* and *A. quitensis*, and the Central American taxa *A. cruentus* and *A. hypochondriacus*.

### *A. hypochondriacus* L.

*Synonym: leucocarpus* Wats.
Gautli, bledo, huautl
$2n = 32$

This venerable grain amarant of Indian tribes in Central America was once the staple food of the Aztecs. It was harvested yearly by the thousands of tons, until the crop was banned and its owners persecuted by the Spanish invaders. Its probable wild form is the diploid species *A. hybridus* (which has black seeds). One of the first steps during Indian domestication was to select clear-coloured seeds (= "leucocarpus") (Agogino 1957).

The inflorescences have stiff spikes and the flowers are clustered into dense, bracteate glomerules. The seeds contain 63% starch and 13–15% protein. Some botanists have tried to separate existing cultivars into "Aztec grain type", "Mixteco grain type", "Guatemala grain type" and finally "Nepal types", which have been selected from Central American introductions and have gained wide extension in Asiatic mountain regions. All these accessions were part of a worldwide "Save the gene pools" project, sponsored by different academic institutions. In Mexico, Ciamex at Chapingo led in this work. In Guatemala, for many years Bressani at INCAP has been active (Bressani 1985) in the improvement of the nutritional values of amarant grains, and recently the Rodale Research Center at Kutztown, Pa. took a leading role in the propagation of this ancient crop. Several North American State Universities like Colorado, Iowa, Kansas, Michigan, Montana und Utah, and Research Units of the USDA are studying the genetic variability and improving the yields, which have already doubled over the last few years (Lorenz 1981, Joshi 1981, Becker et al. 1981).

### *A. cruentus* L.

Purple amaranth, queue de renard, Fuchsschwanz
$2n = 34$

Of Central American origin, but less important than the Mexican "gautli". Plants are very bushy and branched near the ground. Instead of big inflorescenses on the main stalk as in other *Amaranthus* species, this plant develops many small seed heads on side branches. which show severe seed-shattering. Its supposed wild form is *A. powellii*. The grain types have white seeds, whilst the betacyan-rich vegetable forms have dark seeds; the latter type is used to extract red dye for food-colouring (= var. *"sanguineus"*).

Archaeological finds from Central Mexico (MacNeish 1967) (e.g. Tehuacan caves, recovered there from different levels, the oldest being 4000 B.P.), confirm the early use of *A. cruentus*. In the same region it is still grown on a small scale by natives. For breeding purposes it is considered as the most adaptable of all amarants and flowers even under very different daylengths. Furthermore its growing cycle is considerably shorter than that of *A. hypochondriacus*.

### A. quitensis H.B.K

Huisquelite
2 n = 32

This is a semi-weed and probably gave rise to the cultivated "sangorache" amarant in Ecuador. There it is appreciated as an ornamental plant and used for dyeing food and drinks. Sangorache means blood-red and its inflorescenses are brilliant red. The seeds are shiny black (Heiser 1964).

### A. caudatus L

Kiwicha, achita, chaquilla
2 n = 32

Its areal includes North Argentina, Bolivia and Peru. It is easy to recognize by its tail-like, hanging inflorescenses. The plants are sensitive to photoperiodical changes and produce grains only under shortday conditions. For this reason they do not suit the European longday environment (Coons 1982). It is interesting that the early European *"Kreutter-Bücher"* (Herbals) depict plants of *A. caudatus* (e.g. Gessner p. 246,1561, *Hortus Germanicus*). From Europe this grain amarant soon reached the Himalayas and the highland of Abissinia.

According to Hunziker, *A. caudatus* has been used for at least 5000 years, which is testified from North Argentinian grave sites. There are also other finds from 4000–year-old Andean tombs.

*A. caudatus* is a typical shortday plant, well adapted to 2000–3600 m altitude. It has the advantage of producing saponin free seeds, but unfortunately the grains are not held tightly in the ripening inflorescenses, many shattering to the ground. Therefore the yields are rather low, calculated in experimental plots between 1000–3000 kg/ha. To improve this antique cultivar, Peru has now, since 1986, 100 ha in experimental multiplication, less than the USA, where 300 ha

were sown in 1987. The National Academy of Science in the USA, thanks to the efforts of N. Vietmeyer, is paying its special attention to this promising grain crop for marginal regions.

### A. mantegazzianus Pass.

Synonym: *A. edulis* Speg.
Millmi, chaquillon
2 n = 32

We mention here an interesting local selection of the Calchaqui Indians, a higher developed peasant tribe of North Argentina, cruelly persecuted and dispersed by the Spanish conquerers. They had their own regional selection of grain amarant ("trigo de Inka") which is actually nearly exterminated (Hunziker 1952). We studied this species in our plantations of the provinces of San Luis and Mendoza under semi-arid climate conditions and observed its high green-fodder value (Fig. II.1).

Feeding experiments have confirmed the high nutritive value of *A. edulis*, both in

**Fig. II.1.** *Amaranthus edulis* selected by a farmer from the author's seeds

leaves and grains. The grain protein is rich in essential amino acids (like lysine), important for human diet; 100 g protein contain 6 g lysine. This value exceeds that of milk.

There have been extended discussions about the species status of *A. edulis*, often refuting it and subordinating the taxa under *A. caudatus*. Since Pal and Khoshoo (1977) demonstrated that the offspring of crossings between typical *A. caudatus* and *A. edulis* were often abnormal, with considerable seedling mortality, retarded growth and virus-like deformations, we are inclined to maintain the binome *A. edulis* Speg. or *A. mantegazzianus* as distinct good species. It seems that genome and plasmon of the two taxa underwent a certain differentiation. The same authors, with the help of colchicine, produced tetraploids. They reported a gigas effect, which did not disappear even in the tenth generation. These plants were rather stout, did not lodge, had 88% fertility and bigger seeds (2.5 times heavier than diploid seed grains).

A disadvantage for cultivation and especially its mechanical harvesting is its shattering propensity. Seeds drop to the soil when they are ripe. One of the most important tasks for *Amaranthus* breeders is to select non-shattering biotypes and combine this factor with the production of larger seed kernels, and reduced branching of the stalk.

The Chilenian explorer Claude Gay (1849) was the last to have ovserved *B. mango* in cultivation. He met small cultivated plots on the island of Chiloé, near Castro, in the year 1836. From this place there still exists a unique exsiccatum (Nr. 138, typus), now at the Herbarium in Paris. The earliest Chilenean botanist, Molina, had mentioned "mango" in his book published in Italy in 1782, but he stated there that he had never seen this cereal, which the natives called "mangu" or "teca" and the bread from it "covque". Obviously already at that time the Araucanos were abandoning its cultivation.

The oldest citation is from the year 1633 in the book *Novus Orbis* of J. de Laet, (published in Batavia, libr. XII, chap. II, p. 482) about the agriculture of Chile...: "The natives plant there another cereal, which they call 'teca'. Its leaves are only little different from barley. Its culms reach 'media braza española' height. Its seed size is somewhat less than Secale. The Indians roasted these grains and ground them to a sort of flour". The search for a possible wild ancestor in meridional Chile and Argentina is complicated by the huge quantity of wild-growing *Bromus* species. Parodi opined that the best-indicated for emergency human nutrition would be *B. fonckii* Phil. and *B. stamineus* Desv; but that the real *B. mango* survives among them is highly doubtful (Parodi and Hernandez 1964).

---

GRAMINEAE

---

## 2. *Bromus mango* Desv.

Mango, teca, (exterminated in the last century)

This primitive cereal, which Araucanian Indians may have selected from wild-growing *Bromus* species in the most Southern part of the American continent, is, of course, no "neotropical crop", but we include it here to rectify some tendentious assertions that this definitively exterminated Indian cereal still exists and deserves and needs high priority for FAO collections (Esquina-Alcazar in an IPPGR Report 1985). In reality *B. mango* is as "dead as a dodo".

---

CHENOPODIACEAE

---

## 3. *Chenopodium* spp.

The genus *Chenopodium* consists of more than 250 species, mostly "weeds", which grow on all continents. In plant systematics it is closely related to the genus *Amaranthus* (Sherma and Day 1967). They share some morphological characters (e.g. the curved embryo), and also the synthesis of betacyanin (instead of anthocyan). Sometimes the two genera are combined under a practical aspect as pseudo-cereals, in view of their millenarian use in the neotropical mountain

isolated cultivation

frequent cultivation

**Fig. II.2.** Present distribution of *Chenopodium quinoa* in South America

region, before the Eurasian cereals arrived. Three *Chenopodium* have been cultivated: *C. pallidicaule* and *C. quinoa* in South America and *C. nuttaliae* in Mexico (Fig. II.2).

### *C. quinoa* Willd.

Quinoa, quinoa, Reismelde

$2n = 36$

NAME AND ORIGIN

The genetical origin of this widespread Andine crop is uncertain. Simmonds (1971) observed a disomic inheritance and concluded that *C.quinoa* must have an alloploid origin; furthermore he stated that the erratic sex expression is caused by a cytoplasmatic control of male sterility.

The wild-growing *C. hircinum* Schrad. has to be excluded from the phylogeny of quinoa, since Crawford and Wilson (1977) demonstrated the contrary with isozyme eyperiments. Possible wild progenitors may be certain primitive forms which the Kechua Indians call "alko-quinoa", or the Aymaras "isualla" and "ashpa". Such plants have small deciduous fruits and a hard seed testa. Heiser and Nelson (1974) consider this var. *melanospermum* as the most probable wild progenitor of cultivated quinoa.

There is abundant archaeological proof of the early existence of *C.quinoa* south of the equator. For example, Uhle discovered seeds in graves of the Peruvian coast at Tarapaca and Arica. It should be mentioned that today no cultivation of Quinoa is undertaken in this southern part of Peru.

*C. quinoa* is a very ancient crop plant, which ensured the survival of millions of natives in South America in pre-Columbian times. It must be stressed here that it was not only the main grain crop in the Altiplano of the present-day Bolivia – Peru – Ecuador, but certain lowland races called "quingua" have been sown along the western slopes and coastal regions of Chile, as far south as the humid island of Chiloé, by Araucan and Mapuche tribes. We observed the rare existence of such southern races of "quingua" on the island of Chiloé. The English traveller Stevenson found it there frequently in 1832. Several Chilenean botanists, beginning with Molina (1782), until Reiche and Looser, reported domestic use of *C. quinoa* in Curico, Arauco, Cautin, Osorno, south of Rio Bio-Bio and Rio Maule. From this we conclude that the Araucanian Indians had developed their own lowland races of quinoa, physiologically quite different from the high-Andine cultivars. Nevertheless, they are conspecific, as recent isozyme investigations have shown.

The disappearing Chile cultivars of *C. quinoa* should be urgently saved from extermination because they include valuable germplasm for future breeding of quinoa for the hot plains. For such purposes the high-Andine representatives of this species are, of course, not well adapted.

MORPHOLOGY

*C. quinoa* has an extremely plastic habit, which changes according to the environmental factors. We must keep this in mind

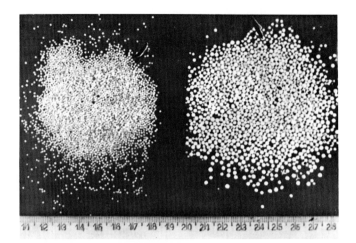

**Fig. II.3.** Progress in domestication in *Chenopodium quinoa* from the Altiplano, Bolivia and Peru. *Left* original landrace with small seeds. *Right* large-seeded selection

when describing this species. In poor soils and at high altitude, plants flower and set fruits when only a few centimeters tall, whilst in humid lowland and fertilized soils the same cultivar may reach 250 cm and have profuse seed production (Fig. II.3).

Plants bushy annual herbs. Stems 100–200 cm tall, branched, stout (1–2 cm diam). Leaves rhomboid mostly asymetric (12–15 cm long), at the top of the plants: small-elliptic, underside covered with glassy shiny papilla. Inflorescences 20–30 cm long. Flowers are concealed between small leaves, they are hermaphrodite and cleistogamous. Fruits glomerules with rich seed production, with fragile pericarp.

Seeds covered with perigone, which becomes detached when ripe. Seed size 1–2.7 mm); a thin episperm covers the embryo, which is typically curved. The seed coat may be thin in selected cvs., like Blanca real, or thick and dark as in the primitive form keitu, with high content of saponin. This glucoside produces a bitter taste and is the main obstacle to more general use of quinoa. It is interesting that in Peru a region exists (between Ayaviri and Sicuani) where natives maintain old landraces with surprisingly low-grade saponin and white testa. A Bolivian Experimental Station has selected a seemingly saponin-free variety called Sayama.

Seeds are very nutritious, with 13% protein on average, but may reach 22% in exceptional cases (Wilson 1980); 58% starch and 4% fat, and a small amount (3%) of sugar. Nutrition experiments concluded that quinoa albumen compares favourably to that of milk.

If we can trust North American newspaper reports, the NASA has recently included also quinoa grains as high energy food for astronauts. It seems that NASA is now emulating with its "space-crackers" the widely used "kispinas" of the mountain Indians, with unlimited conservation. In fact, quinoa grains have a high concentration of essential amino acids, with a lysine content twice that of wheat grains. Therefore it was a good decision of the Bolivian Government to order that all flour products must contain at least 5% of quinoa meal.

*C. quinoa* produces a huge leaf mass which has been analyzed as follows: 52% carbohydrate, 22% raw protein and 15% ash. Natives have long used it as good animal feed, but also here the high saponin percentage is a nuisance, which could be overcome by the selection of saponin-free mutants, which are, of course, genetically recessive and would become supressed in open pollinated quinoa fields. Polyploidy breeding appears promising in view of the fact that the genus has a tendency to genome multiplication, e.g. octoploid spontaneous biotypes of *C. album* exist in Eurasia.

### *C. pallidicaule* Aellen

Kanigua, kanahua
$2n = 18$

It must be considered as only a minor grain crop for marginal regions of the High

Andes, producing at least some yield even at the 4500-m altitude of the Puna. Regular planting occurs in 3300–4000 m, where in well-worked soils kanigua may yield 2000 kg/ha. There are precocious strains which ripen in 95 days, whilst others need 150 days for ripening. Unfortunately, the seeds shed freely, when the plants dry off (contrary to *C. quinoa*, which retains the seeds). Testa colour varies from green, red, brown to black. Seeds have several weeks of dormancy. The leaves have a rhomboid form and are 2–3 cm long. Their lamina is grey-green, with rough texture and covered with globular cells. Their saponin content is considerably higher than in *C. quinoa*. The flowers are hidden between dense leaf clusters.

The cultivar resembles the wild species *C. canigua* Cook, which can be found occasionally in the Altiplano of North Argentina and Bolivia-Peru, called by the natives in the Kechua language "kita canigua" (Tapia-Vargas 1976).

*C. pallidicaule* cultivation has not spread beyond the Altiplano; its main centre is north of Lake Titicaca (Dept. Puno, Peru), which covers over 90% of the actual 6500 ha cultivated in the High Andes.

As scarcely developed and primitive cultivars, kanigua plants do not all mature at the same time, so the harvest extends in the Altiplano from March to May. Plants must be pulled out before seeds have completely matured, to avoid major shattering. All this is laborious work for these long-enduring natives, who live at the very limit of human survival. The resulting minute seeds are dried in the sun and used as roasted "pito" or cooked in porridge. Even the remaining plant residues, after threshing, have their importance in the Altiplano. After burning, it produces an alkaline-rich ash that is in great demand among the native chewers of coca. If they use coca together with this "llipta", the stimulating effect is maximal. Gade (1970), who studied the ethnobotany of *C. pallidicaule*, declares about this rare Andine foodplant: "Perhaps no other crop is so resistant to the combination of frost, drought, salty soil and pests, or requires such little care in its cultivation. At the same time, few grains have as high a protein content as does canihua (13.8%)...".

With the aim of improving this primitive "pseudo-cereal" of the Puna, two Bolivian agronomists (Gandarillas and Gutierez 1973) induced artificial polyploidy. The authors indicate that the original size, which is very small (1 mm diam), was almost doubled after colchicine treatment. Furthermore, some increase in leaf size and total plant weight has been reported.

## *C. nuttaliae* Stafford

Huauzontle, quelite, chia, huah-zontli

$2n = 36$

NAME AND ORIGIN

This quinoa of Mesoamerica is not directly related with the domesticates south of the Equator, and it is not, as some botanists supposed earlier, a pre-Columbian introduction to the Highlands of Mexico (Heiser and Nelson 1974). On the contrary, the most probable ancestor is *C. berlandieri* Moq, wild-growing in North America from Arkansas to Missouri. Perhaps the Ozark Bluff dwellers spread it to Mexico, but it has not been recorded archaeologically.

At present *C. nuttaliae* is still sown in the states of Michoacan, Oaxaca and Tamaulipas, but over a continuously declining area. It seems that it is less used for grains, but more as a vegetable (a spinach-like herb called "quelite" with edible flower clusters). The main area is situated between 1200 and 3000 m altitude. The cv. Chia is still used as a grain crop. The fruits are relatively large (1.5–2.5 mm diam), whilst those of vegetable huauzontle have only 1.0–1.7 mm diameter. Their seeds have a very thin outer seed coat or lack it even completely. This facilitates quick, uniform germination, and consequently also equal maturation of the plants.

Hybrids between *C. nuttaliae* and *C. quinoa* are difficult to obtain, in spite of the equal chromosome number of $2n = 36$. This supports a separate di-phyletic origin, north and south of the Equator. In contrast to the southern varieties of *Chenopodium*, which offer certain breeding possibilities, it seems to us that improvement of *C. nuttaliae* has no great chances (Wilson and Heiser 1979).

## 4. *Zea mays* L.

Indian Corn, maize, milko, maiz, mais, Mais

$2n = 20$

### NAME, ORIGIN AND DISPERSION

In honour of the decisive contributions of several Indian tribes in the selection of wild forms and the final domestication and creation of this marvellous grain crop, the well-considered expression *Indian Corn* should more often be used. The common name in different European languages is derived from Mahiz, which is an Arawak word. The Aztek denomination was "cintli", related to the deity for fertility, Cinteotl. The maize plant was the object of manifold veneration in the peasant religions of Mesoamerican Indians. Uncounted ceramic objects in the Aztek and Maya cultures are adorned with maize cobs. The ornamentation of deities included maize, mostly in their head dresses, but also the funerary urns of zapotec dignitaries depicted maize ears. In this case it was even possible to determine them as the ancient Mexican corn race "Nal-tel" (Fig. II.4). In the secondary centre of maize domestication, in the high valleys of the South American Andes, uncounted artistic reproductions of *Zea mays* also exist. Such phytomorphic ceramics excel in their exactitude and make it easy to recognize the distinct Peruvian varieties.

Maize was totally unknown outside of America before the arrival of Columbus. All reports about "pre-Columbian" maize in Africa or in Asia (Jeffries 1955) or recently by Ji-Geng (1980), claiming its origin in China, must be considered as sheer phantasy. On the other hand, exact descriptions fortunately exist of the circumstances under which the first Europeans came to know this cereal. Columbus himself noted in his diaries that on the 5th of November 1492 his men found on the island of Cuba great fields of a strange new crop. They came back with the samples and the notice: "The Indians have a sort of grain which they call maize and make it into flour".

Columbus took seed samples back to Spain, where already in 1494 the first description of

**Fig. II.4.** Mythological representation of a corn god from Central America

this exotic appeared. It was "Flint maize", then common in the Carribbean region, with hard seeds and quick-growing; otherwise this Indian corn would not have ripened in the temperate climate and rather long-day photoperiod of the Iberian Peninsula. Quite different from the slow initial dissemination of other New World crops, *Zea mays* was spread quickly through the Mediterranean region and was soon recognized as a valuable cereal. Spanish and Portuguese seafarers spread it to Africa and Asia. Maize was being consumed on the Cape Verde Islands in 1535, on the North African coast (Tripoli) in 1550, and in 1560 in Egypt. From Japan it was reported in 1565 on the Island of Kyushu. In China, maize was planted in the Southern provinces, according to Li-Shi-Chen, in 1573; introduced possibly by land from Burma or Indochina. Portuguese ships brought maize to Zanzibar in 1643, to Reunion in 1690.

The earliest illustrated *Herbals* depicted maize plants (Bock 1539, Fuchs 1542 in Germany), sometimes under erroneous names like "Turkish corn". At this time *Zea mays* had already spread in hundreds of distinct varieties through North and South America. Indians who lived far north at the mouth of the St. Lawrence River (the boundary between Canada and the USA) knew maize as well as the Mapuche Indians at the River

**Fig. II.5.** *Zea mays* different landraces from South and Central America. (Science, Vol. 222)

Bio-Bio in South Chile. There is no other cereal in the world, selected and domesticated exclusively by "primitives", with a differentiation and expansion similar to the Indian corn. The creation and domestication of *Zea mays* by American Indians is the most unique plantbreeding event in the history of mankind (Fig. II.5).

It can be assumed for sure that "zapalote" maize moved from the northern part of Central America slowly southward, being adapted (i.e. selected) to other photoperiodical and climatical environments and perhaps introgressed by other Maydeae. Once arrived at the Isthmus of Panama, sea-travelling Indians probably spread the American cereal to the South American subcontinent, whilst others took seeds with them when they crossed the mountain gap of Darién. In any case, an important centre of different maize races developed early in Northwest Columbia.

According to archaeological finds, maize reached Ecuador in 3000 B.P. and approximately 500 years later some irrigated lands of coastal Peru. From there it was introduced in the Andine valleys, where it later formed a remarkable "secondary centre" of genetical diversity. Unquestionably the "sugary" and "tunicate" variety, also the "giant soft maize" were developed there. The question remains open, how the South American Indians achieved such striking seed characters, which had never arisen before in the primary gene centre of Mexico. From the Peruvian Highland *Zea mays* was disseminated to mid-Chile. However, we must question certain maize remains which had been declared to be as old as 4000–5000 years. According to Bird (1980), they have a fresh aspect and look more like modern maize. In any case, maize crossed the Rio Bio-Bio southward and reached the Island of Chiloé (43°S) on the Pacific coast, and Mendoza (33°S) on the East-Andan side, during the maximum expansion of the Inka Empire.

Strangely enough, there was also in pre-Columbian times a reverse movement of southern maize (mostly Peruvian "sweet-

**Fig. II.6.** Teosinte growing wild (probably introduced) in Paraguay and used as forage plant

corn" and "pop-corn") to Central America, where it was found together with ceramic remains and woven clothing.

Fortunately there exist many archaeological records of maize in different regions of America. The oldest material belongs to the Tehuacan caves. (Mac Neish 1967), and consists of small 2–cm–long cobs, with small glumes, dated at 7200 years B.P. The Bat caves (southern USA) also yielded many cobs of reduced size, more than 4000 years old. The Bat caves were occupied by natives for several thousand years. At the beginning they were plant-gatherers, but the later dwellers also practised some primitive agriculture. Their debris and the garbage of plant material was accumulated to a depth of nearly 2 m, and allowed the recovery of a great quantity of maize (Randolph 1976).

In comparision with thousands of remains in Mexico/USA, South America is rather poor in fossils of maize plants, but there is a great wealth of pottery and ceramics from pre-Columbian times, depicting maize cobs. The "Ur-Mais", i.e. the wild-growing ancestor of the cultivated corn, remains a great mystery. Three different hypotheses, inspired by leading authorities in *Zea mays*, have dominated the battle field for decades, and no near victory is in sight. As an expert ethnobotanist recently declared: ... "volumes of words are still being created, concerning the origin of maize"... (Smith 1986).

We took our stand in our book *Tropische Nutzpflanzen* (1977, pp. 36–50) and still favour the maize grass = *Zea mays* subspecies *mexicana* Schrader as direct ancestor. We must, however, admit that cytogenetical and biochemical proof exists for a phylogenetical relationship with another "maize grass", *Tripsacum* (Smith and Lester 1980). Both are members – together with corn – of the American Maydeae.

As Rao and Galinat (1974 and 1976) have demonstrated, the *Tripsacum* chromosomes Tr-7, Tr-9, Tr-12 and Tr-13 have homologues to the maize chromosomes Nr. 4 and Nr. 2. The cytogenetic dates have shown that, for example, *Tr. dactyloides* and *Tr. floridianum* has at least six loci in common with maize chromosome Nr. 2. Is this not sufficient proof that the genus *Tripsacum* and the genus *Zea* evolved from a common ancestor, confirming cytologically the opinion of Mangelsdorf and Reeves, who as early as 1939 stated this, based on morphological reasons.

Mangelsdorf (1947, 1965, 1974) dedicated his whole life to elucidating the origin and evolution of Indian Corn. According to his earlier "Tripartite Hypothesis" *Teosinte* was under no circumstances ancestral to maize, but instead the result of crossings of a (now extinct!) "wild maize" with a species of genus *Tripsacum*. Beadle (1972) presented a series of strong arguments in favour of the old "teosinte hypothesis", which dates back to the last century but had been discarded under the influence of the Mangelsdorf school of thought. In fact the "tripartite hypothesis" has for several decades dominated the discussions on the origin and evolution of Indian corn. "It is repeated in encyclopaedias, in compendia and in many textbooks, sometimes being transformed from an hypothesis to an established fact" (Beadle 1980) (Fig. II.7).

Recently "the catastrophic sexual transmutation theory" (Iltis 1983) appeared, which

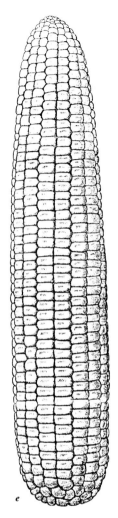

**Fig. II.7.** Corn evolution according to Beadle (1980). Domestication of corn from the wild grass teosinte by a few millenia of human selection has been reconstructed by breeding experiments undertaken by the author. All the specimens are drawn approx. half life-size. The "spike" of teosinte **(a)** is equivalent to the ear of corn and consists of a single row of kernels in hard, shell-like fruit cases. On ripening, the spike shatters, scattering the seeds. Crosses between teosinte and corn yield a modified teosinte **(b)**, which may be similar to an early transitional form. A single mutation in teosinte can also give rise to a tonicate variety **(c)**, in which the hard fruit cases have been converted into soft, husklike glumes from which kernels can be threshed with ease. This mutation may well have been a crucial step in the domestication of teosinte. Crosses between teosinte and modern corn give rise to small, primitive ears **(d)**, which are similar to 7,000-year-old archaeological specimens found in the southwestern USA and Mexico. Modern corn **(e)** is a biological monstrosity created by prolonged domestication. It is well adapted for production of grain but is unable to survive under natural conditions

ignited the decade-long controversy anew (Galinat and Mangelsdorf 1984). For this reason we prefer to postpone the debate.

Iltis (1983) asserts that some 8000 years ago a profound morphological change occurred in the maize plant under human selection, i.e. by native hunter-gatherers. Iltis believes that they observed abnormalities in the flowering habit and favoured a tighter packing of the grains produced by sexual disturbances. Ilties call this a sexual switch from male to female in the tassel flowers ... "since the corn ear is always produced at the tip of a branch, and in exactly the same position as a teosinte tassel spike, it must have evolved from a teosinte tassel spike, its morphological homologue..." Doebley (1984) discussed germplasm interchange between maize and teosinte and supports this hypothesis with isozyme evidence for reciprocal genetic introgression between maize and annual teosinte, but Galinat (1985) is critical.

## MORPHOLOGY

*Zea mays* is an annual, monoecious Gramineae, which surpasses all other annual grasses in size and biomass production. The uniqueness of maize among other cereals lies in its inflorescenses. In contrast to wheat, rice or sorgum, maize has a sharp functional separation of male ("tassel") and female ("spike") flowers. The tassels are formed at the end of the stalk and the female ears in the middle section of the plant.

The stems vary in length from 70 cm to 7 m (in certain tropical biotypes even more) with conspicuous nodes (6–12), from which sometimes adventive raizes develop. The stalks are strong, with a diameter between 2 and 4 cm.

The leaves (linear-long, 20–40 cm) are borne alternatively on the nodes, and their sheaths (with a 5-mm-long ligula) partially enclose the stem. Tropical varieties have leaves 100–140 cm long.

Inflorescences: In maize the male and female organs are borne separately on the same plant. The staminate flowers are situated at the top of the central stem. These "tassels" are composed of numerous lateral branchlets, which bear many spikelets. They contain male flowers which shed the pollen, but sometimes also have a rudimentary gynoeceum. The quantity of pollen is enormously high: per plant 2–4,000,000 grains. The functional pistillate flowers are located in dense ears on a short lateral branch, which emanates from a node axil, in general at the middle of the stem. This female organ is covered by several transformed leaves (= husks). The flowers are produced in spikelets, which enclose two florets. Whilst the lower is sterile, the upper is pistillate. Between a short lemma and palea emerges a long (20–40–60 cm) thread-like style, which receives the pollen. In general there is no auto-fecundation, because the flowering is protandrous. At the base of each spikelet are two glumes without flowers (= empty glumes). Once fecundated, the female inflorescences develop large "cobs", which contain between 600 and 1000 seed kernels, tightly enclosed in several layers of protective husks. Once dry-ripe, a situation unique in the whole plant kingdom exists: the seeds sit so tightly on the cob that they cannot be disseminated on their own. In fact, the maize plant has lost the ability of natural dispersal. In the case of a whole ear falling, hundreds of seedlings would emerge on a space a few centimeters square, should they come in contact with moist soil. Indian corn with its monstrously built ears relies on the help of man for propagation. Once domesticated by

Indians, their main grain crop needed constant intervention in harvest, seed cleaning, storing and sowing: it had become a typical "cultigen".

There are many well-established races of *Zea mays*. The most primitive are the "chapalote" (= zapalote) with isodiametric kernels. They still exist in the Mexican lowlands (100–600 m alt). In the coastal region of Chiapas "zapalote chico" is still planted as a local cultivar. The plants are quick-growing and remain short. They have small cobs, which are covered with thick husks, with strong anthozyan content.

"Zapalote grande" has its major area in the hill region (500–600 m alt). In Mexico four ancient Indian corn races still exist today. They have maintained their identity since time immemorial, in spite of the overall existing open pollinization, possibly by geographical isolation and different flowering time. According to Mangelsdorf (1974) these are: Arocillo amarillo, chapalote, naltel and palomero toluqueño. They belong to the "pop-corn" group.

It is traditional to classify the maize races in accordance with their agricultural use and kernel morphology. While considering this of little taxonomical value, we nevertheless follow this scheme and separate the main cultivars in:

Flint Maize (= *indurata*)
The ears are cylindrical with few tassel branches and a flag leaf. The seed kernels are yellow, very hard, and contain in their interior very little soft starch. The "Northern Flint" race dominated the Indian fields of North America (Eastern part of USA) for several hundred years before its discovery by Europeans. Its place of selection may be in Guatemala. Its chromosomes are knobless.

Dent Maize (= *indentata*)
Characterized by a "dent"-like depression in the crown of the seed. The kernels are white, large with a central core of floury endosperm which shrinks when drying. The sides are hard corneous starch. Its outer colour is white or yellow. Its supposed origin is from a hybridization between Indian maizes from Mexico. Dent corn is the traditional maize of the North American corn belt. "Southern dent" was introduced from Mex-

ico. It has tapering ears, many tassel branches, no flag leaf. Its chromosomes possess 5–12 knobs (some local selections have 1–8 knobs).

It has various practical advantages in comparison with flint maize: higher digestibility for animals, more suitable for production of glucose.

### Floury Maize ( = *amylaceous*)

In general large, soft kernels, which can be broken easily. This is the common maize of the Andine natives, who selected in pre-Columbian times certain curious forms with "giant" kernels, called "pescoc runtum" in the valleys of Cuzco. They have a long growing period. Several schemes to cultivate this variety in other parts of America have failed. Hybrids with "amylaceous" corn have aquired industrial interest, because their starch consists of 80% amylose.

### Sweet Corn ( = *saccharata*)

This variety has a South American origin. It is characterized by translucent horny seed structure, which becomes wrinkled when dry. This "saccharata" biotype originated in a recessive gene mutation on the 4th chromosome. The factor "su" prevents the conversion of sugar into starch. The immature grains have a translucent, glossy appearance. In the last decade sweet corn has become a very popular product for fresh consumption and canning.

### Pod Corn ( = *tunicata*)

The kernels of this Indian corn are covered with glumae. This variety exists only in restricted regions of South America, mostly in Bolivia and Paraguay. The Guarani Indians use it even today and their witchdoctors call it "avati-guaicuru" and plant it in hidden places.

### Pop Corn ( = *everta*)

A very ancient variety of Mexican origin. It has small ears with small seeds and very hard endosperm, covered with a glossy exine. When put into the fire the seeds explode, popping up to 20 times their original volume. In this state they could have been eaten by natives, long before the fabrication of vessels or cooking ceramics had been invented.

### Waxy Corn ( = *ceratina*)

Instead of the usual starch, it produces amylo-dextrin in its seed kernels. As this curious variety was first reported from China, it was thought that it must have an east-Asiatic origin, even starting the claim that the whole maize genus may be of Asiatic origin (Stoner and Anderson 1949).

Today we have certain proof that this "waxy mutation" occurred also in different places in America. According to Brieger, the gene "waxy" exists even in the Paraguayan Indian corn. The gene "waxy" has been localized in the chromosome No. 9.

For nutritional purposes *ceratina* maize may be preferred for its easy digestibility and has recently aroused the interest of the health food industry.

Recently has been discovered in the "Sierra de Manantlan" (Mexico) the most primitive wild relative of Indian Corn, designated by taxonomists (Iltis, Doebley, Guzman and Pazy 1979) as Zea diploperennis. The species has 2 n = 20 chromosomes, the same number as Indian Corn, but perennial rhizomes and is tolerant to seven virus which attack maize fields. Intensive hybridization in Northamerica was undertaken to create "perennial maize". Actually the hybrid plants are 3–4 m tall, very promising as silage forrage.

### ECOLOGY

True to its origin in the hill region of Mexico, *Zea mays* is by nature not a tropical crop, but the Indians had patiently selected landraces for the humid tropical lowlands of Yucatan-Guatemala, expanding them as far as the Isthmus of Darién. On the other hand, mountain tribes of Ecuador and Peru created Indian corn for the High-Andine valleys. Such an early differentiation of *Zea mays* in opposing ecological and photoperiodical directions underlines its astonishing genetic plasticity and adaptability. In fact, maize can be cultivated in more divergent climates than any other cereal. Optimal conditions are mean temperatures of 25° C, warm soils during germination, with sufficient moisture. Maize can tolerate high temperatures (38° C), but is very susceptible to light night-frosts. Exceptions are reported from Alti-

**Fig. II.8.** Corn from Guayana, a selection several metres high grown by the author for forage in Trinidad

plano maize and European selections from Sweden and Russia. Our collection of Sikkim maize (high valleys of the pre-Himalayan Tista region) contained cvs. which ripened when only 70 cm high already in 70–80 days, at the same time showing a fair frost tolerance under −2°C. Similar features have been reported from the Russian variety Viatka.

During flowering, all maize races are very sensitive to water deficit, which may damage pollen formation and reduce the silk development (Goldsworthy et al. 1984).

Furthermore, one should bear in mind that maize, especially the tropical races, is very

sensitive to photoperiodical influences and needs shortday conditions as flowering incentive. In Guiana I observed local Indian corn which needs more than one year (14–15 months!) to produce ripe cobs (Fig. II.8) reaching 5 m in height.

In contrast, European plant breeders created quick-ripening, cold-tolerant varieties, like Velox, Iskra etc., which produce short plants with fair yields in 150 days. The tropical regions contribute at present approximately 50 mio t/year. The Central American countries (Mexico, Guatemala, Honduras) have a high maize consumption, 140 kg yearly per inhabitant, whilst Europeans consume only 11 kg per capita.

There is a principal difference in the natives' use of maize. The Central Americans use the ground flour and make flat cakes from it ("tortillas, tamales") whilst the Andine population prefer the cooked or roasted kernels ("choclo cocido", huminta, mote). The usual cooking of maize kernels in water with ash (using its alkaline reaction to soften the pericarp) is a very ancient custom in Bolivia.

## PRODUCTION AND ECONOMY

*Zea mays* cultivation circles the globe between 58° lat. N. to 45° lat. S., growing equally well at sea level and at 3800 m altitude in the Andes, extending its areal each year, thanks to newly developed strains for marginal climate. Maize has now acquired the second position in world grain production in the year 1987 with 535 mio t; followed by rice with approximately 466 mio t. The spectacular increase during the last decade is demonstrated in Table 6.

At present the USA produce almost half of the world production of Indian corn, more than 210 million t, which represents a value of 25,000,000,000 US $. In comparison, the leading country in South America, Argentina, produced in 1982 9 mio and in 1983 7.5 mio with decreasing output during the past years (Table 6).

**Table 6.** World maize production (in millions of tons)

| 1971 | 1972 | 1973 | 1979 | 1980 | 1982 | 1983 | 1984 | 1985 | 1987 |
|------|------|------|------|------|------|------|------|------|------|
| 293  | 310  | 313  | 423  | 404  | 438  | 446  | 456  | 490  | 492  |

Comparing the areas sown with Indian corn in the whole of America, it is astonishing that the surface in the USA (approx. 25,000,000 ha) is slightly inferior to that of Latin America (26,000,000 ha), but the yield is nearly four times larger.

Table 7 (adapted from FAO production yearbook 1981) lists the ten leading countries in maize production, comparing their yields and areas.

**Table 7.** The ten leading countries in maize production, comparing their yields and areas (FAO 1981)

| Country | Yield in 1000 t | Area in 1000 ha | % of maize world crop (1981) |
|---|---|---|---|
| USA | 208 314 | 30 200 | 46.1 |
| China Mld. | 61 601 | 20 537 | 13.6 |
| Brasil | 21 098 | 11 491 | 4.7 |
| Mexico | 14 766 | 8 150 | 3.3 |
| South Africa | 14 650 | 7 000 | 3.2 |
| Argentina | 13 500 | 3 500 | 3.0 |
| Romania | 11 200 | 3 150 | 2.5 |
| USSR | 8 000 | 3 545 | 1.8 |
| India | 7 000 | 5 800 | 1.5 |
| Philippines | 3 176 | 3 319 | 0.7 |

In South and Central America the major part of the maize yield (with the exception of the mechanized production in big estancias of Argentina and Brazil) is assigned to human nutrition, quite differently from in North America, where only 10% enters the human food chain. Therefore the breeding directions are fundamentally different.

The yield improvements in irrigation fields of North America are really impressive. Some years ago (1970), 6000 kg/ha under optimal conditions of field technique and artificial manure were considered maximum, but only 10 years later breeders from New Jersey reported 22,700 kg/ha, with the maize hybrid Os Gold SX 5509. High fertilization and plant density of 92,000 plants/ha in irrigated fields were the prerequisites, but recently we have been informed that the genetical potential of corn hybrids in the USA would allow even 31 t/ha, given optimal fertilization and irrigation.

## GENETICAL IMPROVEMENT AND RESISTANCE BREEDING

*Zea mays* has a tremendous yield potential, if we compare the 3-cm-long ears of archaeological Natel maize with the 30-cm-long cobs of modern maize selections or hybrids. Especially the creation of "hybrid corn" has revolutionized the whole cereal production. It is scarcely known that the creation of "hybrid corn", considered today as a unique result of efficient plant breeding in the USA, goes back to Charles Darwin. Between 1870 and 1880 he performed controlled crossings between different maize races, and compared the results with those of self-pollinated plants. Darwin was the first biologist to observe that under identical environment conditions crossings between unrelated varieties of *Zea mays* produced hybrid vigour. He concluded correctly that increased yield occurred only when he combined "diverse heredities" (Darwin 1877: *The Effect of Cross- and Self-Fertilization in the Vegetable Kingdom*.

The practical application of the heterosis effect in the 1930s by Wallace in the USA doubled the corn yields between 1930 to 1950. Quite rightly this was called the most far-reaching development of this century in the field of applied biology (Fig. II.9). Outstanding geneticists like Shull, Jones, East, Hayes and McClintock prepared the experimental bases for this achievement. Ironically, at the same time another large maize-producing country, the Soviet Union, drastically diminished the corn yields. The spiritual father of this disaster was Lyssenko, who forbade the application of "hybrid vigour" as a product of "western capitalism and morganism", as long as he presided over the Academy of Agricultural Science in the USSR.

In spite of all its undisputed advantages, "hybrid corn" has one drawback. It should not be sown in impoverished soils or under lack of sufficient precipitation. This is often the case in under-developed countries. There it is better to use the locally adapted varieties which still yield under poor environmental conditions (Hoffmann et al., 1985).

The maize kernel has a rather low (6–8%) protein content. There are different ways to

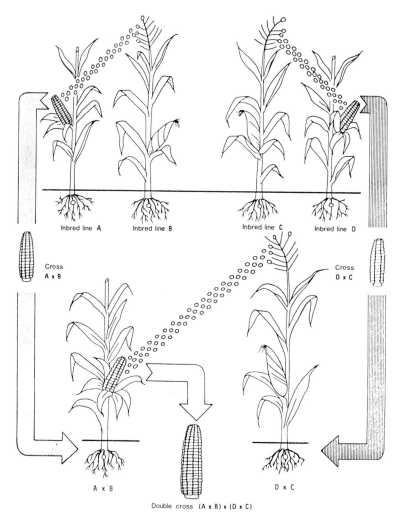

**Fig. II.9.** Scheme showing how maize hybrids with higher yield are produced; double cross $(A \times B) \times (D \times C)$ causes considerable heterosis effect in the resulting hybrid plants. This effect is drastically reduced if a planter uses the seeds from such ears for further multiplication

change this. The theoretical possibility of a re-construction of the maize caryopse has been demonstrated in a 70-years-long experiment in the USA, with recurrent selections in local varieties. In one direction, 22% protein was achieved, in the reciprocal case it was lowered to 3%. The reason may be found in increase of embryonal tissues, which are rich in amino acids. For example, in the var. Doce Cuban 47, the portion of protein-rich embryo is 22% of the whole kernel.

Another way to improve the protein content of the maize caryopse is to increase the aleuron layers which surround the kernel. The conventional commercial varieties have only one cell layer. Rather by chance, a Bolivian local Indian corn has been found, called "Coroico", which has five layers of aleuron cells. Crude protein in such cells varies from 35 to 38%. This must be a casual mutation, preserved by Indians of the Eastern Andine valleys, because until today additional search in several countries has not resulted in any similar anatomical character. It seems that "Coroico" is the sole source of multiple aleuron layers in *Zea mays*.

A quite different approach to the improvement of the protein components in the maize kernel was undertaken by geneticists of Purdue University. With multimillion funds, they have worked since 1964 on the incorporation of the "opaque-2" gene in high-yielding corn varieties. "Opaque-2" produces high lysine content, often 70% more than in common maize strains and double that of triptophane. Later experiments combined the factor "opaque" with the genes "floury 2" and "sugary 2", resulting in new biotypes with better digestibility. The combination with the cultivar Illinois High Protein (which has 22% protein) finally produced breeding lines with 17% protein and 3.6% lysine; however, some problems still exist in the practical field.

The maize plant produces in general one or two huge cobs. This morphological character is traditionally – since Indian times – considered as advantageous for practical reasons during the harvest.

Certain Indian corn races have, however, always been known with many small ears. Such "prolific varieties" have generated recently special interest among maize breeders. Quantitative yield studies have shown that there is a high correlation between ear number (5–7) and higher yield. Such plants survive better under climatic stress and are considered as an insurance against losses by drought because in general they have a larger root system and a higher photosynthetic efficiency.

Another breeding project which may give unexpected results in a still distant future are the "additional bastards" between *Zea mays* and *Sorghum*. In view of their origin on different continents and their profound genome differences, most scientists considered it impossible to obtain viable offspring. Nevertheless, some plant breeders at CIMMYT in Mexico reported the first positive results in 1976. From more than 25,000 crossings between different maize and sorgum strains, they obtained one sole hybrid plant which ripened. Due to severe pairing difficulties in diacinesis, no viable pollen was produced in the terminal inflorescence; but from backcrossings with pollen of maize some progeny developed.

Maize planters are sometimes confronted with a strange phenomenon, which seems to contradict the well-known Mendelian segregation of seed colours in simple crossings. The expected colour segregation occurs already in the ear of the mother plant. The reason is that the endosperm of *Zea mays* has a triploid constitution. This is the result of a "secondary fertilization" of the polar nuclei of the embryo sac which receive the second nucleus of the pollen tube. From this union the endosperm develops. Dominant genes transferred by the pollen are visibly manifested in the endosperm tissue. For example, yellow or violet is dominant over white, "sugary-crinkled" over smooth seed surface.

Kernels on the ear of the mother plant show the dominant factors of the father plant. Already in 1882, Focke introduced the term xenia to describe this curious immediate effect of foreign pollen on another individual.

## Breeding for Resistance Against Pests and Diseases

As one of the biggest monocultures in the world, it was inevitable that Indian corn plantations were "punished" with innumerable parasites, virus, bacterias, fungus etc. They became cosmopolites, the same as the large agrichemical industries which contribute to and live from combating them. To reduce the huge costs of the "chemical warfare" against "army-worms" (*Agrotis, Spodoptera,, Pyrausta* and *Heliothis)* and "blight" (e.g. *Helminthosporium, Cercospora* and *Diplodia*) it would, of course, be more convenient to build up genetical resistance; but this needs time, and a profound change in cultivation methods.

Warnings had long been issued, but the institutions and personalities with the "power to listen" did not do so, until "blight" suddenly struck the United States in 1970, reducing the maize crop of the southern states by half, causing millions of dollars damage. Whilst the old Romans, long before the Christian era, implored their god Robigus to protect their cereal fields from "rust" diseases, probably the Maya peasants cursed their gods, who were unable to save the maize fields, when a completely new disease (according to Brewbaker 1979, Mosaic Virus) decimated the traditional maize yields. The consequences were the total collapse in 900 A.D. of the Maya civilization in Central America.

### DISEASES CAUSED BY VIRUS

a) *Maize streak virus*
   The symptoms are severe chlorosis, arranged along the veins. There are many host plants. The vectors are *Cicadulina* leafhoppers. Damage is especially high in moist climate.

b) *Corn stunt virus*
   The infected plants have a stunted, bushy appearance, due to this virus, which causes abnormal growth of shoots from axillary buds. The vectors are leafhoppers. The ears develop rudimentarily.

c) *Corn mosaic*
   The symptoms are yellow irregular mottle on darker green leaves. Vectors are aphids, which live on many host plants. Losses are severe.

d) *Striped dwarf* (= enanismos rayado in Venezuela)
   This virus appears to be endemic in tropical maize fields of South America with a very low incidence (0.5–4%) in small fields of the natives. However, it increases in dangerous form (50%–60% losses) in extended monocultures of large farms, where it is transmitted by sucking insects, like *Peregrinus maydis*. The disease differs from the common "corn stunt", but has some similarity with the "white stripe virus" reported from Cuba and Hawaii. During my maize breeding work at the Island of Trinidad, I observed heavy infections in 1971 in our experimental fields in Chaguaramas.

### BACTERIA

Between the different bacterial diseases which affect *Zea mays*, the most dangerous is "bacterial wilt" caused by *Bacterium stewartii*. It attacks the different corn races, but sweet corn suffers most. At one time bacteri-

al wilt paralyzed the canning industry for sweet corn in the USA, until tolerant cultivars had been found, which are based on dominant resistance genes. Minor damage is caused by *Xanthomonas* and *Pseudomonas*, which prosper in humid environment.

### FUNGUS

*Ustilago maydis* ("smut")
This severe disease is distributed worldwide and is well known for its monstrous deformation of the ears. The disease is not carried over on the grain, as in other cereal smuts; therefore chemical seed treatment is of little value. The spores overwinter in the soil. Breeding of resistant lines has been quite successful. As a curiosity it may be mentioned that in certain Asiatic countries *Ustilago*-caused shoot deformations (full of black spores) are appreciated as vegetable and commonly sold on the market places. We have observed a similar custom also in native markets in Bolivia-Peru.

*Helminthosporium* spp.

Several distinct species of this fungus cause lesions on maize leaves and, worse, they attack the ears during development. Its spores survive in the soil and enter all parts of plants, under humid warm climate conditions, especially during morning fogs. Under such favourable conditions the "blight" spreads incredibly quickly over long distances, because the fungus completes its life cycle in only 2–5 days, and its spores are spread by wind.
This happened in the North American corn belt, when in 1970 and 1971 a new aggressive biotype of *H. maydis* raced across the fields.

*H. maydis* ("Southern leaf blight")

This disease kills the green tissue of leaves and produces typical lesions which are 3 cm long and 0.5 cm wide, of straw yellow colour, by reducing the assimilatory surface. This has long been known in North America; but suddenly in the year 1969 there appeared a new, very aggressive race, perhaps introduced from Asia, which attacked especially the TMS lines. The biological reason behind this epidemic was that the farmers of the USA at this time used nearly exclusively commercial corn hybrids built up on the so-called Texas male sterility cytoplasm. This proved highly susceptible to the new race of Southern leaf blight, whilst the older local maize varieties had a natural resistance to race-T-blight.

In comparison with *Helminthosporium*, other fungus diseases seem to be of minor importance. We mention here: *Diplodia zeae, Fusarium, Cercospora* and *Sclerospora sorghi*. The last-mentioned, called in Venezuela "mildiu lanoso", also affects different *Sorghum* species, which act as host plants. Over the past decade this disease has reached alarming proportions in tropical maize plantations. We mention here a personal observation from North Argentina. In the hot humid climate of the province Tucuman, the imported maize varieties and especially hybrid corn from the USA were dying out, but the local Indian corn variety called Perla resists infection by *Helminthosporium* very well.

INSECTS

Army worms

*Spodoptera frugiperda* ("barredor del maiz" in America)
*Spodoptera exempta*
Synonym: *Laphigma frugiperda*
*Prodernia eridania* (Afrika)

The adults of the army worms are small (2–3 cm) moths with a pronounced migratory tendency. They migrate over tens and hundreds of kilometres, and are preferably active at night, depositing eggs on young maize plants. The larvae devour the young maize shoots for 3 weeks, then creep down to the soil, to change into "pupas". After only 2 weeks the new moths appear and infect the plantations, which in this way suffer from various generations of "army worms". The result can be disastrous in extended monocultures but not in the small fields, hidden in forests, where many birds and other predators of larvae exercise a certain control. Big farms rely exclusively on expensive fumigations with strong insecticides (which of course also destroy the natural enemies of *Spodoptera* ,like a dozen *Dipterous* and *Hymenopterous* parasites, which would normal-

ly keep the pest in check). We observed, for example, in Trinidad that the larvae of a small wasp (*Euplectrus plathypenae*) live as ectoparasites on "army worms" and destroy them.

To complete the possibilities of biological methods to overcome the damage of "army worms" we mention the genetic resistance to the corn earworm (*Heliothis*) found in an indigenous primitive maize from Mexico. This is the Indian corn "Zapalote chico", which seemingly possesses some chemical factors that act as growth inhibitors to the earworm. Scientists of the Missouri Agric. Exp. Station at Columbia produced top crosses of high-yielding cultivars with Zapalote chico, and reported less larval penetration.

Other damaging insects are the "stemborers" *Diatraea saccharalis, Pyrausta, Ostrinia nubilalis, Agrotis ypsilon*, and *Euetheola bidentata*, a beetle which perforates the stem, and several phytophagous *Thysanoptera* (e.g. *Frankliniella*).

**References**

Aellen P (1967) Amaranthaceae. Illustr Flora Mitteleuropa, Hegi III:461–1532
Agogino GA (1957) Pigweed seeds dated oldest US food grain. Science Newsl Washingt 72:345
Anonymous (1984) Amaranth, modern prospects for an ancient crop. Natl Acad Res Counc Natl Acad Press, Washington
Beadle GW (1972) The mystery of maize. Field Mus Nat Hist Bull 43:1–11
Beadle GW (1977) The origin of *Zea mays*. In: Reed (ed) Origins of agriculture. Mouton, The Hague, pp 615–635
Beadle GW (1980) The ancestry of corn. Sci Am 242:112–119
Becker R, Wheeler E, Lorenz K et al. (1981) A compositional study of amaranth grain. J of Food Sci 46:1175–1180
Bird RMcK (1980) Maize evolution from 500 B.C. to the present. Biotropica 12:30–41
Bock H (1539) Neues Kräuterbuch. Strassburg
Bressani R (1986) Amaranth Newsletter 1–5, Arch Latinoam Nutr INCAP Guatemala
Brewbaker J (1979) Diseases of maize in the wet lowland tropics and the collapse of the classic Maya civilisation. Econ Bot 33:101–118

Brieger F (1968) Die Indianer Maisrassen des südamerikanischen Tieflands und ihre Bedeutung für die Züchtung. Die Kulturpfl 16:159–173

Cardenas M (1969) Manual de plantas economicas de Bolivia . Imprenta Icthus, Cochacamba

Coons MP (1982) Relationships of *Amaranthus caudatus*. Econ Bot 36:129–146

Crawford D, Wilson HD (1977) Allozyme variation in Chenopodium fremontii Syst Bot 2:180–190

Doebley JF (1984) Maize introgression into teosinte, a reappraisal. Ann Missouri Bot Gard 71:1100–1113

Doebley JF, Goodman MM, Stuber CW (1984) Isozyme evidence for reciprocal introgression between maize and Mexican annual teosinte. Econ Bot 9:203

Downton WJS (1973) *Amaranthus edulis*, a high lysine grain. World Crops 25:20

Gade D (1970) Ethnobotany of canihua (*Chenopodium pallidicaule*), rustic seed crop of the Altiplano. Econ Bot 24:55–61

Galinat WC (1977) The origin of corn. In: GF Sprague (ed) Corn and corn improvement. Agron 18. Amer Soc Agr Madison Wisc

Galinat WC (1978) The inheritance of some traits essential to maize and teosinte. In: Walden (ed) Maize breeding. Wiley, New York

Galinat WC, Pasupuleti (1982) *Zea diploperennis*. II. A review on its significance and potential value for maize improvement. Maydica 27:213–220

Galinat WC (1984) The origin of maize. (a refutation) Science 225:1093–1094

Galinat WC (1985) The missing links between teosinte and maize, a reviews. Mayidica 30:137–160

Gay C (1849) Historia fisica y politica de Chile. Vol otanica, Santiago, Chile

Gandarillas S, Gutierrez J (1973) Polyploidy induced in Canahua (*Chenopodim palidicaule* with colchicine. Bol Genet Castelar Arg 8:13–16

Grant WF (1959) Cytogenetic studies in *Amaranthus*. III. Chromosome number and phylogenetic aspect. Can J Genet & Cytol 1:313–328

Goldsworthy PR, Fisher NM (eds) (1984) The physiology of tropical field crops. Wiley, Chichester, England

Grubben GJ (1976) The cultivation of amaranth as a tropical leaf vegetable. Comm 67, Royal Trop Inst Amsterdam

Grubben GJ, Van Slooten D (1981) Genetic resources of amaranths. IBPGR Secr FAO, Rome

Hauptli H, Jain K (1984) Genetic structure of landrace populations of the New World grain amaranths. Euphytica 33:857–884

Heiser C (1964) Sangorache and amaranth used ceremonially in Ecuador. Am Anthropol 66:136–140

Heiser CB, Nelson DC (1974) On the origin of the cultivated chenopods (Chenopodium). Genetics 78:503–505

Hoffmann W, Mudra A, Plarre W (1985) Lehrbuch der Züchtung landwirtschaftlicher Kulturpflanzen. II. ed. Paul Parey Vlg. 434 pp

Hunziker AT (1952) Los Pseudocereales de la agricultura indigena de America. Acme Agency, Buenos Airs, p 103

Iltis HH (1983) From teosinte to maize, the catastrophic sexual transmutation. Science 222:886–894

Iltis HH, Doebley JF (1980) Taxonomy of *Zea* II. Subspecific categories in the *Zea mays* complex and generic synopsis. Amer J Bot 67:994–1004

Iltis HH, Doebley JF, Guzmann R, Pazy B (1979) Zea diploperennis, a new teosink from Mexico. Science 203:186–188

INCAP (1986) Amaranth Newsletter. Office Archivos Latinam Nutr Guatemala

Jain SK, Kulakov PA, Peters I (1984) Genetics and breeding of grain amaranth. In: Proceed III. Amaranth Conf Rodale, USA

Ji-Jeng (1980) Isozyme studies on the origin of cultivated corn. Acta Genet Sin 7:223–230

Joshi BD (1981) Catalogue of amaranth germ plasm. Natl Bur Plant Genet Resources Simla, India, p 42

Khoshoo TN, Pal M (1972) Cytogenetic patterns in amaranthus. Chromosomes Today 3:259–267

Kulakov PA, Hauptli H, Jain SK (1985) Genetics of grain amaranth. J Hered 76:27–30

Lorenz K (1981) *Amaranthus hypochondriacus*, characteristics of the starch and baking potential of the flour. Staerke 33:149–153

MacNeish RS (1967) A summary of subsistence. In: DS Byers (ed) The prehistory of the Tehuacan valley. Univ Texas Press Austin 290–309

Mangelsdorf P (1965) The evolution of maize. In: Hutchinson (ed) Essays on crop plant evolution. Cambridge Univ Press

Mangelsdorf PC (1974) Corn, origin, evolution and improvement. Belknap Press of Harvard Univ Press, Cambridge, Mass

Martin FW (ed) 1984) Handbook of tropical food crops. CRC Press boca Aaron fl pp296

McClintock B (1929) Chromosome morphology of Zea Mays. Science 69:629–30

McClintock B et al (1981) Chromosome constitution of races of maize. Colegio de Postgraduados, Chapingo, Mexico

National Academy of Sciences (1975) Quinoa. In: Underexploited tropical plants with economic value. Washington DC, pp 20–23

Pal M, Khoshoo T (1977) Evolution and improvement of cultivated amaranths VIII. Induced autotetraploidy in grain types. Z Pflanzenzücht 78:135–148

Pal M, Pandey R, Khoskoo TN (1982) Evolution and improvement of cultivated amaranths. IX. J Hered 73:353–356

Parodi L, Hernandez JC (1964) El mango, cereal extinguido en cultivo, sobrevive en estado salvaje. Cienc Invest 20:543–549

Pearsall DM (1978) Early movement of maize between Mesoamerica and Southamerica. J Steward Anthrop Soc 9:41–75

Randolph LF (1976) Contributions of wild relatives of maize to the evolutionary history of domesticated maize. Econ Bot 30:321–345

Rao BGS, Galinat WC (1974) The evolution of the American Maydeae. J Heredity 65:335–340

Sauer JD (1967) The grain amaranths and their relatives. A revised taxonomic and geographic survey. Ann Mo Bot Gard 43:102–137

Sauer JD (1969) Identity of archeologic grain amaranths from the valley of Tehuacan, Puebla, Mexico. Amer Antiqu 34:80–81

Senft J, Kaufman C, Bailey N (1982) A comprehensive bibliography, with 2500 entries. Rodale Research Center, Rodale, USA

Sharma AK, Dey D (1967) A comprehensive cytotaxonomic study on the family Chenopodiaceae. J Cytol Genet India 2:114–127

Simmonds NW (1971) The breedings system of *Chenopodium quinoa*. I. Male sterility. Heredity 27:223–235

Simmonds NW (1976) Quinoa and relatives. In: Evolution of crop plants. Longman, New York

Smith JSC, Lester RN (1980) Biochemical systematicas and evolution of *Zea tripsacum* and related genera. Econ Bot 34:201–218

Smith CE (1986) Importance of palaeoethnobotanical facts. Econ Bot 40:267-278

Stoner C, Anderson E (1949) Maize among the hill peoples of Assam. Ann Miss Bot Gard 36:355–404

Tapia-Vargas W (1976) La quinoa. Academica Nac Cienc Bolivia, La Paz

Tapia M, Gandarillas H (1979) Quinua y Kaniwa, cultivos andinos. CIID, Bogota pp 288

Vietmeyer ND (1982) Amaranth: return to the Aztec mystery crop. In: Yearbook of science in the future. Encycl Britannica, Chicago Ill, USA

Walden DB (ed) (1978) Maize breeding and genetics. Wiley, New York, p 794

Weber L, Hubbard E, Putnam D, Nelson L, Lehmann J (1988) Amaranth grain production guide. Rodale Press Inc, Emmaus, PA and American Amaranth Institute, Bricelyn, MN

Wildenow CL (1790) Historia Amarantharum. Zurich

Wilkes HG (1977) Hybridization of maize and teosinte in Mexico and Guatemala. Econ Bot 31:254–293

Wilkes HG, Mangelsdorf PC (1979) *Zea diploperennis*, the "missing link" in corn's genealogy. 12 Ann Meet Soc Econ Bot Raleigh N C

Wilson HD (1980) Artificial hybridization among species of *Chenopodium*. Syst Bot 5:253–263

Wilson HD (1988) Quinoa-Biosystematics I. Domesticated populations. Econ Bot 42:461–477

Wilson HD, Heiser CB (1979) The origin and evolutionary relationships of Huazontle (*Ch. nuttalliae* Safford) domesticated chenopod of Mexico. Am J Bot 66:198–206

Yacovleff E, Herrera FL (1934) El mundo vegetal de los antiguos peruanos. Rev Mus Nac (Lima) 3:243–322

# III. Protein Plants

## Introduction

In world nutrition there is a pronounced protein deficiency. With the exception of a few fortunate countries in the subtropical latitudes, like Argentina, South Africa and Australia, which still produce an excess of animal protein (e.g. Argentina had until 1986 per capita beef consumption of 80 kg/year), the majority of the inhabitants of the tropical belt (and these are many hundred millions) have no – or only very limited – access to animal protein and often not even to vegetal protein, which in general is more abundant and easier to produce. The world shortage of proteins is increasing each year as a consequence of the irrational and explosive population growth in the underdeveloped countries, whilst yields decrease, for example on the African continent (with the exception of a few nations like the Republic of South Africa and Nigeria) caused there by dramatic effects of soil erosion, deforestation, rural exodus and the chronic inability of the natives and their governments to produce better yields. In spite of these maladies, many scientists around the world and non-profit organizations such as CIAT, CIMMYT, CIP, IITA and ICRISAT are sparing no efforts to improve the protein situation of the starving masses.

Their paramount goal is to improve the Leguminosae as protein-producing plants. This refers essentially to the genera *Vigna, Phaseolus, Lupinus, Canavalia, Pachyrrhizus* (also *Arachis* and *Glycine*) in the huge botanical family of Leguminosae with over 13,000 species, subdivided in the three subfamilies: Mimosoideae, Caesalpinioideae and Papilionoideae. The last-named contain what are commonly called "pulses".

The majority of these "useful" pulses are of neotropical origin. For a layman they are rather difficult to identify and even in official statistics they are cast together. Their yield on a world scale is registered between 4,700,000 to 5,000,000 t/year, not including soybean, for example, with approximately 10,000,000 t/year; but in spite of their high protein content, soya seeds are industrialized as an oilcrop.

In general, "pulses" are low yielders in comparison with maize, rice and wheat. Especially now, after expansion of the so-called high yield varieties created by CIMMYT in Mexico, legume production has diminished throughout the world. This negative effect on pulses has been recognized by Borlaug. The trend must become reversed, especially in view of the importance of bean cultivation for an enormous quantity of smallholders in underdeveloped countries and the biologically important beneficial effect on soils. All legumes are symbionts of nitrogen-fixing bacteria and considerably improve the soil substrate.

LEGUMINOSAE

## 1. *Canavalia* spp.

This pantropical genus is represented with 50 species on both hemispheres, according to Sauer's (1964) exhaustive taxonomic work. Its worldwide distribution was facilitated by its buoyant seeds, which resist extremely long transport by ocean currents and may have been washed up on tropical sea coasts for millions of years, e.g. *C. maritima*. Most of the species are slender, climbing vines, or lianas. The flowers have a typical and peculiar bilabiate calyx. In general they are woody perennials with the exception of the few which are cultivated as annuals and therefore of especial interest for us, leaving the trailing lianas aside. From the neotropical taxa three have been used by Indians.

### *C. plagiosperma* Piper

Giant bean, poroto gigante, pallar de los gentiles, Riesenbohne

$2n = 22$

NAME AND ORIGIN

This is an ancient Indian cultivar, which apparently no longer exists in the wild form. According to Kaplan (1980) it appeared in the coastal region of Southwest-Ecuador already in pre-ceramic times and pre-dated the use of *Phaseolus* beans. Archaeological finds from coastal Peru date back to at least 2000 B.P. and consist of well-preserved pods and seeds. The oldest material is pre-ceramic from Huaca Prieta and includes several dozens of pod fragments and seeds. Abundant remains of *C. plagiosperma*, more than 250 seeds, have been recovered in the Paracas Necropolis (Yacovleff and Herrera 1934). From the widespread presence of seeds at different places of the Peruvian coast, we conclude that this pulse was a common crop together with *Cucurbita* and *Phaseolus* beans in the arid, irrigated fields inhabited by fishermen-peasants. As a last vestige of this millenarian domestication, we may mention that even today *Canavalia* is still cultivated in reduced quantities at the Oasis of Ica, in South Peru, under the name "pallares de los gentiles".

MORPHOLOGY

Plants bushy, or short trailing vines, annual or in favourable climate perennial.
Leaves tripartite on long petioles. Leaflets ovate, 12–15 cm long, pubescent on surface. Flowers lilac with 13-mm-long calyx. Fruits compressed, 25 cm long and 4 cm broad, each valve with sutural ribs. Seeds with brown blotches or entirely white, $26 \times 18 \times 10$ mm size, oblong and moderately compressed with a long (11-mm) hilium(Fig. III.1).

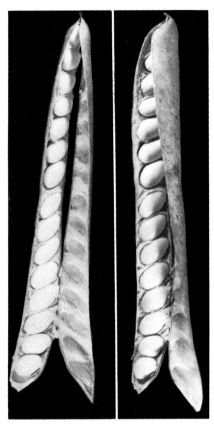

**Fig. III.1.** *Canavalia plagiosperma,* ripe pods twisting for seed release

### *C. ensiformis* (L) DC.

Jack bean, pois sabre, frijol del burro

$2n = 22$

NAME AND ORIGIN

We could not find a typical Indian denomination for this millenarian cultivar, but there are no doubts about its use as a food plant. In caves near Tehuacan (Puebla, Mexico), domesticated *Canavalia* seeds have been found in levels dated between 5200 and 3400 B.P., and from the Abejas phase beween 3400 and 2300 B.P. (Mac Neish 1967). Much more recent is the abundant material from the Oaxaca valley, associated with Monte Alban IV pottery (900 A.D.). All the seeds have the typical ivory-white testa. It is remarkable that more recent finds from Arizona (1400 A.D.) revealed a genetic variation which included yellow, dark spotted, brown and dark brown variants. The seeds have been found near dwellings of Salado Indians, who used irrigation. Obviously these Indians maintained very primitive biotypes, remains from early domestication when the dark colours of wild *Canavalia* dominated. These variations in colour and shape are unknown in modern *C. ensiformis* populations. As the wild ancestor is probably *C. brasiliensis*, with dark-coloured testa, we suppose that the Indians maintained some archaic types in their *Canavalia* beans. A similar case has been reported by Sauer and Kaplan (1969) from Yucatan. A larger find of *Canavalia* seeds at Dzibil-chaltun in Yucatan, which was radiocarbon-dated at 320 B.P., yielded seeds of varying size, with smaller ones similar to *C. brasiliensis*, which is native in Yucatan. This indicates the possibility of phylogenetical relations between the two species. We believe that the widespread *C. brasiliensis*, diffused from Paraguay, Brasil, Colombia, Panama and the whole of Central America to Mexico was the ancestral species from which the Indians selected *C. ensiformis* (Fig. III.2).

## Morphology

Bushy, semi-erect plants, mostly cultivated as an annual. When grown in favourable tropical climate, acquire perennial habit, with metre-long climbing vines and stems woody at the base. Roots deeply penetrating and well-nodulated.

Leaves trifoliate on long petioles (10–15 cm). The leaflets are ovate-elliptic, entire, leathery and 8–15 cm long, with pubescense. Inflorescences axillary racemes with strong peduncles, which bear many (30–50) flowers, of purple-red colour. The standard petal is erect with a white throat. Flowers are generally self-pollinated. Fruits are large, flat and firm (20–35 cm long and 2 cm broad) and contain 12–20 seeds. The first ripe pods are produced after 4 months. Seed size 20 × 15 × 10 mm oblong compressed, ivory-white, with a 9-mm-long hilum (the hilum of *C. gladiata* is much longer and covers the whole seed length).

Seeds are edible, also the green immature pods make an excellent vegetable. We consumed cooked, peeled seeds, changing the cooking water twice, and found them very tasty, with mealy texture.

The average yield of dry beans varies according to planting conditions. In Paraguay we obtained 1500 kg/ha, but there are reports from an Experimental Station at Puerto Rico with 5000 kg/ha. When grown as green manure, or forage, the harvest of biomass is 30–50 t/ha. With its dense leaves

**Fig. III.2.** *Canavalia ensiformis* in Paraguay

it is also an excellent soil cover in regions with laminar erosion.

The nutritional value of *Canavalia* seeds is high: 50% carbohydrates, 24–26% proteins and 2–3% fat. Milled seeds could be used in protein concentrates for human food in underdeveloped regions. *Canavalia* has the advantage that it can be produced on most soils and even tolerates the heavy, acid soils of the humid tropics, as we observed along the occasionally flooded river banks of the Amazonas and Orinoco.

Future genetic improvement should be directed to eliminating the toxic substances which are present in all wild and in most of the cultivated species. Chemical analysis confirmed the presence of alkaloids, the poisonous amino acid canavalin and certain hemagglutinins (Con-canavalin A). The latter, for example, have a negative effect on growth in domestic animals, reducing the body's ability to absorb nutrients (Liener 1972). Some of the anti-metabolic substances can be eliminated relatively easily by soaking the dry seeds for several days in cold water (to avoid sprouting, Indians place them in mountain rivers). Another possibility is to boil them in two or three changes of water.

We must include two *Canavalia* taxa, which are of Old World origin, but so widely diffused in the neotropics that some authors considered them as native to America. These are the wild-growing *C. maritima* and the edible *C. gladiata*.

### *C. maritima* (Aubl.) Thouars
$2n = 22$

The buoyant seeds, which maintain impermeability to water for over 1 year, are the biological reason for the worldwide distribution. It exists on the seashores of Australia, Malaysia, China, India, Africa and South- and Central America, where it is presumed that *C. maritima* has grown for millions of years. Together with *Ipomoea pes-cabris*, it covers the sandy beaches of the Caribbean Islands and provides there efficient protection against erosion from the action of maritime surf on the loamy coasts. The stems are

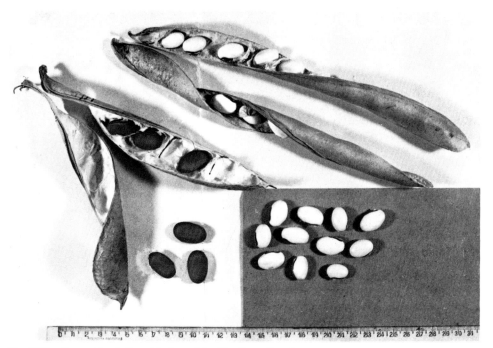

**Fig. III.3.** *Canavalia gladiata* (dark seeds) with *C. ensiformis*

creeping, several metres long. The pods are oblong (15 cm × 2.5 cm), with many brown seeds (18 × 13 × 10 mm). In the case of shipwreck they have been used as emergency food.

## C. gladiata (Jacq.) DC.

Sword bean, poroto sable, haricot sabre, Riesenbohne
2 n = 22

The vegetative aspect of the plant is very similar to *C. ensiformis*, with which it crosses easily. It is extensively used in India, China and Malaysia as vegetable (green pods) and as grain food (with due care as to its slightly toxic effect). This perennial climbing species is also a valuable cover crop and green manure, and has therefore long been introduced also in the New World. The seeds have different colour (scarlet red) and shape (a long hilum which extends almost the entire length of the seed) than *C. ensiformis* (Fig. III.3). Hybridization between these two similar species should be stimulated as well as mutation breeding to eliminating the genetic factors for toxicity.

Besides its multiple use for human food and animal pasture, we may call attention to the "fungicide" action of *Canavalia* leaves. Lampard (1979) described a decade ago the existence of demethyl homopterocarpin in the plants. When *Atta* ants cut the leaves and transport them to their "fungus gardens", this substance paralyzes the procreation of larvae. It has been observed that the cessation of activity of leaf-cutting ants lasts for several months. Therefore it is recommended to place freshly cut *Canavalia* leaves on top of the *Atta* mounds, as a cheap biological control method.

LEGUMINOSAE

## 2. Lupinus mutabilis Sweet

Synonyms: *L. cruickshanksii* Hooker, *L. tauris* Hooker
Tawri, chocho, ullu, tarhui, pearl lupin, Indian lupin
2 n = 48

The large-seeded, domesticated *Lupinus mutabilis* and *L. montanus* belong to a worldwide distributed genus, with a strong background in the New World, where 200 taxa have been established. From the Old World only a dozen species are known, the most important being *L. luteus* and *L. albus*, both selected by modern plant breeders as nonshattering, low-alkaloid cultivars. The creation of so-called sweet lupins is connected with the name of Reinhold v. Sengbusch, whose creations achieved paramount importance as a new protein plant in Germany during the food restrictions of the Second World War. Cytogenetic barriers exist between the Euro-Asiatic group with 2 n = 50 and 52 chromosomes and the American lupins, which in general have 2 n = 48 chromosomes. For this reason the genes for low alkaloid of selected European varieties cannot be transferred by simple hybridization to the otherwise useful American lupin species.

### NAME AND ORIGIN

The taxonomy of this Andean lupin needs further research. Sweet (1825) used a garden exemplar in England, grown from seeds which had been sent from Colombia, as "typus" for its nomenclature. Definitively, this was not a wild species, which still has to be recovered. Probably the term *"L. tauris"* would be more appropriate.

This protein and oil-rich pulse was selected in praehistoric times by Indian tribes who inhabited the high valleys of Bolivia and Peru; we presume that it was derived from real wild-growing lupins of Andine origin, which merit urgent investigation to save the gene pool of this important crop.

The Ayamara Indians call this plant "Tarhui Tawri", the Kechua use the name "Chocho" or "Ullu" and consider it as an important grain food despite metabolic problems. Its early use is documented by seed remains in grave sites of ancient Peru and also by depiction in pre-Columbian ceramics.

### MORPHOLOGY

Annual herbs with a strong mainstem (80– 180 cm tall), which develops continuously lateral branches. The growing period extends over 5 months. The branches bear flowers, leaves and pods in different states of maturation, thus giving the impression that *L. mutabilis* has an indefinitive growth habit. Leaves digitate-palmate with five or more leaflets, covered with fine short hairs.

Inflorescences are many-flowered on terminal racemes, decorative showy, exuding a pleasant scent to attract pollinating insects. In contrast to the Afro-European *Lupinus* species, *L. mutabilis* is not self-pollinating. Flowers are multi-coloured (yellow, white, blue, purple) changing colour during ripen-

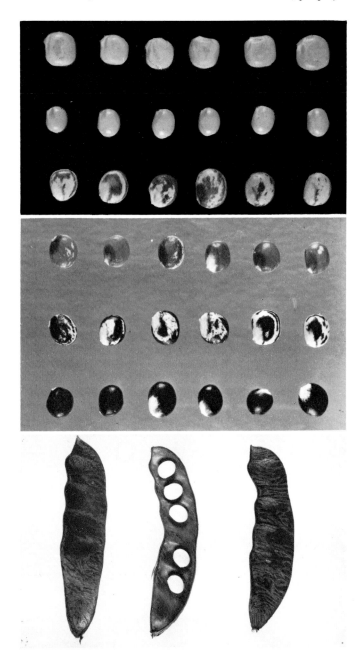

**Fig. III.4.** Seeds from South American *Lupinus* cv. with the European species *Lupinus albus* (in the first row)

ing, for which reason they are called "mutabilis".

Fruits. The pods are 8 cm long and 2 cm broad and contain 5–6 large, bean-like seeds of different colour; according to different landraces, they are black brown, white, mottled or speckled with varying colours and patterns (Fig. III.4). Very important is the fact that *L. mutabilis* pods are non-shattering and maintain the seeds inside the valves, quite differently from the Old World wild lupins which have a pronounced tendency to twist and split and shed the ripe seeds onto the soil. This characteristic is a considerable advantage – due to intentional or unintentional selection by early Indian planters – when they harvested ripe *L. mutabilis*. To achieve this non-shattering, several selection steps had to be taken: elimination of the natural tendency to split on the suture, aquisition of leathery, rather than brittle valves, firm joining of the ripening seeds to the pod. Thus the Indians selected and combined useful factors in their South American lupin, an advantage which was only recently achieved in the European lupins by modern plant genetics.

The optimal area for *L. mutabilis* and its related *L. montanus* is situated between 8° N and 22° S of the Andes, preferentially in the high valleys and cooler portions of tropical mountains at 1500 m to 2500 m in altitude (Fig. III.5). However, we observed in the Al-

tiplano of Bolivia and Peru "tawri" fields also near the 4000-m agriculture limit.

This legume might appear to be an ideal source of protein and fat for the undemanding inhabitants of the American mountains and high valleys, did it not contain bitter-tasting and even toxic substances in its seeds.

Since time immemorial Indians have tried to leach out the alkaloids by watering seed-filled bags in ice-cold mountain streams (to avoid premature germination), but this is a troublesome and lengthy process and insoluble alkaloids always remain in the material. These complications have discouraged the inhabitants of the more wealthy nations, like Colombia and Venezuela, from sowing and consuming their local lupin strains, losing valuable landraces by this negligence (Brücher 1985), whilst the cultivation and consumption of "tawri-chocho" has never ceased in remote valleys of the Bolivan and Peruvian Andes. There is an urgent necessity to save the gene funds of such local lupin strains. The challenge for plant collectors persists by screening huge quantities of lupin seeds on the spot with elaborated "quick tests" (described by Brücher 1970), to discover and select immediately the few low-alkaloid biotypes which are hidden in the huge quantity of usually toxic seeds.

Biochemical investigations of seeds and plants of different lupin species have shown

**Fig. III.5.** A native of the Peruvian highland with a fresh harvested lupin plant (*Lupinus mutabilis*)

that they contain several quinolizidine alkaloids, mostly sparteine and lupanine, which
not only produce a bitter taste when consumed by humans, but may have also toxic
effects on livestock, if eaten raw, with the
exception of sheep.

This problem can be resolved either by seek-
ing low-alkaloid biotypes which seldom occur in free nature or by creating them on a
larger scale by mutation breeding. Possession of alkaloids is a dominant factor which
protects wild-growing lupin plants from extermination by grazing animals.

With the intention of creating for the dispos-
sessed and starving mountain people of
Venezuela and Colombia a useful vegetable,
free of toxic substances, we began radiation
experiments in 1965 at our Dept. of Genetics
at the Universidad Central de Venezuela,
with financial support from Germany and
the International Atomic Agency in collabo-
ration with the Atomic Reactor at IVIC-
Caracas. We obtained promising low-alkaloid mutants from the Andine species *L.
mutabilis-tauris* and *L. montanus* described
by Brücher (1968, 1970), using gamma-radi-
ation.

Hardly had we successfully made the first
steps in our mutagenic experiments, when
misguided students and fanatic "Red
Hordes" of Caracas occupied the University, closed my Genetics Dept. and obliged me
to emigrate. Thus an altruistic endeavour
was destroyed. The challenge remains – after
many lost years – to repeat such experiments
under more favourable conditions.

In the meantime, v. Baer and collaborators
(1977) performed successful work in selecting *L. mutabilis* strains with as little as
0.05% alkaloids. These selections came
from natural populations. It remains to
combine high-yielding domesticates with
these low alkaloid sibs. We need threshold
values at which human beings cannot taste
the bitter substances in lupins (Fig. III.6).

Once liberated from its anti-metabolic substances, *L. mutabilis* – like Old World lupins
– could find its place in the ever-growing
market of pulse crops. Similarly to soybean,
protein and oil make up more than half the
seed's weight. The protein component is on
the average 45%, with a maximum of 49%
of the seed's volume. The chemical composi-

**Fig. III.6.** A native women on a Peruvian market
place offers cooked lupin seeds, free of bitter substances by special treatment

tion is favourable; it is rich in cystine and
has a fair part of lysine, and other essential
amino acids, but is deficient in methionine
and other sulphur-containing amino acids.
When mixed with wheat flour, which is
lysine-deficient, the tarwi grains represent a
valuable supplement for native bread.

The fatty substances of *L. mutabilis* seeds
consist of 14% on average, but may reach
21% in selected favourable strains. The oil is
rich in unsaturated acids, especially in linoleic acid, has a light yellow colour and a
pleasant taste.

Due to its symbiosis with *Rhizobium* bacterias, *L. mutabilis* can fix as much as 400 kg
nitrogen per hectar. It also produces a huge
biomass, calculated at 40–50 t/ha of green
mass, containing 1,750 kg of raw protein per
hectare. This considerably exceeds the vegetable matter produced by European lupin
species.

*L. mutabilis* exceeds Old World lupins in its
frost-hardiness and tolerance of acid soils.

In spite of its origin from a tropical shortday region, it possesses good photoperiodical adaptability. *L. mutabilis* produces flowers and ripe pods in the longer summer days of temperate zones as was shown in experimental plots in South Africa (Pakendorf 1974) and Australia.

It has further been observed that certain strains of *L. mutabilis* have genetic resistance to fungus diseases caused by *Erysiphe, Fusarium* and *Rhizoctonia* and also *Phomopsis*, a highly toxic fungus which infects lupin plants and may kill livestock feeding on lupin pastures.

Plant geneticists and development helpers, who have not lived and worked personally long enough in Latin America to have suffered malevolent treatment, persecution and even destruction of their altruistic endeavours and help projects for the poor and hungry, may still feel enthusiastic about the recent recommendations of the US National Academy of Science to develop *Lupinus mutabilis* into a new world crop. One of its leading personalities, N. Vietmeyer, expressed this in the following personal communication (Dec. 1987): "Tarwi is almost unknown outside the Andes ... tarwi is outstandingly nutritious. Mixing tarwi and cereals makes a food that, in its balance of amino acids, is almost ideal for humans... Uncertainty over the amount of insoluble alkaloids has previously hindered tarwi's introduction to areas outside the Andes. Alkaloid-free, sweet tasting types, could make tarwi into a major crop for cool, tropical highlands".

---

LEGUMINOSAE

## 3. *Pachyrrhizus* spp.
$2n = 22$

This neotropical genus consists of several species useful for Indo-American natives and deserves intense breeding work, supported by modern plant geneticists, to liberate it from certain toxic characters. *Pachyrrizus* plants have double use: the starchy tubers are appreciated by natives both in Bo-

livia-Peru-Ecuador and in Mexico. Also protein-rich seeds are used, after having been cooked abundantly, to eliminate certain toxic substances, $C_{23} H_{22} O_6$, considered as belonging to the rotenone group.

We recognize the following six species, four of Central American origin and two from South America: *P. panamensis* Clausen, *P. vernalis* Clausen extending from Panama to Nicaragua-Mexico and *P. strigosus* Clausen confined to the Mexican-Guatemalan border region. These are wild-growing species, occasionally used by natives. From these wild forms Indian tribes of the Aztek and Maya cultures developed the high-yielding *P. erosus* (L.) Urban. These "jicamas" were cultivated for their large roots (20–30 kg) and abundant protein-rich seeds, from Mexico southward through Panama-Colombia as far as the Coast of Ecuador, as was reported by Cieza in the 16th century (Cieza de Leon, reprinted in 1924 *Cronica general de Peru, Lima*). There it came together with another *Pachyrrhizus* species, favoured by Amazonian Indians for its starch-rich roots, *P. tuberosus*, known under the native denomination in Brazil "bacuzu", in Venezuela "nupé" and generally also called "xicama".

Finally there is the mountain species *P. ahipa*, domesticated by Aymara and Kechua Indians in the high valleys of Bolivia and Peru, called "huitoto" and "ajipa" and found also rarely in North Argentina.

## *P. erosus (L.)* Urban

Yam-bean, iguama, yicama, patate-cochon, xicama, macucu

$2n = 22$

### NAME AND ORIGIN

With the Aztec denomination "xicama" Hernandez described this tuberous crop in the 16th century in his *Historia de las plantas de Nueva España*, considered by the Spanish conquerors of such high value that they took the roots to Spain and later to Africa and Asia. Actually it is cultivated in the Philippine Islands and in China to such a great extent that it was claimed to be of Asiatic origin. (Soerensen 1985). The real origin is

**Fig. III.7. a** *Pachyrrhyzus erosus* from Central America **b** *Pachyrrhizus tuberosus* from Amazonia, with edible roots

undoubtedly in Central America, where several biotypes still occur wild, as Williams stated from Guatemala Fig. III.7). He wrote in 1981: "The roots of wild plants are considered poisonous in some regions... nursing women are warned that damage to the child may result from eating the root..."

MORPHOLOGY

Quick-growing vines (5 m long when flowering). Leaves with three leaflets which are entire and serrate; but some biotypes have lobate-palmate leaflets. Terminal leaflet larger and broader than the laterals.
Inflorescences erect, many-flowered and up to 50 cm long. Many of the white flowers do not develop ripe fruits. Pods flat, 7–12 cm long, covered with short dense hairs, with 6–12 seeds. The kernels are rectangular, flat, with a hard testa of brown colour, 6–10 mm diameter.
Tubers are abundant and develop after 4 months to a size of 30 cm diameter, with a weight between 1 and 2 kg. Their pulp is white, succulent and sweet-tasting. Besides 80% water tubers contain 10% starch, 5%

sugar and 1–2% raw protein. They are not bitter or poisonous and can be eaten raw or cooked. I have seen them offered and sold as refreshment at native Mexican markets. They resembled white turnips.
Pods and seeds are said to contain dangerous quantities of rotenone. We tested selections from Mexico personally (cv. Cristalina and Agua dulce provided by J. Salinas-Gonzalez, Inst. Nac. Investig. Agricolas, in Celaya) and found them good. To avoid adverse effects, it is recommended to change the cooking water. I was informed that at present 4000 ha are planted with *P. erosus*, especially in the states Michoacan, Guanajuato and Morello, and one can only hope that this breeding work receives due support.

**P. tuberosus (Lam.) Sprengel**
Potato bean, pois patate, cacara, jicama
$2n = 22$

NAME AND ORIGIN

Similar in use and habit to the above species. Its origin has been traced back to the humid tropical valleys of the Upper Amazonas region. Its areal extended in pre-Columbian times through Brazil, Venezuela to the Caribbean Islands northwards, but also reached the nearly tropical region of East Paraguay. There Guarani Indios planted it both as tuber crop and medicinal plant, according to Montenegro (1710). He reported that this "mbacucu" was sown each year, that its roots were used as a conserve prepared with salt. The Paraguayan botanist Bertoni several decades ago saw plantations of *P. tuberosus* at the margin of the Parana River and mentioned that their weight was 6–8 pounds and "that the Indians consumed the tuberous roots like potatoes". During our explorations of the flora of Paraguay (1970–1973), we tried in vain to recover this interesting species, which most probably has been exterminated in Eastern Paraguay by the violent advance of modern agriculture.

### MORPHOLOGY

Stems long (5–7 m) climbing on bushes and trees, at the base lignified.
Leaves tripartite, with leaflets of different size and form. The terminal leaflets are rhomboid, whilst the laterals are asymmetric-triangular.
Inflorescences many-flowered with violet-coloured flowers (seldom white).
Fruits large (15–30 cm long) with rather large seeds (11–14 mm diam) sometimes covered with irritant hairs.
Tubers large (50 cm diam), watery, with fleshy edible pulp, 20–30 kg in weight.
It is noteworthy that the plants do not show insect damage, even in environments which are heavily infested by all sorts of pests and parasites. This may be the result of the slight natural content of rotenone, which is well known for its insecticide effect (Clausen 1945).

### *P. ahipa* (Wedd.) Parodi
Aymara: huitoto, kechua: ajipa
$2n = 22$

### NAME AND ORIGIN

Oviedo (1535) gave the first authentic description, comparing it with turnips "dulce que se deshacen en la boca ...". While studying ancient Peruvian plant descriptions and painted ceramics of the Mochica culture, we came across stylized depictions of *Pachyrrhizus*, that indicated that this species must have played an important role in the daily life of Andine Indians. They probably selected it from regional wild forms in the "Ceja de Montaña", which are now exterminated. *P. ahipa* has a double utility; both its tubers and seeds are consumed. Unfortunately, however, its use is rapidly decreasing. Occasionally one can find some examples in house gardens at 1000–2500 m altitude in Peru and Bolivia. The sites in the North Argentinian province Jujuy, which Parodi (1935) reported in his taxonomic description, may have become exterminated.

### MORPHOLOGY

The plants have a compact growth, with short (30–90 cm) semi-erect stems.
Leaves have extremely assymetric lateral leaflets. The short terminal leaflets are broader than long.
Inflorescences with few flowers, mostly violet, some white or lavender; 4 cm long.
Fruits are short (8–10 cm long) with 4–5 seeds. Seed colour is brown-black, their diameter is 10 mm. Their raw protein content is approx. 20%.
The roots produce tubers 10–15 cm in size (500–800 g in weight) with white, succulent, sweet pulp and yellow skin. They may be consumed raw or also boiled.
Burkart (1952) considers *P. ahipa* as most indicated for future genetic improvement and rational cultivation, due to its compact habit and comparatively rapid growth. Sown in December, it can be harvested – both seeds and tubers – already in April. In spite of its origin in tropical shortday regions, it adapts well to other environmental conditions. Vietmeyer, US Nat Acad Sci (personal communication) ascribes great potential for future commercial cultivation to *ahipa*.
Once improved genetically, *P. ahipa* could make a considerable contribution to the diversification of protein supply in tropical countries. In view of the near extinction of ancient landraces in the Altiplano, the exist-

ing vestiges of a formerly rich gene pool should be collected as soon as possible and made available to advanced plant breeding institutions for further improvement.

Fig. III.8. *Phaseolus acutifolius,* a promising "rediscovered" pulse for arid regions

---

LEGUMINOSAE

## 4. *Phaseolus* spp.

Introductory Remark:
"Beans" have been since time inmemorial a basic food and cash crop for the American small farmers. They provide in Brazil, Colombia, Ecuador and Peru probably more than one-third of total protein intake and offer – due to their low price – for millions of poor people the main nutrition, with a per capita consumption of 50 kg/per year. We estimate that more than four million tons of shell beans are produced annually in Latin America, but these figures do not appear in world statistics because they represent the product of innumerable small farms and house gardens. The yields are generally low, perhaps only 400 – 500 kg/ha, whilst the commercial bean production on big farms is much higher and may range from 2000 – 3000 t/ha.

According to FAO statistics of 1981, the world yield of "beans" (several botanical species counted together) may be estimated between 12 – 14 mio tons, a proof of its global importance as basic food. There exists also considerable consumption of "snap beans" as green vegetable.

### *P. acutifolius*
with two varieties:latifolius, tenuifolius
Tepary bean, Texan bean, escomite, frijol de colima
2 n = 22

ORIGIN AND HISTORY

This hardy bean originated in the semi-arid regions of Central America. Wild forms still exist in Guatemala, Arizona and Mexico, especially in Sonora, Jalisco and Guadalajara, climbing there on desert bushes with long vines. Indians of the Sonora Desert, probably belonging to the tribes of Basketmaker III, used to collect wild-growing escomite seeds and domesticated this bean plant in remote times (5000 years B.P.), even before they knew *P. vulgaris*. Seeds have been recovered in residues of the Tehuacan caves and at Durango USA. For the Papago Indians tepary beans were still a basic food in the year 1700 and they continued to harvest wild-growing plants until 1940. Under the Spanish domination tepary beans had a considerable commercial importance along the Pacific Coasts of California and Northern Mexico (Fig. III.8).

MORPHOLOGY

Annual, short-lived, glabrous herbs, often with indeterminate growth habit, sub-erect, small bushes (30 cm height), but some primitive biotypes with 2-m-long trailing vines. Leaves trifoliate, with pointed ovate leaflets on long petioles.

Inflorescenses axillary, with 3 – 5 small cleistogamous, self-fertilizing flowers. According to variety they are white, pink or pale lilac, flowering generally under shortday conditions, whilst some are day-neutral.

Fruits (5–8 cm long and 1 cm wide) covered with long silky hairs. They maintain the wild character of pod dehiscense which complicates the harvest, and contain 5–7 small, round-oblong seeds. The colour of the seeds is very variable, dull matt, white, yellow, brown speckled or violet. They possess a high protein content (24–32%). The nutritional value compares well with other legumes, but they need a longer cooking time. To 100 g there are 600–800 kernels. Tepary seeds react very quickly when sown in dry land which receives some moisture. Water penetrates easily; the testa wrinkles in humid soils within 5 min. They yielded 2200 kg/ha seeds in water-supplemented fields in the Sonoran Desert, but can produce more than 4000 kg/ha under less extreme conditions. In spite of having such a perfect adaptation for arid farming lands, the tepary bean is faced with extinction, and is in fact no longer cultivated on a commercial scale in North America (Nabhan and Felger 1978).

## P. coccineus L.

Synonym: *P. multiflorus* Willd.
Scarlet runner bean, ayocote, haricot d'Espagna, chamborote, ixcumite, Feuerbohne
$2n = 22$

### ORIGIN AND HISTORY

This bean is without doubt of Central American origin. Wild forms occur at moderate altitudes (1000–1800 m) in the cool humid mountains of Mexico and Guatemala, but no wild form of *P. coccineus* has ever been reported from regions south of the Isthmus of Panama. Archaeological findings from Mexico (Tehuacan) indicate that *P. coccineus* was probably domesticated 2200 years ago. The wild races are perennial and produce a considerable amount of large tuberous roots, which are dug up by natives and used as starchy food. In Central America, exclusively wild-growing, the related species *P. formosus* HBK and *P. obvallatus* Schlecht exist, which produce similar thickened roots rich in fecula, which are eaten boiled by natives.

Williams (1981) stated for Honduras: "Scarlet runner beans are cultivated in all the highland regions of Central America. They are found in all markets of Guatemala and Costa Rica in season as shell beans or green beans"

For the botanist travelling in South America it is an unsolved enigma, why *P. coccineus* never had a place in Indian agri-horticulture. At least, no local South American plant collector reported this species and neither have I observed this bean species during repeated exploration journeys. It seems that South American natives have a traditional preference for their autochthonous bean species *P. vulgaris* and *P. lunatus*, with their abundant local landraces (Fig. III.9).

### MORPHOLOGY

Vigorous twining and climbing stems, in tropical climate perennial, in Europe and North America grown as annuals, sometimes as selected dwarf varieties. Climbing varieties can reach 3–5 m length with flexible stems which in cultivation need support with poles and wires to secure the heavy load of mature pods. Most varieties produce fleshy tuberous roots.

Leaves large (10–12 cm long), trifoliate, with ovate leaflets.

Inflorescences axillary racemes, which produce a considerable quantity (10–15) of bright, scarlet-red flowers (exception some white-flowering mutants).

Flowers (2–2.5 cm long) are borne on long peduncles, and have in general an allogamous fecundation system, which explains why they are frequently visited by pollinating insects.

Fruits are relatively long (30–40 cm) broad-flattened, and often slightly curved and pubescent. The pods contain 5–10 large seeds (25 mm long) of different colours, but mostly purple-black with red mottle. They have hypogaeic germination. Natural crossings with other *Phaseolus* species do not occur, in spite of their identical chromosome numbers.

Lamprecht (1964) noted that sterility in *P. coccineus* x *P. vulgaris* bastards is plasmoncaused. The alternative combination *P. vulgaris* x *P. coccineus* grows rather well in artificial hybrids.

**Fig. III.9.** *Phaseolus mulitflorus*, flowers and seeds (after Bianchini et al. 1976)

To explain the high rate of pollen abortion in the $F_1$ plants of crosses between *P. vulgaris* and *P. coccineus*, Basset et al. (1981) studied the meiotic behaviour in such hybrids and the degree of similarity between the two genomes. Two pairs of chromosomes were differentiated by inversions. The meiosis in the $F_1$ hybrids was anormal. The conclusion is that the chromosomes of the two species are structurally incongruent. The high crossing-over rates are responsible for most of the pollen abortion. We conclude from these cytogenetic analyses that species differentiation between *P. coccineus* in Central America and *P. vulgaris* from South America took place at an early stage of phylogeny.

Reimann-Philipp (1983) also observed the high degree of sterility in the first crossing generations, but $F_6$ populations of backcrosses with commercial varieties of *P. vulgaris* grew well and had a fair tolerance to low germination temperature inherited from *P. coccineus* and they are expected to retain their resistance against CBM virus and antracnosis.

Due to its origin in the cool uplands of subtropical regions, *P. coccineus* does not suffer from low temperatures, either during germination, or during flowering. This may be the reason that the runner bean grows so well in England or in Sweden, where it is popular for its profuse flowering and occasionally high pod-yield in favourable frost-free autumns.

### *P. flavescens* Piper

Synonym: *P. leucanthus* Piper, *P. polyanthus* Greenm.

Murutungo-bean, chaguita

$2n = 22$

ORIGIN AND HISTORY

The systematic position of this Venezuelean/
Colombian species needs further explana-
tion in view of the existence of a seemingly
similar taxon from Mexico, denominated *P.
polyanthus*, and *P. leucanthus* described
from Jamaica. The distance of the original
places of finding the three units is enormous: more than 3000 km between the locus
classicus of *P. flavescens* (Cerro Tama, Cal-
das. US Nat. Herbar 143.511) and *P. polyanthus,* which was originally collected in
Jalapa, Vera Cruz, Gulf of Mexico. There
are also anatomical differences between the
two taxa.

The existence of this wild bean, which is
used regularly by natives of the Andine valleys of Venezuela and Colombia, has been
described only recently by Berglund-
Brücher and Brücher (1974). We discovered
this interesting edible bean during our exploration journeys of the cool-humid tropical mountain forests in the border region of
Colombia/Venezuela, 1965, and cultivated it
for several vegetation periods in our experimental fields at Caracas. There we studied
the seeds and found them free of adverse
metabolic substances and easy to digest.
This lends support to the ancient custom of
mountain natives to harvest and consume *P.
flavescens* regularly in the wild state. The
existence of this notable bean has been overlooked so far by economic botanists and
plant collectors in South America and by
agricultural science in general. With the intention of improving this situation, we submitted this interesting "cultivar in statu
nascendi" to mutation experiments at the
IVIC-Atomic Reactor and began selections
at our Genetics Department of the Central
University of Venezuela (Brücher 1977) and
distributed seed samples to the Faculté des
Sciences Agronomiques Gembloux, where
Prof. Marechal believes that *P. flavescens* is
related to *P. polyanthus.*

We determined the chromosome number as
$2n = 22$ and observed normal meiotic behaviour, excluding the possibility that *P.
flavescens* might perhaps be a bastard plant.
It can be hybridized with runner beans (*P.
coccineus*) as well as with the common haricot bean (*P. vulgaris*), thus representing an

important genetic bridge between these two
distinct bean groups.

MORPHOLOGY

Stems with indeterminate growth (3–5 m
long) twining and climbing on other shrubs,
at base sometimes woody, but without enlarged tuberous rhizomes; covered with
short bristly hairs.

Leaves trifoliate, bigger than in the common
bean, often 30 cm long and 24 cm broad, on
long petioles, leaflets broad ovate, nearly as
broad as long, short acuminate, densely covered with short hairs.

Inflorescences borne on the stem in leaf axils
as lax racemes.

Flowers larger than in *P. vulgaris* (15–
18 mm long), at the beginning white, then
typically changing to creamy-yellow
(= "*flavescens*") which is the colour in
Herbarium exsiccata. Style visibly curved
introrse with an extended stigma. Bracteoles of the calyx narrow and small
(6 mm × 1 mm).

The fruit pods are covered with a faint hair-
indumen (in contrast to the coarse surface of
*P. coccineus* pods). The fruits are 8–10 cm
long, rather broad (20–28 mm), straight,
with a short stout beak. When ripe, they
aquire a weak dehiscence for seed release.

Seeds are 14 mm long, 7 mm thick and
10 mm broad, with a typical gold-brown testa colour, a red hilum and dark caruncula.
Seed germination is epigaeic (as in *P. vulgaris*), not hypogaeic as in *P. coccineus*, to
which some similarity exists in stem and leaf
characters. Besides this systematically very
important germination behaviour, *P. flavescens* differs from *P. coccineus* in one essential flower character: the stigma is not ex-
trorsely inserted on the spiral pistil as is the
case in *P. coccineus*, but quite differently, it
has an introrse position on the pistil coil.
There are further differences to other *Phaseolus* species. *P. flavescens* has a remarkably
slow juvenile growth and needs in total 150
days until flowering begins. Quite frequently
one comes across plants which still bear dry
pods of the previous growing season together with new blossoms on the long twining
stems (Fig. III.10). We cultivated *P. flavescens* for several years under shortday condi-

◄ a                                      b

**Fig. III.10.** *Phaseolus flavescens*, a perennial wildbean, regularly collected by natives as food in Colombia and Venezuela

tions similar to its original habitat at Rio Pamplonita, near Diamante, 2500 m alt. (Colombia) and Rio Motatan, near Timotes (2000 m) province Trujillo (Venezuela). We observed good yield and high resistance to pests (*Empoasca*) and fungal diseases (*Colletotrichum* and *Uromyces*). We came to the conclusion that the genetical improvement of "Murutungo" beans would represent a considerable enrichment for the protein-hungry world of tomorrow, once the adverse factors of perennial growth and shattering of the pods have been eliminated.

### P. lunatus L.

*with two agrotypes: var. microcarpus* = sieva bean, var. *macrocarpus* = Lima bean

Butter bean, Lima bean, pallar, chuvi, haba lima, haricot de Lima, Silva bean, pángoa (Colombia), cubace (Costa Rica), Mondbohne

$2n = 22$

### Name, Origin and History

Spanish travellers reported this Indian bean first from Lima in Peru, but this name should not be related to its genetical or geographical origin. In our opinion, *P. lunatus* was selected independently at very distant places in the Neotropics from similar wild forms, which today still exist spontaneously in Central and South America (Burkart 1952, Mackie 1943, Brücher 1977). According to these authors, the most probable polyphyletic origin and domestication is:

a) Northern group. Hopi Indians selected biotypes with flat small seeds, few HCN, brown or white (but never red)-coloured, at present distributed from Florida, Virginia to the Oklahoma-Appalachian mountains.

b) Caribbean-tropical group. Red-coloured testa, seeds small to medium sized with high HCN content. Perhaps related to

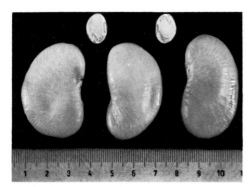

**Fig. III.11.** *Phaseolus lunatus.* Seeds from a wild form from Guatemala together with gigas cultivar from Peru

the small-seeded wild and primitive forms of Guatemala and Mexico. Adapted to humid hot climate.
c) Pallares group. Large seeds with different colours and patterns, selected from Andean wild forms, which grow spontaneously in North Argentina, Bolivia and Peru (Burkart, Brücher, Kaplan). Used and domesticated by South American Indians for nutrition but also as "messenger beans" in the Inka Empire (Fig. III.12).

MORPHOLOGY

Due to its very ample segregation in innumerable landraces and commercial varieties, it is rather difficult to give an obligatory general description. Annual and perennial biotypes exist, also bush forms and long twining (1–4 m) varieties.
Leaves trifoliate, often hairy on the underside. Leaflets ovate (5–11 × 3–8 cm) on long (9–18 cm) petioles, with very small stipulae.
Inflorescenses borne in the stem axils, many-flowered, self-pollinating, with white or greenish flowers.
Fruits large, broad pods (5–10 cm long and 2–3 cm broad), generally curved with a sharp beak and few (3–4) seed kernels.
Seeds are very variable in colour, ranging from pure white to creamy, red, brown, black mottled with striking patterns, mostly flat and rather different from other bean seeds. The biggest "macrocarpus" seeds

may measure 4 × 3 cm(!) and are real gigas selections.
*P. lunatus* is a very ancient crop among American Indians. Material has been recovered in Peru at Ancon, Chillon, Chuquitanta belonging to the pre-Ceramic epoque. Seeds from Chilca have been dated at 5200 B. P. Further, early Indian potteries also exist with attractive artistic "pallares" motives. The Mochica culture is rich in pottery with figures of "bean-warriors". Some may represent shamanistic metamorphosis from plants to animals. The Moche deity "Tusked God" has been painted as a big bean seed. There is proof that pallares with different patterns have been used for transmission of secret messages. On one of our exploration trips in Northern Columbia we were surprised to observe that different families reproduced around their houses distinct *P. lunatus* cultivars. The seeds had expressive colours and patterns and were used as "playing chips", and considered as a typical "house mark" signature (Fig. III.12).
Plants of *P. lunatus* have epigaeal germination, i.e. as in *P. vulgaris*. In general, they possess a high grade of resistance against disease and pests, and thrive rather well in humid areas of the tropics, where common beans (*P. vulgaris*) suffer from the climate. This is the reason why *P. lunatus* was spread rather early after the conquest of America through the tropics, like India, Burma, Nigeria and Madagascar. Viney types are preferred in tropical regions, with long growing (5–9 months) season. They produce there an impressive biomass and protect the soils against erosion and leaching during heavy rainfall. In the Parana region of Paraguay we selected from perennial indigenous *P. lunatus* strains our own cultivars which yielded 2500–3000 kg/ha seeds annually. In crossings we observed the dominance of the primitive factors, like twining, dark flower colour over white, and mottled and coloured seed testa dominant over white; indeterminate growth dominates over bushy.
Similarly to the haricot bean (*P. vulgaris*) the Lima bean is also used in two commercial ways: as green pods for vegetable cooking and as the usual dried bean. In both cases, Lima beans are a valuable and nutri-

**Fig. III.12.** *Phaseolus lunatus,* colour samples of local Andine varieties which are still used as "playing chips" by natives. *Below* similar samples from excavations of the Mochica culture

tious foodstuff. If consumed as immature green bean pods the major part is water (66%), the rest of the edible part consists of 8% protein, 23% carbohydrate and 0.7% fat. The mineral content shows a fair amount of phosphorus: 111 mg in 100 g and potassium: 747 mg in 100 g, also vitamins A, B and C.

Shelled dry Lima beans have 20% protein, 64% carbohydrate, 1% fat and 2% sugars, 3% ashes (calcium 84 mg in 100 g and 5% iron in 100 g). As with most *Phaseolus* beans, the protein content has a rather low

digestibility, which can be overcome by long cooking procedures.

Under certain circumstances *P. lunatus* seeds provoke toxic effects. This is mainly a question of the variety. Certain Caribbean strains are described in literature as rather poisonous. Such plants contain major quantities of the glucoside phaseolunatine and the enzyme linamarase. This combination can liberate under hydrolysis the very poisonous hydrocyanic acid (HCN). Such toxic effects are less when cooked dry bean seeds are consumed instead of cooked green pods.

## *P. vulgaris* L.

Haricot bean, kidney bean, snap bean, poroto, caraota, frijol, haricot commune, Gartenbohne

$2n = 22$

ORIGIN AND HISTORY

The discovery by Burkart and Brücher (1953) of wild-growing beans in the Andes of South America (North Argentina, Bolivia, Peru) initiated long-lasting and controversial discussions as to the real homeland of the garden bean. The wild bean may be crossed easily with *P. vulgaris*, but on the other hand possesses so many independent genetic factors which distinguish it from the cultivated bean, that it was designated as a separate species in its own right: *Phaseolus aborigineus*.

The presence of a similar wild bean has been reported by Gentry (1969) for Mexico. Earlier, McBryde (1947), Burkart (1952) described it from Guatemala and Honduras, Brücher found *P. aborigineus* in Costa Rica and in the Chiriqui mountains of NW Panama. Considering its presence in both Middle and South America, the wild bean covers a distribution arc of 10,000 km.

We believe that the different cultivars of the *P. vulgaris* group have a polyphyletic origin in the neotropics, i.e. they have been developed and domesticated by different Indian tribes in quite distant regions, for example in the Aztec area of Western Guatemala-Mexico, in the Talamanca civilization of ancient Costa Rica/North Panama (Chiriqui mountains), in the Colombia-Venezuela Andes and finally in the highly developed agriculture of the Kechua-Aymara Indians south of the equator and even south of Capricorn, where Diaguites and Huarpes excelled as planters.

In my opinion, the gathering of beans from the wild state and the gradual domestication of primitive cultivars in house gardens began around 10,000 years B.P. in the intercordillerian valleys with temperate climate and easy irrigation. Such regions were, for example, the "callejon de Huaylas" of Central Peru, and the valley of the Urubamba River in the Depts. of Cuzco and Vilcanota; the Yunga valleys and the region of Cocha-bamba in Bolivia, and finally the highly developed Diaguita irrigation schemes of North Argentina, which extended southward along the fertile Aconquija mountain chain in the present provinces of Tucuman and Catamarca.

## *P. aborigineus* Burkart

Research into the phylogeny of useful plants for mankind presents only few cases where the origin, distribution and domestication of an important foodcrop can be traced back so completely to its wild ancestor as in the case of the common bean, *P. vulgaris* (see also Gepts 1988).

The wild bean of South America is an authentic member of undisturbed natural plant communities of the Montana district (Cabrera 1971) in the Cordillera de los Andes and the Sierras Preandinas. Its habitat is limited to a rather narrow belt (1000–2500 m altitude) in the mesothermic forest, often represented as a "cloud forest" in the higher regions. Here, trees like *Juglans australis, Alnus acuminata* and *Sambucus peruviana* abound, whilst the understorey of the dense vegetation is composed of different species of the genera *Clematis, Geranium, Datura, Madia, Rumex, Salvia* etc. In these sometimes impenetrable thickets thrives *P. aborigineus*, climbing on shrubs and trees, its vines easily reaching 3–6 m length.

The geographical extension of *P. aborigineus* in South America is enormous. Its distribution arc (Brücher 1968) begins at 32°2′S.lat. in the mountains of Comechingones (prov. San Luis-Cordoba), and covers all the Northwest Argentinian provinces of Catamarca, Tucuman, Salta and Jujuy. We discovered it in Bolivia in high mountain valleys of Tarija and Cochabamba, but are convinced that *P. aborigineus* grows also in many other places on the Eastern slopes of the Andes, provided that the original *Alnus-Juglans*-forests have not been destroyed, as has unfortunately been the case for several decades. From Peru there are several reports, dating back to Weberbauer (1945), of the spontaneously growing wild beans. Vargas (Herbar Nr. 7110), Smith (pers. commun) (Fig. III.13).

**Fig. III.13.** *Phaseolus aborigineus*, the wild bean from South America

Even if we have no data from Ecuador, we are convinced that *P. aborigineus* exists there in the high valleys of the cloud forest region. From Colombia and Venezuela we reported its existence from many places (Berglund-Brücher 1968), beyond latitude 10° north.

In all these countries the age-old custom of primitive mountain dwellers survives of gathering the ripe pods of the wild bean in the autumn. The seeds are not only an emergency food, as we observed at Mitisu in Venezuela, but are considered as a valuable supplement of the native diet in the poorer regions of northwest Argentina, where I observed that the daily "locro" consisted of cooked meat, together with seed kernels of maize and wildbeans. I still consider it as a

privilege for a foreigner to have been invited by the shy and retiring natives of the remote Chabarilla valley in the mountains of Catamarca to share with them this archaic wild bean dish (Fig. III.15).

It should be mentioned here that the domestication of the haricot bean from the South American wild bean *P. aborigineus* is one of the few instructive models in crop selection and evolution with a clearcut separation between the still-existing wild-growing ancestors and the selected cultivars. We mention this fact with particular emphasis to refute opinions like Gentry's statement (1969) that *P. aborigineus* is a "weed-bean" or an "escape biotype" formed after a spontaneous crossing between wild and cultivated biotypes, or that of Flannery (1973) when he says on p. 302... "the Andean wild form *P. aborigineus* is so similar to *P. vulgaris* that one could conceivably lump them all as races of one species... Indeed, some varieties of *aborigineus* are less impermeable than wild *vulgaris*... and may possibly be escapes derived from cultivated *vulgaris*...".

Both authors never explored the beans of South America, nor do they cite the places where the supposed spontaneous crossings of wild and cultivated beans might have been observed. On the contrary, these have never been reported. During our repeated exploration journeys and collecting expeditions from North Argentina to Venezuela along the distribution arc of *P. aborigineus*, we never found spontaneous hybrids, or "introgressed populations" between wild and cultivated beans. We may recall that outcrossing in *P. vulgaris* is practically inhibited for floral-biological reasons. Self-pollination is predominant and experienced beanbreeders consider spontaneous hybridization to be negligible in their experimental plots.

North American ethnobotanists needed an unusually long time to understand these facts and to accept the South American origin of the kidney bean. The first sign of change and comprehension is due to the notable archaeological bean record from the Guitarrero Cave (Central Peru), described by Kaplan et al. (1973). Kaplan declared in 1980: "... The Guitarrero gives strong sup-

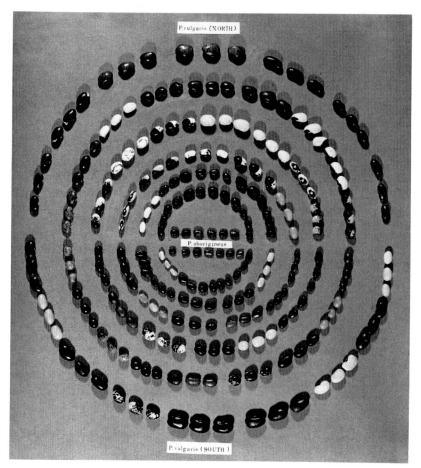

**Fig. III.14.** Samples of primitive bean cultivars from Argentina, Bolivia, Ecuador and Peru in comparison with seeds of the wild bean *Phaseolus aborigineus* in the centre (Brücher 1987)

port to the contention of Burkart and Brücher (1953) and Berglund and Brücher (1976) that at least one group of *P. vulgaris* domesticates had its origin in South America..." In addition to this very old find from Peru, we know also some few pre-Columbian archaeological bean remains from North Argentina. We mention the finding of the Huachichocana Cave in the Quebrada de Purmamarca (province Jujuy), the Inca Cave from province Salta, and bean samples from province San Juan, described by Roig (1977) in a local review of limited distribution (Fig. III.15).

These are unmistakable proofs of early bean cultivation (*P. vulgaris*, of course) in NW

Argentina, where *P. aborigineus* grows profusely, providing genetic sources for domestication by Indian plant growers south of the equator. At the other end of Amerindian bean domestication, between 15° and 25° of northern latitude, the indigenous inhabitants of Guatemala and SW Mexico performed similar work. Based on the Central American biotypes of *P. aborigineus*, they created numerous local landraces of different testa colour and plant habits. Archaeological remains have been reported from Eastern Mexico, dated at 4300 B.P. and others even 7000 years B.P. (Smith 1965). McBryde (1947) found wild, primitive *Phaseolus* beans in Guatemala and claimed

a

**Fig. III.15.** Phaseolus aborigineus, the ancestor of the common bean is still collected and used by natives in NW-Argentina for food

b

that this country "is the origin of the common kidney bean". More southward, the hitherto earliest discovered pre-historic proofs of primitive *P. vulgaris* originated from Rio Chiriqui shelters in NW-Panama, near the border with Costa Rica. Carbonized plant remains from excavations at Cerro Punta, also contain beans, dated 4800–500 B.P. These findings received the following comment from Smith (1980): "...common bean remains from the Cerro Punta site are clearly of no special morphological type". Their measurements are the following: length = 7–10 mm, width = 4–7 mm, thickness = 2–3.5 mm, based on cotyledons". These primitive beans from

Chiriqui (NW-Panama) may be considered as a continuity from Mesoamerican beans, Mexico, Guatemala, Honduras, Nicaragua and Costa Rica. We discard the possibility that they entered from the South (Colombia) because the climatical and ecological conditions of the hot-humid Panama Isthmus are quite adverse for its natural dispersion.

MORPHOLOGY

The habit of the *P. vulgaris* group with its hundreds of primitive landraces and highly selected commercial cultivars is so heterogenous that it seems impossible to bring their structure to a common denominator. The first-named landraces are in general high-climbing pole-beans, whilst the second grow as so-called bush-beans. All bean plants have in common well-developed taproots which penetrate the soil rapidly to 100 cm depth, and lateral roots in the upper part which bear large quantities of *Rhizobium* nodules.

Stalks are ribbed and covered with rough hairs. In the case of climbing pole beans, the stems are rather weak, twisted, reaching 2–4 m length, supported by other cultivated plants, like maize or sunflower on native fields, or poles and wires in modern plantations. They develop many nodes and inflorescenses and have an indetermined growth habit.

The stems of bushy types have only few (6–8) nodes, stop their growth at 20–50 cm height, are rather stout, producing generally only one inflorescence with a great quantity of flowers. Modern industrial varieties based on selections of mutations ripen at the same time, which allows mechanical harvesting.

Leaves are alternate, trifoliate with long petioles and typical enlarged parenchyme tissues (pulvinus) at the base, which allow quick torsions. The leaflets are ovate-acuminate (12–16 cm long, 4–11 cm wide).

Inflorescences are lax racemes; the flowers are inserted on short pedicels and have pronounced self-fertilization. Flower colour is pink or violet in wild and primitive biotypes, in cv. often white or creamy-yellow.

Fruit pods are straight or slightly curved with a prominent beak, 5–12 cm long and 1 cm broad (but certain fancy varieties of snap beans are 28 cm long and 3 cm broad). Seeds are very variable in size and colour: white, yellow, pink, red, purple, brown, black, with innumerable patterns, spots and mottle. Primitive varieties have kidney-shaped seed forms, but modern selections may be round, oblong, or compressed, varying also in shape, size and weight.

According to Dierbach (1836) the garden bean (*P. vulgaris*) was considered during the first century after its introduction to northern Europe from the New World mostly as an ornament for fences and garden houses. Its seed production in Central and North Europe was so low (probably for photoperiodical reasons) that farmers had no use for it. This was, however, different in the Mediterranean region, Spain, Italy, Greece and Turkey, where *P. vulgaris* was useful as a field crop. This is manifested in the famous *Herbal* of L. Fuchs (New Kreutterbuch, Basel, 1543), which describes the "Welsch Bone" as a bean from the Mediterranean region. Fuchs and also Bock (1560, Strassburg) indicated its high frost susceptibility, which inhibits its maturation if sown North of the Alps.

The USA received *P. vulgaris* from Captain Gosnold in 1602, who planted it on the coast of Massachusetts; from there it spread to New York in 1644. It seems that bean cultivation in Virginia is older, perhaps introduced there by Indian tribes who received bean seeds in cultural exchange from Mexico.

**Fig. III.16.** Precious picture of the first cultivated beans in Europe, painted by Oellinger 1553. It is remarkable that the ripe pods split spontaneously

**Fig. III.17.** Comparison with a modern bean cultivar, testifying a dramatic genetical and morphological improvement in scarcely 400 years (Courtesy of seed grower Wagner, Heidelberg)

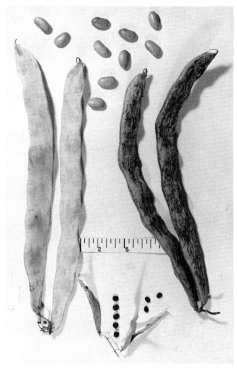

**Fig. III.18.** The gigas effect in bean breeding. The wild bean together with some Argentinian local cultivars

The progress of bean breeding during the last centuries in Europe and later in the USA is well documented. After its introduction from Indo-America in the 16th century, *P. vulgaris* was mainly a curiosity in the show gardens of scholars. Fortunately there exists a precious hand-coloured picture of a bean plant in *Oellingers Herbal*, which was composed in 1553 at Nürnberg. It clearly depicts strongly climbing vines with shattering pods and small, dark red seeds (Fig. III.16). Harvesting of such a crop was not easy, of course, and farmers had not much use for this exotic pulse.

The first selection step must have been the elimination of the undesirable shattering factor, which depends on the anatomical structure of the dorsal and ventral suture fibres. This aim was achieved at the end of the 18th century, as we conclude from the descriptions by the Italian authors Zuccagni (1809) and Savi (1825). The seeds were,

however, still small and mostly dark-coloured. The pods, of course, were not tender. Stringless bean fruits, which today represent a large share of the bean trade, did not exist at this time (Fig. III.17). The beautifully coloured book of Martens (1860) depicts many black-seeded varieties like "var.*nigerrimus*", "var. *gonospermus*" and "var. *atrofuscus*", some of them still with shattering pods. "...the two valves spring open separately, curling and rolling like shrews, vehemently ejecting the seeds ..." stated the author, whilst many others in Marten's book are rather advanced selections.

On the other hand there are reliable descriptions (Krüger 1852) of impressive "gigas varieties", like var. *romanus* with foot-long pods and 10–12 seeds inside, probably selected in Germany because they are only recorded there. Figure III.18 shows such a biotype with 280 mm long and 27 mm wide fruits and considerable seed size of 13 mm × 8 mm.

We conclude that the major aims of empirical improvement of *P. vulgaris* were achieved within only 300 years after its introduction from America. Notably, all this was realized before Mendel discovered the basic laws of heredity, and before scientific plant breeding was established. Recent improvements, like bush types, the elimination of excessive vine length and indeterminate growth, stringless soft pods, devoid of the coarse inner parchment layers, the selection of artificial mutations in seed and pod colour, are the result of worldwide commercial seed production. *P. vulgaris* achieved in the meantime the widest geographical extension of all edible legumes, covering not only the tropics, but also the temperate latitudes from Scandinavia (Uppsala) to Patagonia (Ushuaia).

It has long been known that the ingestion of raw bean seeds can produce grave health disorders. Because of some strange physiological effects, the eating of green pods or unripe seeds causes intoxication, sometimes even with fatal consequences. There is a thermolabile substance, phyto-haemagglutinine (PHA), in *Phaseolus* which is especially dangerous for human beings and mammals, because of its destructive effects on red blood particles. Still it is not quite clear (according to Jaffé et al. 1972) if PHA itself is toxic, or if it is only a secondary effects of other toxic substances in bean seeds. It was found, for example, that *Pisum* agglutinin has no toxic effect. In free nature, the toxic glycoproteids may have a protective role against being eaten by animals. Feeding laboratory animals such as mice and rats exclusively with bean meal, they lost weight and finally died (Liener 1972). With respect to human nutrition there are still doubts if bean meal as food for children causes severe disorders if only baked (together with maize) and not thoroughly cooked. In view of this potential danger of bean nutrition in primitive populations, our discovery seems of exceptional importance (Berglund Brücher 1969) of certain indigenous *P. vulgaris* strains which are free of toxic substances.

The nutritional value of bean seeds compares well in its amino acids with soybean, for example, as is shown in Table 8.

**Table 8.** Comparative nutritional value of bean seeds of different types

|  | Common bean | Lima bean | Soybean |
|---|---|---|---|
| Alanine | 4.2 | 4.3 | 4.4 |
| Arginine | 6.4 | 4.3 | 6.9 |
| Asparagin acid | 12.7 | 17.5 | 12.4 |
| Cystine | 1.1 | 1.5 | 1.4 |
| Glutamin acid | 16.6 | 14.8 | 19.8 |
| Glycine | 3.9 | 5.3 | 4.5 |
| Histidine | 2.7 | 2.8 | 2.7 |
| Isoleucine | 4.7 | 3.6 | 4.5 |
| Leucine | 8.3 | 7.5 | 7.9 |
| Lysine | 7.1 | 6.6 | 6.6 |
| Methionine | 1.1 | 0.9 | 1.2 |
| Phenylalanine | 5.4 | 4.7 | 4.9 |
| Proline | 3.9 | 3.7 | 5.3 |
| Serine | 5.8 | 7.5 | 5.3 |
| Threonine | 4.2 | 3.4 | 3.6 |
| Tryptophan | 1.2 |  | 1.4 |
| Tyrosine | 3.2 | 3.6 | 3.4 |
| Valine | 5.1 | 3.8 | 4.9 |

## Breeding Against Diseases and Pests

### Fungus

*Colletotrichum lindemuthianum* ("anthracnose") is the most common and destructive bean disease, especially in temperate, humid climate and elevated (1000 m alt.) regions. The disease is seed-borne and easily transmitted by spores. The symptoms appear on leaves as rust-brown or purplish lesions, on stems as dark spots; at the beginning the pods show pink lesions which develop in circular depressions with black borders. Heavily infested plants cannot survive. Resistance breeding was begun several decades ago based on immune *P. aborigineus* collected and provided by us from the Aconquija-Mountains (Brücher 1954, Hübbeling 1957). In the meantime *Antracnosis*-resistant cultivars have been distributed worldwide. This fact contradicts the opinion of Gentry (1969) that "wild beans have not been used for crop improvement for several thousand years".

*Uromyces phaseoli* (bean rust) is distributed worldwide and is of considerable economic importance, because it can cause total losses of the bean crop in tropical environments. Its symptoms are rust-coloured pustules which are full of spores which disseminate the disease rapidly. The infected pods show red spots on their surface and may become malformed. Certain resistant *Phaseolus* varieties exist, but breeding and selection of high-yielding resistant strains is difficult due to the great number of rust races (in the USA alone: 35 races).

*Sclerotium rolfsii* and *S. sclerotiorum* (wilt) occur especially in tropical countries and attack stems, leaves and especially green ripe pods, causing a soft watery rot. The disease causes heavy losses (50%) during transportation of haricot beans for green consumption. Several other fungus diseases may be mentioned, e.g. *Ascochyta, Fusarium, Botrytis cinerea, Erysiphe polygoni, Entyloma* and *Alternaria*, which limit production in tropical environments. Genetical resistance against *Ascochyta* – a pathogen prevailing in cool-humid Andean climate – has been observed in *Phaseolus flavescens* from Colombia, combined with certain tolerance against effects of cold.

Very important are selections which do not suffer from *Fusarium oxysporum*, called "podredumbre radicular", which attacks especially stems and roots and in Peru sometimes causes 60% losses in commercial bean plantations.

Among the bacterial diseases we mention only the blight-causing *Pseudomonas phaseolicola* and *Xanthomonas phaseoli*. Infections occur especially at cool, humid temperatures.

**Viruses**

The *common mosaic* (MC) is a worldwide problem in bean plantations. Transmitted by aphids and seeds, it causes leaf distortion and at higher temperature necrosis and loss of leaves. Losses may be 40% of the normal yield. *Rugose mosaic* (MR) provokes symptoms similar to MC virus. Its transmission occurs by coleopteros and mechanical contacts.

The worst virus infections are caused by *Virus Dorado* (BGMV) which is native to Central America. Its main vector is the white fly, *Bemisia*. The symptoms are easy to recognize, because the leaves take on a yellow-golden aspect. Drastic damage has been reported from several tropical countries, which obliged farmers to abandon bean cultivation. Therefore it is promising that after 10 years of research work, CIAT-Colombia released the var. ITA-Quetzal (DOR 41) and Negro Huasteco, which are resistant and give yields of 1300 kg/ha in comparison with the usual 550 kg/ha in tropical latitudes. *Brown streak virus* is a new disease which causes especially high losses in Africa (in East Africa 60–95%) and has been introduced recently also to India and Java. The International Centre for bean breeding are at present screening thousands of accessions of American *Phaseolus* to discover genetical virus resistance.

Pests

A quantity of insects attack bean plants and their stored seeds. Resistance breeding, e.g. against caterpillars, beetles and weevils, which live mostly underneath the plants in the soil, is rather difficult.

In this context it may be interesting that the dark-coated wild bean *P. aborigineus* is less often attacked by the bean weevil *Acanthescelides,* a destructive pest of dry beans in storage. It is not uncommon to observe that native bean farmers in the Andes lose most of the yield, when it is stored under primitive housing conditions; but also modern processing plants in the USA and Mexico suffer from invasions of *Acanthoscelides* in their silos. Therefore resistance breeding against all sorts of bean weevils is of the outmost importance.

While living in Venezuela, we observed an interesting resistance against the *Bruchus* beetle in "caraotas negras", local black-seeded varieties. When the seeds of black beans had been mixed with other commercial varieties, only the light-coloured kernels suffered infestation.

*Epilachna varivesta* is considered as the worst parasite of bean cultivation in Mexico and the USA, but efforts to discover resistance genes have failed so far.

To combat such pests and parasites, of course, many chemical insecticides exist, but their application is limited in bean seeds destined for human consumption. Therefore the search for biological immunity is of great importance.

In this context we mention the efforts of the international CIAT Institute in Colombia and its germplasm bank of *Phaseolus* species. More than 30,000 specimens have now been collected and evaluated for favourable reactions to diseases and pests.

# References

Acosta J (1590) Historia Natural y Moral de las Indias. Madrid

Atabekova AI (1962) The karyological system in the genus *Lupinus* (in Russian) Trans Moscow Soc Nat 5:238–246

Baer E v, Gross R (1977) Auslese bitterstoffarmer Formen von *Lupinus mutabilis*. Zt f Pflanz Züchtg 79:52–58

Basset MJ, Cheng SS, Quesenberry KH (1981) Cytogeneic analyisis of interspecific hybrids between common bean and runner bean. Crop Sci 21:75–79

Baudet JC(1977) Origine et classification des espèces cultivées du genre *Phaseolus*. Bul Soc Roy Bot Belg 110:65–76

Berglund-Brücher O (1968) Wildbohnenfunde in Südamerika. Naturwiss 54:466–468

Berglund-Brücher O (1969) Absence of phytohemagglutinin in wild and cultivated beans from South America. Am Soc Hort Sci Tropic Region 12:68–85

Berglund-Brücher O, Brücher H (1974) Murutungo, eine semi-domestizierte Wildbohne *Phaseous flavescens* aus den tropischen Gebirgen Südamerikas. Angew Bot 48:209–220

Berglund-Brücher O, Brücher H (1974) The South American wildbean (*Phaseolus aborigineus*) as ancestor of the common bean. Econ Bot 30:257–272

Berglund-Brücher O, Wecksler M, Levy A, Palozzo A, Jaffé W Comparison of phytohemagglutinins in wild beans (*Ph. aborigineus*) and in common beans *(Ph. vulgaris)* and their inheritance. Phytochemistry 8:1739–11754

Bianchini F (1976) Fruits and vegetables. Crown Publ Inc, New York

Brücher H (1954) Argentinien, Urheimat unserer Bohnen. Umsch Wiss Tech 54:14–15

Brücher H (1968) Genetische Reserve Südamerikas für die Kulturpflanzenzüchtung. Theor Appl Genet 38:9–22

Brücher H (1969a) Gibt es Genzentren? Naturwissenschaften 56:77–84

Brücher H (1969b) Die Evolution der Gartenbohne *P. vulgaris* aus der südamerikanischen Wildbohne *P. aborigineus*. Angew Bot 42:119–128

Brücher H (1970) Beitrag zur Domestikation proteinreicher und alkaloidarmer Lupinen in Südamerika. Angew Bot 44:7–27

Brücher H (1977) Tropische Nutzpflanzen, Ursprung, Evolution und Domestikation. Springer Berlin Heidelberg New York 529 pp

Brücher H (1988) The wild ancestor of *Phaseolus vulgaris* in South America. In: Gepts P (ed) Genetic resources, domestication and evolution of *Phaseolus* beans. M Nijhoff, Dortrecht, Netherlands

Burkart A (1952) Las leguminosas argentinas. II ed Acme Buenos Aires 570 pp

Burkart A, Brücher H (1953) *Phaseolus aborigineus*, die mutmaßliche Stammform der Kulturbohne. Züchter 23:65–72

Cabrera A (1971) Fiotgeografia de la Republica Argentina. Bol Soc Argent Bot 14:1–42

Cardenas M (1969) Manual de las plantas economicas de Bolivia. Cochabamba, Icthus

Clausen RT (1945) A botanical study of the yam beans (*Pachyrrhizus*) Mem Cornell Univ Agric Exp Stat 264:1–38

Ducke JA (1981) Handbook of legumes of world economic importance. Plenum, New York

Flannery T (1973) Mesoamerican food plants archeology. In: The origins of agriculture. Reed (ed) Mouton Publ Paris, 1013 p

Gentry HS (1969) Origin of the common bean (*Phaseolus vulgaris*) Econ Bot 23:55–68

Gepts P, Osborn TC, Rashka K, Bliss FA (1986) *Phaseolin* protein variability in wild forms and landraces of the common bean as evidence for multiple centers of domestication. Econ Bot 40 451–468

Gepts P (1988) Genetic resources, domestication and evolution of *Phaseolus* beans. M Nijhoff, Dortrecht, Netherlands

Hammons RO (1982) Disease-resistant groundnut released. FAO, Plant Genet Resource Lett 51:12–14

Harlan J (1971) Agricultural origins, centers and non-centers. Science 174:468–474

Hondelmann W (1984) The lupin, ancient and modern crop plant. Theor Appl Genet 68:1–9

Hübbeling N (1957) News aspects of breeding for disease resistance in beans. Euphytica 6:111–141

Jaffé WG, Berglund-Brücher O (1972) Toxicidad y especifidad de diferentes fitohemaglutininas de frijoles (*P. vulgaris*). Arch Latinoam Nutr 22:267

Jaffé WG, Brücher O, Palozzo A (1972) Detection of four types of specific phyto-hemagglutinins in different lines of beans (*P. vulgaris*). Zt Immunitäts Forsch 142:439

Jaffé W, Hannig K (1965) Fractionation of proteins from kidney beans *(Ph. vulgaris)*. Arch Biochem 109:80–91

Kaplan L, Lynch T, Smith CE (1973) Early cultivated beans (*P. vulgaris*) from an intermontane pruvian valley. Science 179:76–78

Kaplan L (1980) Early man in the Andes. Acad Press, New York, pp 145–158

Kay DE (1979) Food legumes, crop and product digest nr 3. Tropical Research Institute London

Lampard C (1974) The use of *Canavalia ensiformis* as a biological control for leafcutting ants. Biotropica 11(4) 1979:313–314

Lamprecht H (1959) Der Artbegriff, eine experimentelle Klarlegung. Agri Hort Genet 17:3–105

Lamprecht H (1964) Species concept and the origin of species. Agri Hort Genet 22:272–280

Liener I (1972) Antitryptic and other antinutritional factors in legumes. PAG Sympos UN New York 239–270

Macbride J (1943) Flora of Peru (Leguminosae) Field Mus Nat Hist 13:1–507

Mackie W (1943) Origin, dispersal and variability of the lima bean. Hilgardia 15:1–24

Macneish R (1967) The prehistory of the Tehuacan valley. Byers D (ed) Texas Univ Press Vol I

Marechal R (1971) Observations sur quelques hybrides dans le genre *Phaseolus*. Bull Rech Agron Gembloux 6:461–489

Marechal R, Mascherpa JM, Stainer F (1978) Etude taxonomique d'un groupe d'éspèces des genres Phaseolus et Vigna sur la base de données morphologiques et polliniques, traitées par lánalyse informatique. Boissiera 28:1–273

Martens GV (1860) Die Gartenbohnen, ihre Verbreitung und Kultur. Ebner, Stuttgart

Mc Bryde FW (1947) Cultural and historical geography of Southwest Guatemala. Smithsonian Inst Publ no 4,75–76,134–135

Nabhan GP, Felger R (1978) Teparies in SW Northamerica. A biogeographical study of *P. acutifolius*. Econ Bot 32:2–19

National Academy of Sciences (1979) Tropical legumes, resources for the future. Natl Acad Sci, Washington DC

Oviedo Y Valdez (1535) Historia General de los Indias. Sevilla

Pakendorf KW (1974) Studies on the use of mutagenic agents in *Lupinus*. Zt Pflanzenzücht 72:152–159

Parodi L (1935) Agricultura Prehispanica. Acad Nac Agr (Buenos Aires) I:115–167

Pinchinat AM (1977) The role of legumes in tropical America. In: Eploiting the legume. Rhyzobium Symbiosis. Hawaii University Press

Piper V (1926) Studies in American phaseolinae. Contrib US Natl Herb 22:663–701

Reimann-Philipp R (1983) Notwendigkeit, Probleme und Ergebnisse von Artkreuzungen zwischen Busch- und Feuerbohnen. Zt Pflanzenzücht 3:181–189

Roig FA (1977) Frutos y semillas arqueologicos de Calingasta (San Juan) In: La Cultura de Ansilta, San Juan, 217–250 (Argentina)

Sauer J (1964) Revision of *Canavalia*. Brittonia 16:106–181

Sauer J, Kaplan L (1969) *Canavalia* beans in American prehistory. Am Antiquity 34:417–424

Schwartz HF, Galvez GE (1980) Problemas de la produccion de porotos, (*Phaseolus vulgaris*) Centro Internac Agric Trop Cali, p 424

Smartt J (1980) Evolution and evolutionary problems in food legumes. Econ Bot 34:219–235

Smith CE Jr (1969) Archeological record of cultivated crops. Econ Bot 19:322–334

Soerensen M (1985) A taxonomic revision of the genus *Pachyrrhizus* (Inst f System Bot Kopenhagen, unpubl)

Verdcourt B (1980) The classification of Dolichos, Lablab, *Phaseolus, Vigna* and their allies. In: Adv in Legume Science. Summerfield & Bunting, Royal Bot Garden, Kew

Voysest O (1983) Variedades de frijol en America latina y su origen. CIAT Colombia, p 87

Weberbauer A (1945) Die Pflanzenwelt der peruanischen Anden. Min Agric Lima

Williams LO (1981) The useful plants of Central America. Tegucicalpa, Honduras. Ceiba 24:1–243

Wittmack L (1888) Die Heimat der Bohnen und der Kürbisse. Ber Dtsch Bot Ges Suppl 6:374–380

Wynne JC, Gregory WC (1881) Peanut breeding. Adv Agron 34:39–72

Yakovlef E, Herrera F (1934) El mundo vegetal de los antiguos peruanos. Rev del Museo Nac Lima

Vanderborght T (1983) Evaluation of *P. vulgaris* wild types and weedy forms. Plant Gen Res Newslett FAO 54:18–25

# IV. Oil Plants

LEGUMINOSAE

## 1. *Arachis hypogaea* L.

Groundnut, peanut, mani, cacahuete, arachide, amendoim, inchis, Erdnuss

$2n = 40$

NAME, ORIGIN AND HISTORY

The English name groundnut refers to the striking phenomenon of geocarpy, i.e. the fruits develop underground, a rare feature in the whole plant kingdom.

South American natives have numerous local names for *A. hypogaea*, for example: inchis (Kechua), choccopa (Aymara), choccopan (Araucano), tagalatic (Toba), mandobi (Chiriguano), amendoim, (Tupi, Brazil), mani, mandubi (Guarani), kupekana (Arawak). In Central America, where the cultivar has been introduced from the south, it is commonly denominated by the Aztec word cacahuatl.

Based on these linguistic differences, some ethnologists have tried to trace the origin and migration of peanut through the American continent. Under the assumption that the cradle of cultivars may be the South Bolivian Paraguayan Chaco, they claimed that it was the Arawak Indians who took groundnut seeds with them on their wanderings from Paraguay and South Brazil towards the Atlantic coast and even to the Caribbean Islands. I, personally, am more inclined to believe that the early Guarani played a decisive role in the origin and selection of the groundnut. First of all, we should not forget that *A. hypogaea* is an alloploid. Its genome of 40 chromosomes is composed of two wild diploid species of the Guarani region, with consecutive chromosome doubling. It is difficult for me to believe that the genome synthesis occurred completely spontaneously in the wild and that the offspring of this species hybrid survived on their own and were vigorous enough to superimpose themselves in the dense natural Chaco vegetation. It seems more probable that this notable event took place in a primitive house garden of Guarani planters. They ate and collected wild *Arachis* seeds and most probably planted them for convenience near their homes. I doubt that the Indians crossed *Arachis* flowers, which is not so easy, as for example in Solanaceous species, but they may have observed the vigorous growth of polyploid offspring from occasional natural crosses (Fig. IV.1). It is noteworthy that Hoehne already in 1944 found irrefutable proof that South American Indians successfully selected wild *Arachis* and domesticated a form which he designated *A. villosulicarpa*; another received the name *A. nambyquare*. But his contributions – written in Portuguese – remained nearly unknown. Even if no archaeological remains of Indian preagricultural activities in the Matto Grosso Paraguay region exist, we would not exclude the possibility that the natives knew how to select potentially useful plants; besides *Arachis* we have the examples of *Ananas*, *Capsicum* and *Manihot* (Fig. IV.2).

*A. villosulicarpa* is a diploid species, belonging to the section Extranervosae, which includes some other prostrate primitive wild-growing groundnuts, all with $2n = 20$ chromosomes (Ressler and Gregory 1979). There are, however, other sections with tetraploid species and edible seeds, attractive to the Indians. In addition to *A. villosulicarpa*, cultivated even today by natives of the region of Diamantino (Matto Grosso, Brazil), there exists the surprising Guarani selection of Mandovi guaycuru, the largest fruited *Arachis* of the world.

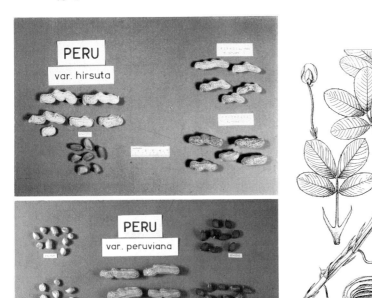

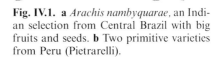

**Fig. IV.1. a** *Arachis nambyquarae*, an Indian selection from Central Brazil with big fruits and seeds. **b** Two primitive varieties from Peru (Pietrarelli).

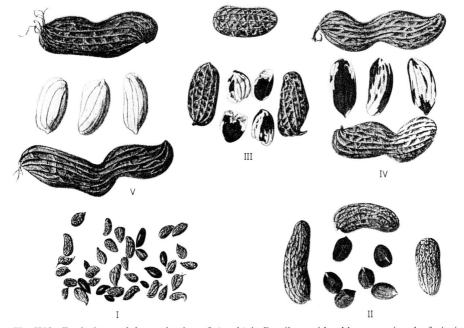

**Fig. IV.2.** Evolution and domestication of *Arachis* in Brazil, considerably surpassing the fruit size of *A. hypogaea*. **I** wild; **II** *A. hypogaea;* **III** *A. macrocarpa*; **IV** other *macrocarpa* biotype; **V** *A. nambyquarae* and white form (Hoehne 1944 )

The genus *Arachis* has a wide geographical distribution in South America, extending in E-W direction from the Atlantic coast to the foothills of the Andes and S-N from the river system of Parana La Plata to the Amazonas. The numerous wild species cover an areal of roughly 3,000,000 km². It is unnecessary to say that such an enormous radiation area in the tropical and subtropical latitudes of South America cannot be called a gene centre in the sense of Russian agrobiologists. If one looks for a sphere of major genetic diversity, probably the most indicated would be the Great Pantanal-Matto Grosso. We mention here only the following taxa:

### *A. batizocoi* Krap. & Greg.

A diploid wild species (annual growing habit) from Bolivia, presumably an ancestral form of the tetraploid cultivated mani-group. An experimentally produced hybrid has been raised by Krapovickas (1974) and called *A. batizogaea*, considered as basic breeding material.

### *A. monticola* Krap. & Rig.

Amphidiploid, annual, from northwest Argentina. This wild-growing species, sometimes considered as direct wild ancestor of *A. hypogaea* L., or perhaps a primitive parallel form. It crosses easily with cultivated peanut cvs.

### *A. glabrata* Benth.

Perennial, tetraploid with utility as forage legume. In the southern part of the USA this perennial peanut is commercialized as "Florigraze".

### *A. repens* Handro and *A. prostrata* Benth.

In Brazil called Amendoim rastreiro and recently used as perennial component of mixed pastures.
Several wild species are useful as forage plants. We mention especially the taxa *A.*

*glabrata, A. prostrata, A. pintoi, A. pusilla,* commonly called sacha-mani in the Argentinian north by natives, who collect and eat its seeds; *A. prostrata* and *A. villosa* are used in a similar way in Paraguay (Prine 1964).
Ecologically, the genus *Arachis* occupies pre-mountain zones, (1500 m alt.) in the Eastern Andes (provinces Jujuy and Salta of North Argentina) and South Bolivia, to the Planalto of Goyas (650 m), to 100 m elevation in Matto Grosso, extending to Bahia at the Atlantic Ocean (Fig. IV.3). Until now, 40 taxa have been described, mostly of diploid genome structure. Krapovickas, a South American botanist, has established (1969) the following botanical sections:

1. Trierectoides (with three leaflets and erect growth)
2. Tetraerectoides (four leaflets, erect growth)
3. Extranervosae (four leaflets, growth prostrate, sometimes erect)
4. Ambinervosae (four leaflets, growth prostrate)
5. Caulorrhizae (four leaflets, growth prostrate)
6. Rhizomatosae (four leaflets, growth prostrate)
7. Arachis (four leaflets, growth mostly prostrate, includes 12 taxa with $2n = 20$ and $2n = 40$ chromosomes)

## NAME AND ORIGIN

The species *A. hypogaea* has long been divided by botanists and taxonomists into different varieties and subspecies. Without discussing their morphogenetic validity, we accept for practical reasons the following, which have a distinct growing habit:

a) ssp. *fastigiata* Waldron (1919) has an upright growth with sequential branching (var. *valencia* is little branched, whilst var. *vulgaris* has more branches). The floral axes are situated on the main stem. The leaf colour is light green. Modern cultivars form upright bunches, and ripen early.

b) ssp. *hypogaea* Krapovickas (1969) has a prostrate growing habit and is characterized by alternate branching. A primary branch produces other smaller branches

AR  – ARACHIS
AM  – AMBINERVOSAE
CA  – CAULORHIZAE
ER  – ERECTOIDES
EX  – EXTRANERVOSAE
RH  – RHIZOMATOSAE
TR  – TRISEMINALE

**Fig. IV.3.** Distribution of the different *Arachis* sections in Brazil (Valls 1983)

at its first and second node. The third and fourth node bear inflorescences, the fifth and sixth nodes again have secondary branches etc. Most cultivars are true "runners" or have a spreading habit. One primitive type, for example, is Peruvian runner; more advanced are Virginia types with short branches. The leaf colour is dark green. Their fruits produce only two seeds per pod. They have a long growing period and ripen after 120–160 days.

Peanuts are herbs with strong, stout, pivotal roots and many side roots. After germinating, the young taproot elongates rapidly downward and has penetrated after 2 months to 150 cm deep, forming a profuse lateral rootsystem which gets covered with many nodules of *Rhizobium* bacteria. This symbiosis produces such a high quantity of nitrogen that fertilization with inorganic N is unnecessary.

Leaves paripinnate, bijugate with two opposite pairs of leaflets.

Inflorescences compressed spikes with zygomorphic flowers, inserted in the leaf axils. It is noteworthy that only a low percentage of the profuse number of flowers develop fruits. Fertilization is in general cleistogamous and only seldom have cases of cross-pollination been observed. The calyx is deep bilabiate, the corolla golden-yellow, the staminal tube has eight to nine anthers. There is a large style in a tubular receptacle (hypantium), similar to a peduncle. After anthesis the ovary elongates geotropically and forms a "peg" (not really a gynophore or carpopodium as it is sometimes called), which penetrates quickly 5–7 cm into the soil with the growing ovary and 2–5 ovules inside. Then the apical region swells and forms fruits. Each plant may produce 20–40 pegs, which develop their pods underground. This geocarpy of *Arachis* is, together with the indehiscency of the pods, a unique feature in the Papilionaceae.

The ripe fruit of *A. hypogaea* has a single cavity, bearing two, in certain cvs., three to five seeds; whilst wild forms produce fruits with separated segments which contain only one seed. These segments are separated by a thin filamentous connective which may vary from a few mm to 10 cm. This typical wild character has been completely eliminated during the process of domestication. The seeds consist of two oil-rich cotyledons and a small embryo, covered with a thin testa of different colours (white, red purple, violet-black). Seeds have no dormancy, so they germinate in humid weather under the soil surface, which causes considerable loss during the harvest of cultivated *Arachis*.

UTILITY AND WORLD PRODUCTION

Groundnuts are a versatile crop. Besides the usual classification as an oil seed, they should also be considered as an important protein source. The composition of dry *Arachis* kernels is the following: 43–53% oil, 25–35% protein, 5–15% carbohydrates. The rest are crude fibres. The levels of lysine and other essential amino acids are rather low.

This analysis confirms that peanuts represent one of the most concentrated human foods, because they contain both sufficient protein and a high fat percentage, and are also easily digestible. Among Indians the seeds had a wide utility; they were eaten raw, boiled and toasted. The latter use persists even in "civilized societies", as an omnipresent snack for cocktails and television time.

One gram of *Arachis* seed provide 25 kJ, compared with 17 kJ in sugar, 15 kJ in rice and 14 kJ in maize.

The principal use of the peanut harvest is for oil extraction. Industrial elaboration includes cooking oil, peanut butter and margarine. The non-drying oil has a good quality, comparable to olive oil. It has a high percentage of oleic acids and moderate quantities of both linoleic and saturated fatty acids.

The residual cake, after pressing, contains 50% of raw protein, which unfortunately is not used for human consumption but serves as a main component in animal feed. We consider it highly disappointing that none of the great world organizations for food and nutrition puts more emphasis on this fact. Underdeveloped regions of Asia and Africa export whole groundnuts to Europe or North America instead of processing them locally. The remaining residual expeller (after oil exportation) could easily satisfy the albumin need of starving people.

A striking example for this inconceivable situation is the state of Senegal, where peanut unfortunately created – since its introduction as monoculture in the last century – "la dictature de l'arachide". Around 40% of the arable land of Senegal is sown with peanut for export to France. For decades, until 1976, production of *Arachis* seeds represented more than 50% of the total export of Senegal. (Only the last few years have shown a certain diversification of commerce). Most of the food for the 6,000,000 population, with a pronounced deficiency in protein and carbohydrates, has to be imported.

Further proof of the incongruency of the world trade with peanuts with respect to the "developing countries" is the following statistic (1977–1979). The homelands of *A. hypogaea*, Brazil and Paraguay, planted only 256,000 ha. Argentina: 396,000 ha. The African countries Senegal (1,000,000 ha), Sudan (1,013,000 ha) and Nigeria (673,000 ha), together with other African nations, contributed with an estimated 3,000,000 ha. In Asia the paramount peanut producer was India with 7,250,000 ha, followed by China with 2,450,000 ha, Indonesia with 512,000 ha and Burma with 552,000 ha. Thus the area planted with peanuts outside of their original home region is 20 times bigger than in South America. From a harvested area in the whole world of 20,000,000 ha, the poor and underdeveloped countries of Asia and Africa contributed over 18,000,000 ha, but the economic benefit remains with the highly developed industrial nations (ICRISAT Newsletters 1987).

Yields in peanut plantations vary considerably. In most tropical developing countries, the mean yield does not surpass 800 kg/ha. But in the USA, the average farmer harvests between 3000–3780 kg/ha. Experimental plots in Georgia gave 5000 kg/ha (Table 9).

composition gives good results. Loose sandy soils favour peg penetration and later the lifting of the plants during the harvest, with their heavy weight of fruits. Like most legumes, *Arachis* depends entirely for its nitrogen requirements on symbiotic bacteria of the *Rhizobium* group, which fix atmospheric nitrogen. Nitrogen production in groundnut fields has been calculated at an average of between 11 and 33 kg/ha, but Williams (1979) recorded much larger values from optimally managed fields in Zimbabwe, with 240 kg/ha.

The actual commercial peanut varieties are the result of crosses between different subspecies. On the market are four types: Valencia, Spanish, Runner and Virginia. The latter two are very similar, with only a slight difference in fruit size. Whilst in the USA in the year 1970 nine cultivars accounted for 95% of the crop, recent reports indicate a

**Table 9.** *Arachis* world area and production compared with area and production (in 1000/ha and 1000/t) in South America (FAO yearbook 1986)

| | Area in 1000 ha | | | | Production in 1000 t | | | | |
|---|---|---|---|---|---|---|---|---|---|
| Year: | 1980 | 1983 | 1984 | 1985 | Year: | 1980 | 1983 | 1984 | 1985 |
| World: | 18.573 | 18.434 | 18.787 | 18.955 | World: | 18.636 | 18.966 | 20.333 | 21.260 |
| South America | 0.643 | 0.401 | 0.381 | 0.415 | South America | 0.969 | 0.589 | 0.660 | 0.651 |

Notwithstanding the impressive world acreage of over 20 mio ha, *Arachis* provides less than 10% of the global edible oil production. Per capita consumption in the USA is 5 kg of peanut and its subproducts annually, whilst in South America it is lower, with only 500 g per capita. Most astonishing is the fact that Argentina has lost its influence on the international groundnut market. Its present (1985) area is only 130,000 ha, whilst in the years 1978–1981 400,000 ha had been sown each year. Also Brazil experienced a sharp decline in *Arachis* production, mainly caused by the spectacular increase of soybean plantation in this country (Weiss 1983).

Good yields of peanut plants are favoured by the availability of calcium. Liming of the generally acid tropical soils with oxysol

strong diminution in the number of varieties.

GENETICAL IMPROVEMENT AND RESISTANCE BREEDING

In hybridization programmes it has to be taken into consideration that *A. hypogaea* is a self-pollinating amphyploid with a complicated genome structure. A further handicap for earlier breeding work was that the germplasm spectrum in North America was rather narrow, as pointed out by Hammons (1977) and Norden (1972). This situation has changed recently with the inclusion of the substantial gene pool of approximately 50 wild *Arachis* species from the South American continent and hundreds of primitive cultivars. These important germplasm resources contain desirable characters.

One of the first positive results of inter-specific hybridization in North America was the new cultivar Spancross (Hammons 1982), an early maturing bunch peanut selected from progenies of crosses between a Georgia variety with the wild annual tetraploid *A. monticola* collected in NW Argentina. A notable resistance against damage from leaf-chewing insects was obvious in the first multiplication trials. At the ICRISAT Institute in India a gene bank of *Arachis* exists which is said to maintain 11,000 different lines and 200 wild forms. This vast breeding programme intends to transfer desirable genes from wild relatives of the groundnut into cultivars adapted for Asia. Recently the new strain Robust 33–1 has been released with "bold pods" and big seeds, useful for confectionery purposes. This was the result of crosses with the wild species *A. batizocoi* from Brazil (Singh 1986).

### DISEASES

Peanut plants suffer especially under tropical conditions from a series of fungal diseases, like leafspot and rust, bacterial wilt and several virus infections (Jackson and Bell 1969).

### Fungus

*Cercospora arachidicola* (tikka disease)
Leafspot is the most common disease and causes some of the worst damage in peanut plantations. Attacks can occur at any time during the growing season and begin with small brown leafspots which increase in size and are often surrounded by yellow halos that give the typical appearance of the infection. After a certain time the damage extends to the whole plant, because the fungus produces a toxin which reduces assimilation. Breeding for leafspot resistance is a major goal for breeders, but has not been successful up to the present. In the world collection some primitive Peruvian landraces exist with certain genetical resistance.

*Puccinia arachidis* Speg.
Rust infection can be distinguished from leafspot, because the latter shows clearly circumscribed spots, whilst the rust produces diffuse pustules with quantities of spores on the lower leaf surface. Early defoliation is its visible damage. Historically interesting is the fact that this fungus was discovered 100 years ago by Spegazzini on groundnuts in Paraguay, and was described as one of the first tropical fungus diseases. It may be that the disease had its origin in this region. In the meantime this rust has expanded more and more on the American continent, entering also the peanut cultivation of the USA. Since 1970, this highly destructive fungus has invaded other continents: Asia, Oceania, Australia and a great part of Africa. Rust does not overwinter from one year to the next in North America, due to the short life of uredospores, but air-borne spores are blown in from subtropical areas, or are transported by airplanes, thus disseminating the disease.

It is remarkable that certain wild and primitive *Arachis* of Bolivia, Peru and Paraguay possess natural resistance, or at least tolerance to *Puccinia arachidis*. For example, *A. glabrata*, *A. monticola*, *A. chacoense*, *A. villosulicarpa* are considered as immune. Primitive cultivars, like Guaykuru, Overo and Tarpoto (Peru) have field resistance. Based on their resistance genes, worldwide breeding work is going on, but until now no rust-resistant cultivar has been released.

*Aspergillus flavus*
Among the pathogenic fungi which affect groundnut yields, the *Aspergillus* group has special importance. This toxigenic fungus produces aflatoxins, when the *Arachis* seeds become wet in storage or are damaged by insects. The deadly consequences for humans and house animals are well known (Coker 1979).

Much has been published during the last decade about the "seed toxicity" of peanuts. Seeds may become toxic from metabolic substances produced by the omnipresent fungus *Aspergillus flavus*, and similar moulds. They cause aflatoxines, which are primarily liver poisons and may even cause cancer. On several occasions, loss of poultry has occurred after feeding with moist peanut seeds. Favourable conditions for mould infection are produced if the ripe peanut pods stay too long on the fields in rain or high air humidity. Quick drying is decisive. Once dried down to 15–10%, the infection danger is reduced. Attention is being concentrated on the utilization of genetic resistance against *Aspergillus* infections, discovered in certain *Arachis* biotypes.

### Bacteria

The omnipresent *Pseudomonas solanacearum*, which causes heavy loss in several important crops for mankind, also attacks peanut. Bacterial wilt is a widespread disease in tropical peanut plantations. It is claimed that the cv. Schwarz 21 showed field resistance to bacterial wilt in Indonesia.

### Viruses

*Spotted wilt* is caused by the tomato wilt virus (TSWV). This virus is now widespread in Texas and transmitted by thrips from weed plants, but is not seed-borne. Infected plants are stunted and have ring spots on the leaves. The typical appearance is of axillary shoots with deformed leaves and discoloured seeds.

*Rosette Virus* (GRV) is a serious problem in African peanut cultures. The virus is transmitted by *Aphis craccivora* from other infected host plants. Infected plants show different symptoms, either dark green or chlorotic leaves and severe stunting. Combatting the virus is difficult, but recently some resistant varieties have been observed in West Africa: Philippine White, Basse, or RG-1 from Malawi and Virginia type 69–101 from Senegal, also RMP 12 and RMB 91.

*Mottle Virus* (PMV) was observed by Herold and Munz (1969) in Venezuela, where 10% of the peanut fields were infected. The symptoms are a vein clearing of young leaves and a weak mottling. It was shown that the virus is transmitted mechanically and by aphids.

*Clump Virus* (PCV) has not been observed in America, but is a dangerous disease in West Africa and India. The main symptom is a severe stunting and a diminution of leaf size. The disease is soil-borne and probably transmitted by nematodes.

### Pests

Among the insects noxious for peanut cultivation are also several aphids. Not only do they transmit viruses, but also cause direct damage by sucking the leaves and flowers. The omnipresent *Aphis craccivora* has pantropical distribution. Until now, no genetic resistance has been found. The situation is similar with many *Coleoptera* and their polyphagous larvae, which destroy roots and stems of peanuts.

---

LECYTHIDACEAE

## 2. *Bertholletia excelsa* Humb. & Bonpl.

Brazil nut, paranut, nuez de Para, noix de Brasil, castanha do para, Paranuss

Widely distributed in humid forests from Peru, Brazil, Panama to Nicaragua. Large trees (30–45 m tall) with erect cylindrical stems· and greyish bark, sometimes with tremendous diameter (200–250 cm).
Leaves oblong, (25–60 cm) simple, with short petioles.
Flowers in panicles with yellow corollas.
Big fruits (20 cm) diam., "pixidios"), globose capsules with a lid which opens when the fruits ripen. Inside the woody shell are 12–20 triangular seed kernels, which need nearly 1 year to ripen. They have high oil content (60–70%) and 13–17% protein. The "nuts" are highly appreciated for their pleasant taste and are used mainly as desert and not for extracting the oil. The white flesh is enclosed by a hard, brown shell.
Harvesting is difficult, because the fruits are heavy (3 kg) and may injure the collectors when they fall down. Therefore the kernels are gathered by contracted laborers on the ground. One tree bears 100–300 fruits per year.

**Fig. IV.4.** *Bertolletia excelsa* tree; often 50–60 m high; Orinoco River, Venezuela (Photo Fouqué)

Nearly the entire world crop of paranuts is harvested in the wild state. It is one of the few cases where a high-priced product comes exclusively from a wild-growing species. Cultivation proved difficult because a tree needs more than 10 years until the first harvest, but in view of its valuable wood and the rapid extermination of wild stands, any form of domestication should be speeded up, especially in Brazil, which at present exports more than 50,000 t yearly from wild-growing trees, which often grow together in dense groves. The paranut tree exists not only in the para-state of Brazil, but is widespread throughout the humid tropics of South America, from the Guyanas in the east, covering the river systems of Jari and Madeira in Brazil, reaching the rainforests of Peru and Bolivia in the west of the sub-continent (Fig IV.4).

### 3. *Eschweilera odorata* Miers.

This rare species deserves mention, because natives of the northern part of South Amer-

ica collect their seeds and consider them a very nutritious and healthy food. Genetical improvement is recommended.

---

CARYOCARACEAE

### 4. *Caryocar* spp.

This neotropical genus includes several species which could be developed as an excellent source for vegetable oil and for future industrial use and exploitation on plantation scale.
Most of them are wild-growing, e.g.: *C. amygdaliferum* Mut. *C. butyrosum* Willd. *C. glabrum* Pers., *C. microcarpum* Ducke, *C. tomentosum* Willd., *C. villosum* Pers., and the most important of all:

### *C. brasiliensis* Camb.

Piqui brava
$2n = 46$

Exists in extended wild stands from Bolivia, Paraguay to central Brazil, where natives collect the fruits, which have fatty mesocarps and oil-rich seeds and are important in their diet, but also for body paint. Its yellow-coloured wood has recently acquired a high value in the craft work, home furniture and ship-building, due to its outstanding resistance to moisture.

### *C. nuciferum* L.

Butternut, noisette indienne, nuez souari, almendra

Natives of Venezuela and adjacent Guyana use the kernels regularly as food, and the pericarp for fish poisoning.
Large trees (up to 30 m tall) with stout trunks (100 cm diam.) of the humid forests (Fig. IV.5).
Leaves large, opposite, trifoliate, on large petioles.
Inflorescenses in clustered racemes, large red flowers, (hermaphrodite), with five large

**Fig. IV.5.** *Caryocar nuciferum* (the "butternut" of Guyana, Photo: Field Mus Chicago)

(7–8 cm) petals and several hundred yellow stamens.

Fruits globose (7–15 cm diam.) with a thin glabrous exicarp, a thick and fleshy pericarp. The mesocarp becomes lignified during ripening and encloses three to four hard seed kernels. These are 5–7 cm long and contain 75% fat and 14% protein. The cotyledons are tasty, can be eaten raw or toasted. From pulp and seed kernel the natives extract cooking oil.

### C. villosum (Aubl.) Pers.

In Brazil commonly known as "pequia" tree, smaller than *C. nuciferum*. Leaves are trifoliate, with large terminal leaflets (12–24 cm long).
Inflorescences are clustered racemes with numerous (300–700) stamens, the white corolla has five lobes. The fruits are globose (7–10 cm diam.). The pericarp is very oil-rich (60–70%) and has a slight smell of rancid butter. Natives use it as cooking oil and for making soap. The cotyledons of the seeds are also edible.

Schultes (1979) considers the whole genus *Caryocar* of great utility: "We may confidently expect that the future will see plantations of one or several species of *Caryocar*... for the betterment of life in the tropics in general..."

EUPHORBIACEAE

### 5. *Caryodendron orinocense* Karst.

Orinoco nut, palo de nuez, noyer de l'Orinoco, taque, cacay, inchi

Native to Colombia and Venezuela, especially at the headwaters of the Orinoco and Amazonas, but also in the humid tropical parts of Brazil and Ecuador. Others taxa have been described: *C. amazonicum* (Ecuador, Colombia), *C. grandifolium* (Brazil), and *C. angustifolium* (Panama).
Large trees (20–40 cm) tall) with dense canopy. Leaves 25 cm long, elliptic, coriaceous. Inflorescenses bear male and female flowers on separate branches and are wind-

**Fig. IV.6.** *Caryodendron orinocense,* leaf and fruits; *below* a hazelnut for comparison (Reckin 1983)

pollinated Fruits are tricoccal, similar to castor beans and contain small (16-mm-long) seed kernels. These nuts are edible and contain a considerable amount of oil (45–54%) and protein (20%). The average weight of the nuts is 10–12 g.

The oil content compares favourably with coconut and cashew nut. It is rich in linoleic acid, with a chemical constitution similar to *Hevea* seed oil. The nuts have an attractive taste, similar to hazel nut, but they have a reduced viability and in the case of plantations they should be sown immediately after becoming ripe (Fig. IV.6).

According to Reckin (1982, 1983), the orinoco nut has a great potential for cultivation as human food. Several attempts have been made in Colombia and in Ecuador to commercialize the seeds. A 10-year-old tree produces 100–250 kg nuts. An annual production of 3500–5000 kg oil/ha has been claimed by some Colombian agronomists (Fig. IV.7).

It seems that the tree can be easily adapted to a wide range of soils and humid tropical environments.

Hybrids between *C. orinocense* and related species may prove valuable for broadening

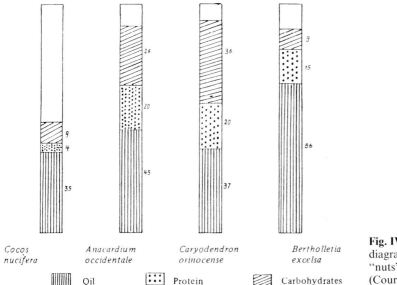

|  |  |  |  |
|---|---|---|---|
| *Cocos nucifera* | *Anacardium occidentale* | *Caryodendron orinocense* | *Bertholletia excelsa* |

**Fig. IV.7.** Comparative diagram of four tropical "nuts".
(Courtesy Dr. Reckin)

Oil            Protein            Carbohydrates

the genetical variability of future cultigens and last but not least for increasing the oil content of selected biotypes.

The wood is used in Venezuela for making furniture, due to its fine texture (red colour) and hardiness.

Natives of Colombia and Ecuador have selected better strains of *Caryodendron* and plant them in their house gardens in the Depts. of Meta and Puttumayo. As wild forms open their fruit shells rather early (which is a disadvantage in domestication), selection must be directed to indehiscent fruits.

---

CUCURBITACEAE

## 6. *Cucurbita foetidissima* H.B.K.

Buffalo gourd, chilicote
$2n = 40$

NAME AND ORIGIN

Its homeland is Northern Mexico and the SW of the United States. Thanks to its low water requirements, this perennial *Cucurbita* grows well in arid lands and disturbed soils.

There it was used in remote times by Indians who collected its oil-rich seeds and excavated its enormous roots, which in three growing seasons may reach weights of 50 kg. Remains of this plant have been recovered from archaeological finds – 7000 years old – associated with the ancient Indian culture of the Greater Southwest (Withaker et al. 1957). The Indians did not cultivate this plant, they most probably only occasionally gathered the seed and roots (Fig. IV.8).

MORPHOLOGY

The plants produce numerous, long runners, which in total may reach a linear length of 200 m, after 5 months growth; they cannot survive frost.

Leaves have an ovate, or sagittate shape, (20–10 cm), with a pronounced midrib and pointed apex. The lamina is dull green with a rough surface.

Flowers are borne at the nodes and are sexually differentiated.

Fruits are round, approximately 5–7 cm diameter, and are striped. They contain a bitter-tasting cucurbitacin, which is water-soluble and can be removed by cooking (natives get rid of the bitter taste with the help of calcium carbonate). The seeds (150–

**Fig. IV.8.** *Cucurbita foeti-dissima*

200 per fruit) have 30–40% edible oil and 30% protein. The oil is similar to sunflower oil and has the following fatty acid composition: 61.5% linoleic, 27% oleic, 3.6% stearic and 7.8% palmitic acid. The high rate of unsaturated fatty acids (over 60%) is especially useful for human food. The cake which remains after seed-pressing contains a considerable quantity of crude protein, which is suitable for animal feed (Curtis 1972).

Fruit per ha is 2000 kg; root yield may reach 8000 kg/ha. Roots are a source of starch and are dug out by inhabitants of deserts. A further advantage of *C. foetidissima* cultivation is the high resistance against diseases and pests, which in general affect other domestic cucurbits. Its deserves special mention that a group of biologists of the University of Arizona under the direction of Drs. Bemis, Curtis and Whitaker are at present working on the improvement of this xerophytic species, for future utilization in arid lands (Bemis et al. 1975, Deveaux and Shultz 1985).

## 7. *Fevillea cordifolia* L.

Abilla, abiria

A climbing herb of the tropical forests, spread from the Amazonian river system through the Isthmus of Panama to Costa Rica, and some Caribbean islands, e.g. Jamaica. Natives know the utility of this nota-

ble liana, but this Cucurbitaceae is still under study by botanists. Pio Correa, Perez Arbelaez (1937) and Ayensu (1981) mentioned it, but only Gentry and Wettach (1986) give a detailed description. They call it "one of the most productive sources of vegetable oil yet discovered in the world" (Fig.IV.9).

### Morphology

Dioecious lianas with several-meter-long climbing stems, which are glabrous, often reddish.

**Fig. IV.9.** *Fevillea cordifolia,* a promising oil fruit from the Amazonas (Wunderlin 1978)

Leaves simple, cordate, 6–16 cm long and 4–12 cm wide. Lamina glabrous, with acuminate apex, the upper and lower surfaces minutely pustulate. Tendrils bifid.
Flowers small, green, yellow, with their stamens inserted on a disk.
Fruits spherical, large (10–15 cm diam.) with a thick exocarp, suitable for water dispersal (Wunderlin 1978); 40–50 fruits produced from one plant. Many seeds in each fruit. Seeds are bitter, large (3–6 cm), with high oil content in the cotyledons. *Fevillea* seeds belong to the largest in the family of Cucurbitaceae. Fifty five % of the cotylodens is extractable fat, a semi-drying oil with a high proportion of stearic acid. There are calculations of a potential yield in wild stands of 800 kg/ha oil. Of course, in plantations this yield would even be improved. There is only one obstacle, that in such future plantations the share of male plants would have to be eradicated by hand or drastically reduced with chemical methods.

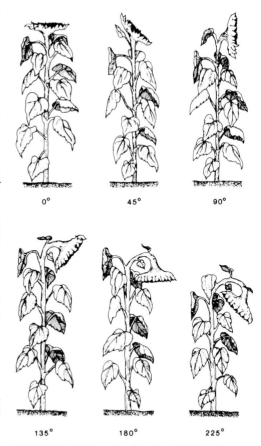

**Fig. IV.10.** Distinct head angle of *Helianthus annuus,* an important charcteristic for selection of tropical sunflower cultivars, protecting the seeds from rain

COMPOSITAE

## 8. *Helianthus annuus* L.

Sunflower, tournesol, mirasol, girasol, Sonnenblume

$2 n = 34$

### Name, Origin and History

The scientific binome refers to its annual status, whilst most of the wild-growing *Helianthus* species are perennial. The Spanish name girasol means that the flowerhead (capitula) follows the movement of the sun. Calderon (1600–1681) the famous Spanish dramatist, for whom this plant was still a newcomer from the New World, expressed this in a poetic, but somewhat exaggerated form: "Aquel Girasol, que está viendo cara á cara al sol. Tras cuyo hermoso arrebol, siempre moviendose va ..." (Fig. IV.10).
In reality, we observed with many experimental accessions of *Helianthus* during our research work in Paraguay, that the heads of sunflower had only a limited phototropic reaction and stayed bent towards the East the whole day.
The geographical origin of the cultivated sunflower is in Central North America, where Indians used the seeds for food long before the arrival of European colonizators. Kernel remains have been found in archaeological sites, which are 2000–3000 years old. Early explorers of North America observed that the natives cultivated sunflower together with beans and maize. Far more important, however, was the regular gathering of seeds from wild-growing *Helianthus* species by Apache and Hopi Indians. Their warriors appreciated the kernels as a remedy against fatigue. Natives of the Grand Canyon regu-

larly collected the achenes of wild *He-lianthus* (Heiser 1965,1976).

The first record of sunflowers in Europe dates back to the year 1510, when Spanish explorers brought seeds from New Mexico to Spain and delivered them to the Botanical Garden in Madrid. From these plants the first botanical descriptions were made. For several centuries sunflowers were grown in Europe only as garden plants at royal courts, where they fascinated by their quick growth and curious height, some reaching 3–4 m on rich soil. The Russian Tsar Peter I. must have seen sunflowers there and introduced them to Russia in 1779, where they were soon developed as an important oil crop. We have thus the curious fact that a useful American plant has been developed in Eastern Europe as a new high-yielding crop. On the American Continent exist many (perhaps 100) *Helianthus* species, extending from Canada, the USA, Mexico, along the Andes to South America, but none of them can be claimed as the true wild ancestor of the cultivated *H. annuus*, var. *macrocarpus*. Heiser (1976), an expert in *Helianthus*, believes that the original wild type may have disappeared, but that it was similar to the subspecies *jaegeri*, which survives in Southern California and Arizona.

Wild sunflowers still cover thousands of hectares in the USA, especially in the Prairies and the Rocky Mountains. The state of Kansas, where it was declared the "state flower of Kansas" included *Helianthus* in its emblem, while in the neighbouring Iowa it is considered as a noxious weed, which invades maize and soybean fields. The Hopi Indians of Arizona made original selections and used them in primitive cultivation, quite different from the actual commercial varieties. These native landraces may be still a valuable genetic source for future breeding aims.

Needless to say, the supposed radiation centre of *Helianthus* variability in the Western part of North America is not connected with the gene-centre Nr. VII proposed by Soviet agrobiologists as the origin of *H. annuus*; we freely admit, however, that the USSR may be called its "secondary centre of variability" (Pustovoit 1975).

MORPHOLOGY

Annual erect herbs, with robust unbranched stems (3–5 cm diam. and 80–300 cm high), with strong tap roots, which penetrate 3 m into the soil and form numerous lateral roots.

Leaves cordate-ovate, large-sized (13–30 cm long and 6–18 cm wide) with stiff hairs on both sides of the coarse lamina. Leaves are numerous, up to 60–70 in a single plant.

Inflorescenses: The small individual flowers are grouped together in a single head, "capitulum", or disk which consists of showy yellow ligulate flowers and inconspicuous disk flowers. The former are sterile and have seemingly the task of attracting insects for pollination. The flower-head is surrounded by several rows of involucral bracts, which are originally modified leaves. The size of such an inflorescence can be enormous, reaching 60 cm in diameter (Fig. IV.11).

The disk flowers are reduced to a minimal size and function. Hundreds, even thousands, of them are arranged in spirals in the centre of the disk (Fig. IV.12). They are composed of five petals which enclose male

**Fig. IV.11.** *Helianthus annuus,* the sunflower

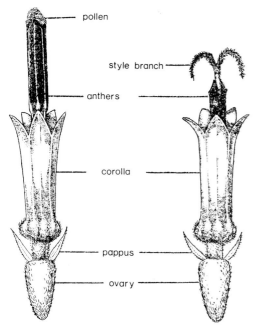

**Fig. IV.12.** Disk flowers of *Helianthus*: flowers with pollen (*left*); with style branches opened (*right*). (Heiser 1976)

**Table 10.** World production of sunflower seeds in mio t

| 1960 | 1970 | 1980 | 1983 | 1984 | 1985 |
|------|------|------|------|------|------|
| 5,400 | 9,872 | 14,396 | 15,584 | 16,425 | 19,078 |

Whilst in the year 1930 *Helianthus* ranked tenth among oil plants, at present it has won the third place, only shortly behind soybean and palm oil.

The leading nations in the world with respect to area sown and annual yield of sunflower are the following (Table 11).

**Table 11.** Distribution of sunflower production

|  | Area in mio ha | Yield in mio t |
|------|------|------|
| Soviet Union | 4.25 | 4.60 |
| USA | 2.00 | 2.10 |
| Argentina | 1.50 | 1.30 |
| China | 0.85 | 1.00 |
| Romania | 0.50 | 0.62 |

and female organs. On either side of the ovaries are two small scales, derived from the sepals. These "pappi" serve for the dispersal of the seeds, which should correctly be called aquenae. The seed kernels are 8–10 mm long, more or less compressed and of different colours and patterns: black, brown, white or white/black-striped. The seeds consists of seed coat, embryo and endosperm. The embryo has two large cotyledons with oil-rich cells, its parenchyma also contains protein particles. Their oil content exceeds 50% in high-yielding cultivars.

ECONOMY

The sunflower is one of the world's most important oil crops. Eastern Europe accounts for 80% of world production, followed by Argentina and the USA (Carter 1978). The world production of sunflower seed and oil is in continuous increase and rose as follows during the last 25 years (Table 10).

These five countries represent 70% of the world area planted with sunflower, closely followed in production by Hungary, South Africa and France, where considerable breeding and hybrid work has been performed, raising the maximum yields to 4–5 t/ha.

The chemical composition of sunflower oil is similar to that of olive oil and has a lower share of linoleic acid than, for example, soybean oil. Selection for oil content resulted in biotypes with more than 50% fat per kernel; additionally there is a considerable part of protein (up to 20%) in the seeds. There are two breeding lines: oil seed and seeds for direct eating, the last-named having big kernels. Most new market varieties are small-sized, oil-rich cultivars.

BREEDING, GENETICAL RESISTANCE AGAINST PESTS AND DISEASES

Plant breeders are making intensive efforts to use cytoplasmatically male sterile biotypes as female partner, with the result that in a few years the commercial *Helianthus* varieties will be hybrids, with the advantage of considerable "hybrid vigour". Often these new hybrids receive genes for resistance against serious diseases like: *Sclerotinia, Sclerotium, Alternaria, Phoma, Phomopsis* and *Puccinia*.

These diseases are especially devastating in humid tropical climates, regions in which *Helianthus* – originally a crop from a temperate climate – is continuously advancing (Weiss 1983).

*H. annuus* varieties suffer to an especially high degree from fungal diseases. Screening for resistance is performed on a worldwide basis but without having achieved final success in most cases.

## Fungus

### Alternaria zinniae
This disease has wide diffusion on the American continent. Its symptoms consist of early necrosis of the young plants. The leaves show brownish-black spots, but worst of all is the infection of the ripening capitulae, which causes the decay of the seed kernels.

### Phomopsis helianthii (Phoma macdonaldii)
Black stem is of American origin and has spread in a catastrophic way to the sunflower-producing countries of the Balkans and also to the Soviet Union. The intensity of the attack depends of the climatic conditions (moist, hot weather). This fungus has an extended (30–45 days) incubation period, therefore the damage is greatest at the beginning of flowering. The stems become black and break. It is not easy to differentiate *Phoma* from *Phomopsis* in the field infections. The last-named is responsible for wrinkled seeds with low weight.

### Downy Mildew
*Plasmopara halstedii*
The degree of infection with this "mildew" is correlated with weather conditions. In our sunflower plantations in Paraguay during cool, rainy periods we observed severe attacks, which inhibited normal seed setting.

### Sunflower Rust
*Puccinia helianthi*
When this disease entered Russia some decades ago, the destruction of sunflower

fields was so severe, that it practically paralyzed the extension of this crop. With the help of rust-resistant wild species it was possible to create new varieties with a certain field resistance to this fungus, which unfortunately exists in different virulent races. A typical symptom are the red-coloured pustules on the leaves.

### Root and Stem Putrefaction
*Sclerotinia sclerotiorum (Sclerotium bataticola)*
The fungus ("microsclerotias") penetrates the central part of the stem or the leaf petioles and destroys the vascular system. The consequences are early leaf drop and half-developed seeds. Resistance breeding is difficult due to the polyphagism of the pathogens. It has been stated that crossings with *H. tuberosus* in Russia gave resistant offsprings.

### Sunflower Withering
*Verticillium dahliae*
Widely distributed in South America. In general, the symptoms appear late, when flowering begins. Then the leaves lose turgescence, show bronze-brown spots and begin withering. Until now resistant varieties have not been developed.

## Pests

Several coleopteras and small butterfly and moth larvae sometimes invade the sunflower plantations with destructive effects. Resistance breeding is very difficult, therefore protection is based on different insecticides, with all their undesirable side-effects.

### "Sunflower Moth"
*Homoeosoma nebulella*
The damage is caused by the small larvae which begin their life cycle in the flowers and penetrate later into the developing seeds, devouring great parts of the kernels. A certain resistance exists in sunflower races which possess phytomelanin pigments as armoured layers of the seed hulls. East European countries have in some years lost the majority of their yields due to these voracious insects.

"Isoca medidora"
*Rachiplusia nu*
The larvae of this small butterfly devour the young leaves. Due to the devastating effect of this "isoca", the sunflower plants lose their foliage and only the petioles may survive. We have observed in Paraguay and Brazil sunflower fields which had lost their leaves to such an extent that assimilation had been interrupted.
A similar defoliating insect is *Chlosyne lacinia* and its black larvae.

*Diabrotica speciosa*
This coleoptera also causes loss of the leaves in an alarming degree and it seems that only chemical products can stop the damage; but such spraying seriously affects the pollinizating insects of the sunflower.

Trees of 18–22 m height, native in the Orinoco and Amazonas basin.
Leaves large (20–50 cm) coriaceous, broadly oblong-oval.
Inflorescences paniculate. Flowers sessile, hermaphrodite with six petals with two much longer than the others. Androeceum covering the ovary with many sterile appendages. Fruits globose, roughly verrucose (10–15 cm diam.), dark brown with a convex margin at the orifice. The seeds have high oil and protein content, but it is not easy to use them, because they are enclosed in a strangely formed fruit case similar to a small pot (10–18 cm) with a lock, which is closed until the fruits ripen. The fruits somewhat resemble Indian pottery. When ripe, the pixids become woody and open their lid like a pot, shedding their seeds. Animals,

---

EUPHORBIACEAE

## 9. *Jatropha curcas* L.

Physic nut, pinhao de purga, pignon d'Inde, Purgier-Nuss

Belongs to a neotropical genus with more than 100 representatives in America. In semi-arid regions of South America several species with castor-like seeds exist. Used by natives of Paraguay and Brazil as purgative (oleum infernale). Other species of the same genus: *J. urens* and *J. multifida*, are also used as ornamentals and hedges and therefore are spread over the whole world tropics, very common, for example, on the Capverde Islands.
The seeds – besides their very toxic substances – contain 40–60% oil, which is used by natives for making soap and candles. Recently it has been tried as substitute for diesel oil. It is claimed that 100 t of seeds may produce 32 t of crude oil.

---

LECYTHIDACEAE

## 10. *Lecythis elliptica* H.B.K.

Monkey pot, coco de mono, tocari, marmite de singe, Affentopf

**Fig. IV.13.** *Lecythis elliptica (L. usitatis).* (Photo: Field Museum, Chicago)

especially monkeys, try to open the hard fruit shells before their maturity (therefore the Spanish name "olla del mono"). The seed kernels have a size from 3–5 cm and enclose small cotyledodons with a big hypocotyl which contains reserve substances for quick germination.

Various similar species exist in the rainforests of Venezuela, Colombia, Brazil and Panama, e.g. *L. tuyrana* Pitt., *L. ollaria*, L. and *L. pisonis* Caub., *L. ampla* Mier *L. elliptica* H.B.K. *L. grandiflora* Aubl. *L.usitata* Miers.

During the past decades several *Lecythis* species have been introduced into tropical Asiatic countries for exploitation of their nutritious seeds. A certain care is recommended when the trees grow on selenium-rich soils, because *Lecythis* seeds accumulate selenium to such a high degree that humans can suffer health damage, even with occasional fatal consequences (Fig. IV.13).

---

CHRYSOBALANACEAE

## 11. *Licania sclerophylla* Mart.

Synonym: *L. rigida* Benth
Oiticia Tree

Grows in the N.E. of Brazil in rather arid environment. Large tree with handsome canopy (20 m tall), appreciated for its intense shadow. Inflorescences racemose with showy flowers. Fruits 4–6 cm long with one single seed kernel. The seeds contain a high quantity of a yellow, dense-flowing oil (58–63%), lican acid similar to elaio-stearin acid, which is quick-drying and industrially used in the same way as Tung oil, for resistant paints and other technical purposes. Natives collect the seeds on the ground, under trees, each of which may produce 200–500 kg annually. Oiticia oil export from Brazil has increased beween 1940 and 1980 tenfold (approximately 50,000 t). The recommendation is made to collect seeds from well-bearing trees in the wild, for future selection and breeding work on industrial plantations.

---

COMPOSITAE

## 12. *Madia sativa* Mol.

Oil-madi, madi, melosa, madivilcun (arauk.), Oelmadie

The origin of this under-exploited oilplant is the Southern Cordillere of Chile and Argentina. It seems to be the only South American species of the 20 Madi taxa. It is easily recognized by its aromatic scent, caused by a dense cover of oil glands (Kunkel 1984). Araucanians developed in this region, where no other vegetable fat sources exist, a humble oilplant. The Chilenean botanist Molina (1782) called the attention of naturalists in Europa to this curious Araucanian cultivar, describing its oil as "identical to the best olive oil". Consequently the plant was introduced to France, where it was cultivated in the 19th century. The cultivation was interrupted on account of difficulties in harvesting the shattering achaenes. During the last World War and due to the great shortage of edible oil in Germany, Brücher and Fischer (Brücher 1977) undertook selections in earlier hybrid populations with considerable success. They combined a better consistency of ripe flower capitulae ("closed heads") with non-shattering achaenae, and a low content of viscocity and glandular trichomes, on stems and leaves, with considerably improved oil content and larger seeds (Fig. IV.14).

Achaene size: 7 × 2 mm, oil content: 35%, iodine number: 121–128, specific weight: 0.92. linoleic acid : 70–78%. Oil quality similar to sunflower oil.

MORPHOLOGY

Annual herbs, with erect stems (30–120 cm high) and covered with numerous trichomes, hairs and glands, which produce sticky viscosity in wild biotypes, representing an efficient protection against being eaten. Leaves numerous lineal, sessile, densely covered with short hairs. Inflorescenses in wild forms small flower heads on long side branches. In cultivars compact flower heads on main stem. Flowers yellow. Fruits without pappus. Achaenes cuneiform, in wild

a                                                              b

**Fig. IV.14.** *Madia sativa* together with the Chilean wild progenitor

forms 3 mm long; in cultivars 7 × 2.5 × 1.5 mm, with greyish colour when ripe. *Madia sativa* might in future be an oil crop for marginal arid zones. Its advantage is good drought-resistance and a short vegetation period. It is protected against insects by dense glands and a sticky leaf surface. Its yields vary according to varietal selection. In 1944 Brücher and Fischer obtained 1700 kg/ha of dry seeds in their advanced breeding material at Lannach (Austria).

LEGUMINOSAE

### 13. *Monopterix uacu* Spruce

Large, attractive trees in the rainforest of the Upper Orinoco-Rio Negro rivers. The pods contain large seeds (3 cm diam.), from which the Indians extract in boiling water a dark-green fatty (melting point 23°C) substance. This oil is locally known under the name "uaku", and is highly appreciated for its

good quality. Further research is recommended.

SAPINDACEAE

### 14. *Sapindus saponaria* L.

Soapberry, parapara, jaboncillo, savonnier, Seifenbaum

This is often mentioned in traveller's tales as the "soap-tree of the Indians": Its saponin-rich fruits are still employed by natives of Venezuela and Brazil to make soap. The reddish fruits contain black-coloured seeds with 28–30% fatty substances with a low melting point (15°C). The evergreen leaves are imparipinnate with an alternate position on the branches. The fruits are spherical (1–2 cm diam.) and have a bitter-tasting viscose mesocarp.
The tree grows in the Orinoco and Amazonas river systems and produces a huge quantity of fruits. The size of the soap tree is

small, its roots have a very stringent effect and are probably poisonous. They are used in Venezuela as a tonic and against snake bites.

---
BUXACEAE
---

## 15. *Simmondsia chinensis* Schneider (Link)

Synonym: *Simmondsia californica* Nutt.
Jojoba

### NAME AND ORIGIN

The origin of this monotypic species is not China, as the unfortunate Latin binome may suggest, but this shrub is indigenous in Mexico and Lower California (Thomson 1985).

The term jojoba has been derived from the word "hohowi" used by the Coahuila Indians of Arizona, where this bush is native. The primitive inhabitants of Lower California and Sonora collect the fruits, which are rich in fat, and use them in different ways. They eat them raw, or grind them for preparing meal which is boiled or consumed as a beverage. This insignificant desert plant would have remained in obscurity, if it were not for some phytologists and chemists who detected some peculiarities in its seed oil.

The confusing name of this American species is due to a mistake by an early botanist (Link 1822), who confounded in his cabinet two seed parcels, one if which was sent from China and the other from California, when he created the taxon. Instead of accepting the appropriate term *S. californica* introduced by Nuttal in 1844, the sometimes too rigid formalists of botanical nomenclature insisted on the conservation of the curious epithet, emended in 1912 by Schneider, and disregarding its geographical inconsistency.

### MORPHOLOGY

The dioecic wild-growing bushes are 2–3 m high, with a rounded canopy. They are long-living (up to 100 years) and grow very slowly, reaching only after 5–6 years an adult stage. They have a deep-rooted root system. Until the adult stage, male and female individuals cannot be distinguished, which is a serious obstacle, as essentially only the female bushes are economically interesting (Fig. IV.15).

Leaves small (2–4 cm long) elliptic-obovate with a leathery lamina, and opposite insertion. Flowers small and nearly imperceptible. Hermaphrodite flowers are very seldom, usually male and female flowers are produced on different bushes.

**Fig. IV.15.** *Simmondsia chinensis*, the "jojoba bush" **(a)**; **b** male and female branches

Male flowers are in clusters, female flowers are solitary. Although insects can be observed on the flowers, pollination occurs by wind. The fruits ripen slowly, have the size of an acorn and contain only one seed, exceptionally two. After maturation the outer husks dry out and expose the soft-skinned, brown seed kernel. Ten-year-old wild types produce 2–4.5 kg seeds per plant.

Domestication and selection of jojoba plants for plantation use is still in its infancy. The actual habitus is mainly a many-branched shrub, whilst the ideal form for mechanical harvest would be a small tree. Handpicking in the remote arid regions, like the Chaco, where jojoba is planted, is made difficult by lack of labour force; harvester machines for jojoba must be improved. As young plants are frost-sensitive, only such arid lands are appropriate for cultivation where there is no winter frost.

It takes more than 3–4 years before jojoba plants bear fruit and only after 10 years are maximum crops achieved. Fifty percent of the offspring in nurseries are male plants, most of which must be eradicated from the plantations and replaced by fruit-bearing plants.

*S. chinensis* is not easy to cultivate, as some enthusiastic investors have learned with losses in the semi-arid regions of Paraguay, Chile and Argentina. I saw spectacular failures there at the beginning of the "jojoba-rush" in the 1970s. Speculations went so high sometimes, that prospectors promised a 50,000 US $ income from 1 ha after 5 years with a calculated fruit harvest of 2,5000 kg/ha (Anonymous 1985).

On the other hand, there are certain advantages to jojoba. The plant is effectively drought-resistant, lives with 200 mm precipitation and can be maintained with a minimum of irrigation, once the young plants have become established. Jojoba has an extraordinarily deep root system, the taproots of mature bushes penetrating in semi-arid land as far as 12 m deep and elevating underground water. Additionally they have a good system of feeding roots for quick uptake of rain water up to 80 cm from the surface. A mycorrhizal fungus, which stimulates the uptake of minerals, has been discovered in natural stands of *Simmondsia*.

The oil – or it should rather be called wax – of jojoba has interesting applications. The pharma-cosmetic industry prefers it, because it does not become rancid. It is similar to spermwhale oil and has remarkable resistance to bacterial degradation. When hydrogenated, it forms a white crystalline wax. It has a very high melting point and suffers no alteration on high (280°C) temperatures. Therefore jojoba oil can be used as lubricant in high speed machines, e.g. in drilling equipment and airplanes. As it seems difficult to make such a special wax-oil synthetically, certain expectations to produce it cheaply from jojoba plantations are sound (Johnson and Hinman 1980). According to the Jojoba Growers' Association, more than 10,000 exist now in California and Arizona alone. The USA produced in 1983 approximately 165 t of jojoba oil, still mostly from wild collection.

Also Iran, Israel, Saudi-Arabia, the Australian Continent and last but not least South American countries like Argentina, Chile and Paraguay with marginal arid land have entered the jojoba euphory. Too much publicity may have damaged the initially well-intentioned work. Prices have already declined. In 1975 1 kg was worth 60 US $, 1983: 30 $, 1985: 10 $ and for the year 1986, it is expected to go down to 4 $. Therefore careful calculations in the case of new plantations are recommended, especially in view of the estimations that the USA alone may produce in 1990 the enormous quantity of 130,000 t jojoba oil.

MYRISTICACEAE

## 16. *Virola sebifera* Aubl

Synonym: *V. surinamensis* Warb.
Ucuba oiltree

This tropical tree has a wide distribution from the Upper Amazonas, Orinoco into the humid forests of Brazil, Venezuela, Guyana and to the Atlantic coast. Its fruits are spherical capsules. On becoming ripe, they release a 3–cm kernel which is surrounded by a red aril (very similar to a nut-meg seed, which belongs to the same family). The greatest part of the seeds (70%) consists of a dense fat of yellow colour. Natives use them to extract oil and make candles. Before using the oil for culinary purposes, a certain resin has to be removed. A *Virola* tree may produce 50 kg of seeds per year. As many of the trees grow on river banks, their fruits fall into the water, where the natives collect them easily with rods. There they use them ornamentally to make floating candlelights for the night, putting several of the fatty seeds on sticks. The oil (mostly myristin acid), which is exclusively extracted from wild-collected fruits, is exported for indus-trial purposes, under the denomination ucuba fat.

# References

Anonymous (1975) Buffalo gourd. In: Underex-ploited tropical plants. Natl Acad Sci, Wash-ington DC

Anonymous (1985) Jojoba, new crop for arid lands. Rep Natl Acad Press, Washington DC, p 102

Ayensu E (1981) Medicinal plants of the West Indies. Reference Publ Algonac, MI

Bemis W, Berry Weber, Whitaker T (1975) The buffalo gourd, a potential crop for the produc-tion of protein, oil and starch in arid lands. Office of Agric Agency Internat Development, Washington DC

Brücher H (1977) Tropische Nutzpflanzen. Springer Berlin Heidelberg New York, 520 p

Carter JF (ed) (1978) Sunflower science and tech-nology. Agronomy 19. Am Soc of Agronomy, Madison USA

Coker RD (1979) Aflatoxin, past, present and fu-ture. Trop Sci 21:143–162

Correa-Pio M (1975) Diccionario das plantas uteis do Brasil e das exoticas cultivadas. 6 vol Minist do Agric Brasil (1926–1975), contin por L de Azerado Penua

Curtis LC (1972) An attempt to domesticate a wild perennial xerophytic gourd *C. foetidissi-ma*. Progr Rep 1, Ford Foundat Beirut

Deveaux JS, Shultz EB (1985) Development of buffalo gourd (*C. foetidissima*) as a semi-arid land starch and oil crop. Econ Bot 39:454–472

Gentry A, Wettach RH (1986) Fevillea, a new oil seed from Amazonian Peru. Econ Bot 40:177–185

Gillier P, Sylvestre P (1969) L'arachide. Maison-neuve + Larose, Paris

Hammons RO (1977) Groundnut rust in the United States and the Caribbean. PANS 23:300–3004

Hammons RO (1982) Registration of peanut germplasm. Crop Sci 22:452–453

Heiser CB (1965) Genetics of colonizing species. In: Baker, Stebbins (eds) Academic Press, New York

Heiser CB (1976) The sunflower. University of Oklahoma Press

Heiser CB (1981) Seeds to civilization, the story of food, 2nd edn Freeman and Co., San Francis-co

Hoehne FC (1940) *Arachis*. Flora Brasilica 25:(122)

Hoehne FC (1944) Duas novas especies de Legu-minosas de Brasil. Arq Bot Estado Sao Paulo 2:15–18

Herold F, Munz K (1969) Peanut mottle virus. Phytopath 59:663–666

Icrisat (1987) International *Arachis* Newsletter. Patancheru, India

Jackson R, Bell DK (1969) Diseases of peanut, caused by fungi. Res Bull 56, Univ of Georgia Agric Exp Station

Johnson JD, Hinman CW (1980) Oils and rubber from arid land plants. Science 208:460–463

Knowles PF (1978) Sunflower, morphology and anatomy. In: Carter (ed) Sunflower science and technology. Am Soc Agron Madison, USA

Krapovickas A (1969) The origin, variability and spread of the groundnut, *Arachis hypogaea*. In: Domestication and exploitation of plants and animals. Ucko & Dimbley, Duckworth, London

Krapovickas A (1973) Evolution of the genus Arachis. In: Agr Genet Jerusalem:135–150

Kunkel G (1984) Plants for human consumption (Checklist of 12,600 species) Koeltz Scientific Books pp 393

Molina J (1782) Saggio sulla storia naturale de Chile. Bologna

National Academy of Science (1975) Underex-ploited tropical plants with promising eco-nomic value. Jojoba. Washington DC

Norden J (1972) Altika, a peanut variety for the tropics. Circ S-215. Agric Exp Stat Gainsville, Florida

Perez Arbalaez (1947) Plantas utiles de Colombia, Publ Cont Gen Republica

Prine G (1964) Forrage possibilities in the genus *Arachis*. Soil Crop Sci Soc Fl Proc 24:187–196

Pustovoit VS (1975) The sunflower (in Russian). Kolos, Moscow

Reckin J (1983) The Orinoco nut, a promising tree crop for the tropics. Int Tree Crops J 2:10–119

Ressler PM, Gregory WC (1979) A cytological study of three diploid species of genus *Arachis*. J Hered 70:13–16

Schultes RE (1979) The Amazonia as a source of new economic plants. Econ Bot 33:258–266

Singh AK (1986) Utilization of wild relatives in the genetic improvement of *Arachis hypogaea*. 8. Synthetic amphiploids and their importance in interspecific breeding. Theor Appl Genet 72:433–439

Thomson PH (ed) (1982) Jojoba handbook, 3rd edn. Bonsall Publ Calif, USA, p 162

Valls J (1983) Collection of Arachis germplasm in Brazil. Plant Genet Res Newsletter, FAO 53:9–14

Weiss EA (1983) Tropical oilseed crops, Longman, London, p 660

Williams JH (1979) The physiology of groundnuts. Nitrogen accumulation and distribution. Rhod J Agric Res 17:49–55

Wunderlin (1978) Cucurbitaceae. In: Flora of Panama, p 315

# V. Palms

## General Remarks

With more than 2000 (perhaps 2500) species, the botanical family of Palmae represents one of the largest taxonomical units of the Monocotyledones. Their economic importance for mankind is second only to the Gramineae. In the tropics, the palms provide many millions of inhabitants with nearly all the necessities for daily existence, i.e. basic food, clothes, wood, shelter, fibres, beverage, wax etc. According to Glassman (1972), 1439 palm species exist on the American continent. The biological and systematic understanding of this huge group still suffers – since Humboldt's and Martius's classical explorations and descriptions – from the extraordinary difficulties in collecting them and making herbarium specimens. Felling a palm tree is always a hard task for a botanist; but without cutting a palm down for study of floral organs it may be impossible to determine the find correctly.

The basis of our knowledge of neotropical palms is still the monumental work of Martius, who over a period of 3 years (1817–1820) investigated a vast number of palms, collecting them in Brazil and describing many as new for science. Later, the Brazilian explorer and botanist Barbosa-Rodrigues studied from 1871–1874 the native palms of his country and published his results in 1903 in a splendid volume with coloured plates. In our century, the work of Dahlgren (*Index of American Palms*, 1959) is of basic importance.

Recently Balick (1979–1981) began a comprehensive survey of useful palms of the American tropics. Uhl and Dransfield (1987) made a classification based on the work of Harold Moore. Lötschert (1985) described ornamental palms.

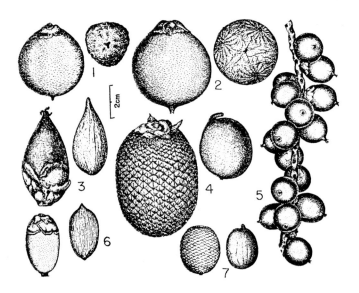

**Fig. V.1.** Palm fruits and seeds (Cavalcante 1977) *1 Acrocomia* sp.; *2 Astrocaryum tucuma; 3 Maximiliana martiana; 4 Mauritia flexuosa; 5 Oenocarpus distichus; 6 Jessenia batava; 7 Mauritia martiana*

Of the South American countries, Brazil and Venezuela are the richest in diversity of palm species, for example nearly 180 species have been notified for Venezuela and for Brazil 230 different taxa (Braun 1970, Pesce 1985) (Fig. V.1).

According to my experience, approximately 50 wild-growing neotropical palm species are used by the natives of South and Central America for oil, starch, vegetable, wax and beverages, besides their use of the subproducts for hunting bows, fibres for fishing nets clothing and mats and leaves for roofing and building of shelters.

Palms are especially rich in fat and oil, which is produced inside the seed kernels, but also in the fruit pulp, in general of different quality and chemical composition, united with high yields.

Palm oils are non-drying oils and stay liquid in a tropical climate, when exposed to air. In general they are tasteless triglycerids, their colour is caused by natural pigments (Hodge 1975).

There is a general world shortage of edible oils. It is in my opinion therefore of paramount importance that the natives of tropical America maintain their self-reliance with palm oils; but native wild stands are more and more endangered by the advances of civilisation (highways and slashing down of rainforests). Increasing demand for vegetable oils and their valuable by-products should be an incentive to plant breeders in the tropics to develop certain New World palms, which have until now been exploited in a rather rudimentary way by native collectors, for example, Babassu palm nut or Pejibey fruits.

## 1. *Acrocomia* spp.

This genus is noteworthy for its climatic tolerance, because it grows well also in less hot regions. It thrives well in the savannas of North Argentina, South Bolivia and Paraguay and survives also in higher altitudes with occasional winter frost.

From the 20 taxa described until now, three acquired especial importance for the American natives. These are *A. mexicana* Karw.

(called there "coyole") with former wide distribution through Guatemala, Honduras, El Salvador and Nicaragua. Due to the custom of felling the trees and producing an alcoholic beverage, the "coyol-palm wine" from the cuttings, the area of this palm has been severely reduced.

In the southern part of Central America *A. panamensis* Mart. is found restricted to the Pacific coast of Panama. I observed there that the natives apreciate its fruits, called "corozo de pacora" and consumed them fresh. Their seed kernels contain an oil-rich endosperm. The stems are rather small (7–9 tall) and bear large (5-m-long) leaves.

More important is *A. sclerocarpa* Mart. It grows in Central Brazil, but also on the Caribbean Islands, where I found it under the denomination "Gee-Gee". The trunks (15–20 m tall) have in general sharp spines, but I observed also on the island of Trinidad mutations which were nearly free from them. The leaves form a rounded crown with the peculiarity that the dead leaves and their sheaths remain attached to the upper

**Fig. V.2.** *Acrocomia sclerocarpa.* This "gi-gi" palmfruit is highly appreciated by inhabitants of the Caribbean Islands

**Fig. V.3.** Harvest of *Acrocomia totai* in Paraguay

part of the trunk, which constitutes a notable characteristic of this palm.

The fruits are yellow-red (4 cm diam.) and have a mucilaginous pulp, with an agreeable sweet taste. The mesocarp adheres strongly to the seed kernel, which is a considerable disadvantage for processing, but it is rich in oil (35%). The black kernels also contain oil (55%) with a low melting point. In view of its rich fat content of quite different composition, future exploitation on a larger industrial scale seems promising (Fig. V.2).

### *A. vinifera* Oerst.

This is present only in Central America. Natives use the fruits, in spite of the spinous trunk which has to be climbed, and the tendency of this palm to extend ripening over a long time. The oil and starch content is good. Prevailing is its use for preparing alcoholic beverages.

### *A. totai* Mart.

Mbocaya, coco paraguayo

The growth habit is similar to a cocos palm. Stems reach 15 m, often covered with spines. Leaves are 3 m long. Inflorescences are separated on the same tree.

It is notable that this palm invades land in Paraguay which has been cleared from the jungle. There it thrives well, often in formerly ploughed terrain. This is the reason why large stands of "coco paraguayo" can be seen round the capital, Asuncion. This astonishing fact attracts the attention of visitors arriving in Paraguay by plane. These are by no means original stands, but a secondary vegetation on formerly cultivated land. In view of its local importance for the Paraguayan oil industry, the Ministry of Agriculture should long ago have initiated a selection programme of this "coco del Paraguay", which still lacks basic domestication. We observed good bearers with eight to ten fruit panicles whilst the average produces only two bunches per year. Due to open pollination, the plus palms must be isolated during the fecundation period. The fruits are relatively small (3–4 cm diam.), dark yellow in colour, assembled in panicles with 200–500 fruits, each weighing 12–14 g before drying. Dried fruits weight 9–11 g. The oil content of the kernel is high (35–60%) whilst the pulp has only 15% fat. The national oil production of 8.800 t/year covers the local demand and the rest is for export (Fig. V.3).

## 2. *Astrocaryum* spp.

This exclusively neotropical genus is differentiated in a dozen species, which belong to the rainforests of Brazil, Colombia, Peru and Venezuela. They are solitary palms with rather large fruits similar to small coconuts and stems always covered with long, sharp spines.

Indians use these palms in numerous ways: fruits are eaten fresh and served as nutritive juices; leaves are used for fibres and trunks for building huts.

### *A. aculeatum* Meyer
Aman palm, waran

The edible fruits contain excellent oil, the "palmhearts" of the young stems provide a vegetable.

### *A. jauari* Mart.
"Awarra" Palm

Distributed in the rainforests of Brazil-Venezuela-East Peru. The red fruits contain 40% oil in their pericarp.

### *A. murmuru* Mart.
"Murmuru" Palm

Native to the Upper Amazonas and appreciated by the natives for its large (3–4 cm diam.) fruits. The seeds contain 40% fat; to obtain it, the hard endocarp must be cracked by force.

### *A. tucuma*, Mart.
"Tucum-assu" palm

Its main area is the Amazonas region in Brazil, where this solitary palm prefers open sites. The erect, 10–15-m-high stems are covered with dense rings of repulsive long black spines. The leaves are very long (3–5 m), densely aculeate with enlarged sheats.

The fruits are produced in long (1 m) clusters, are large (5–6 cm, with 70 g weight) and have an oily pulp which is consumed raw by the Indians.

### *A. vulgare* Mart.
Aiara, curua, chanbira, tucumao

This is the most popular of all *Astrocaryum* species and is generally called "tucumao", by the natives of Para (Brazil), who consume the fruits in great quantities. In contrast to the other species, this palm grows in clusters with rather thin trunks (15–20 cm) which produce small fruits.

Leaves very big (5–7 m long) and always in an erect position. Inflorescences produced in long (150–170 cm) spadix.

The small fruits (2–3 cm diam.) are an important fresh food for natives and also sold on local markets (the kernels are even exported for oil processing). The yellow pulp contains 16% oil, 3% protein and a high quantity of vitamin A, (three times more than *Daucus carota*!). The kernels are rich (33%) in fat of a high melting point, which is valuable for culinary purposes. The total food value of the edible mesocarp is notably higher than in any other tropical fruits and gives 247 calories per 100 g.

In view of this advantage, its consumption should be stimulated, and the breeding of spineless stems for future *A. vulgare* plantations initiated.

## 3. *Attalea* spp.

It is not easy to describe this genus, because it is closely related to the genera *Scheelia* and *Orbignya*.

We mention the following species together with their probable synonyms:

### *A. excelsa* Karst.
Synonym *Scheelia martiana*
"Yagua" palm

### *A. spectabilis*
Synonym *Orbignya spectablis* (Mart.) Burr.

Both abundant in the lower Amazonas region, where they offer a potential yield of thousands of tons of fruits (according to Pesce), which are not exploited.

**Fig. V.4.** A Waika-Indian in
the Upper Orinoco rain-
forest, ready to roast *Attalea*
palm fruits together with soft
termite larva
(Photo: Dr. J. Goetz)

### *A. agrestis* **Barb. Rodr.**

Notable for its acaulescent growth with the
fruit bunches near the ground. This characteristic represents a considerable advantage
for harvesting, (similar are also *A. acaulis*
Burr. and *A. monosperma*).

### *A. humboldtiana* **H.B.K.**

I have seen this tall-growing palm on the
Upper Orinoco, in Venezuela, called by the
natives "yagua", and used in different ways
(Fig. V.4).

### *A. olifera* **Spruce**

With a wide areal from Colombia, Brazil to
Paraguay.

### *A. funifera* **Mart.**

From this palm comes the "piassava" fibre.
The leaf sheaths consist of stiff, wiry bristles,
which are an export article in Brazil. The
fruits have no nutritive value, but buttons
are made from the hard seeds.
This palm has a peculiar habit. It grows during the first years after germination underground. Later it sends its leaves from a sub-
terraneous stem to the surface.

## 4. *Bactris gasipaes* **H.B.K.**

Synonym: *Guilielma gasipaes* (H.B.K.) Bailey,
*Guilielma speciosa* Mart. *G. utilis*, Oersted
Pejibaye, pixba, pifá, chontadura, pupunha

NAME AND ORIGIN

Alone the fact that this useful palm is known
in South and Central America by more than

60 names (from the tupi-guarani: "pupunha" to kechua: "chontadura", to chibcha: "pejiba") underline the extraordinary importance of this species. It was said that this palm is so ancient that it no longer even exists in the wild state; but this is not true. Several botanists and explorers of the tropics like Barbosa, Ducke, Huber, Leon, Plowman and Schultes have met wild *Bactris* in the rainforests of Peru, Ecuador, Colombia and Brazil. They received denominations like: *G. microcarpa* Huber, (from the Ucayali valley in Peru), *G. insignis* Mart. (from East Bolivia) and *G. mattogrossensis* Barbosa (from Amazonia, Brazil). We accept the opinion of the outstanding Brazilian explorer of palms, Barbosa-Rodrigues, who already in 1903 suggested that the wild forms are *Guilielma* ( = *Bactris*) *mattogrossensis*, widely dispersed in the Amazon Valley. Wild-growing examples at the Rio Acre have been recognized as true wild species. Patiño (1958) reported more morphological variability in the western part of Amazonia than in East Brazil.

Seibert (1950) found wild-growing Pejibae palms in typical rainforest areas in eastern Peru, where a century ago Hubert also reported from the Ukayali and Huallaga valleys "pupunha brava", i.e. wild forms of the pejiba palm. We believe that it may have been introduced in semi-domestication at several and widespread places, which explains the present existence of significant differences between South American populations and the Central American varieties (Mora Urpi and Clement 1981).

There are no wild stands in Central America. In prehistoric times, Indians took *Bactris* seeds with them from the southwestern region and the foothills of the Andes and spread them during their search for new land to NW Colombia (e.g. the Rio Atrato), crossing the Isthmus of Darién to Panama – Costa Rica northward. Other tribes migrated with this palm eastward, following the Rio Negro-Orinoco to the Guianas (Fig. V.5).

From the time of the Spanish conquest several reports exist on the wide distribution and the high esteem of "Pejibae-palm" among the Indians. In the Rio Atrato region (North Colombia, bordering on Panama),

**Fig. V.5.** *Bactris gasipaes,* one of the most important palm fruits for Indians who inhabit the tropical rainforest. Observe the climbing implement to avoid the longe spines

there were plentiful *Bactris* plantations when Pedro Cieza de Leon arrived in 1525. Gumilla (1741) reported it from the north coast of Venezuela. Aublet (1775) found it abundant in French Guiana. From Panamá, where the Spanish invaders landed, Pizarro (1525) mentions this palm near Pueblo Quemado. There is a detailed description (repeated by Fernandez 1881) in the history of Costa Rica from 1540. At the south-eastern border with Panamá lies the valley of Sixaola. Highly esteemed by the natives, a Pijba plantation of 30–40,000 palm trees grew there, as a heritage from their Indian forefathers. As the result of internal strife between Spaniards and Indians, this marvelous palm grove was destroyed. The incident went to the High Court of the Spanish Crown in Madrid and therefore the whole history was documented in the Court Records for posterity. This *Bactris* plantation was probably the largest that ever existed in the recently discovered New World.

## Morphology

The slender stems reach 15 m height and have a 50–75-year life cycle. They are covered with spines. Very often various trunks emerge from a common rootstock.

Leaves pinnate, 160–200 cm long and have a spiny rhachis. Inflorescenses are monoeceous and are covered with long spathes, which have male and female flowers together on the same raceme. Pollination is done by small insects; especially small weevils are important agents for fertilization. Flowering occurs twice a year.

The fruit bunches ripen slowly and mature only after 6 months. In general, five to six bunches are produced per tree with an average weight of 10 kg; but there are reports of 12 fruit clusters per plant, each with more than 300 fruits. Fruits are ovoid with a 3–5 cm diameter; several local races exist with different skin colour, which may be pale yellow, deep orange-red, or even green when ripe. The flesh has high carotene content and an attractive gold-yellow colour. The pulp is of mealy consistency. The analysis of the mesocarp revealed the following values: 40% carbohydrates, 7% fat, 2–3% protein, some vitamin C, and a considerable amount of vitamin A and phosphorus.

The seed kernels (15–20 mm long) have a white oil-rich endosperm. They can be easily opened with a hammer and their content eaten raw (Clement and Urpi 1987).

In total, as Schultes (1980) states, the fruits of *B. gasipaes* "are one of the most balanced foods, containing fat, carbohydrates, vitamins and minerals". The natives boil the fruits in salt water after harvesting, remove the skin and consume the fleshy mesocarp immediately. Storing the harvested bunches is limited to 2 weeks; otherwise they must be cooked and dried. After a second brief boiling, the pejibae fruits recover their former consistency. Besides their fruits, the palm is used also in many other ways; the leaves are good for thatching huts, the spinous trunks for palisades and defences against wild animals and the hard wood is good for manufacturing bows. Finally it should be mentioned that younger shoots are used as

**Fig. V.6.** A fresh harvested bunch of fruits from *Bactris gasipaes* in Panama, 1/10 nat. size (Photo Hansen, Balboa)

"palmito" vegetable. As nearly always with palms, the natives know how to use them for preparing alcoholic beverages. In Panama and also in Costa Rica "pijba" fruits are a popular food of the lower classes, sold at local markets and street corners already boiled, but at an incredibly high price (1 dozen for one dollar in Panama!) (Fig. V.6). Future breeding should be directed to selecting the casually occurring mutations with spineless stems. There are also individuals which produce fruits without hard-covered seeds. They could be multiplied in a clonal way, and would be useful to produce fruits for milling purposes (Johannessen 1967). We hope that the Governments of the Central American states will take more interest in this historic food plant of their native populations and stimulate plantations of selected *Bactris gasipaes* on a commercial scale.

## 5. *Elaeis oleifera* H.B.K.

Synonym: *Corozo oleifera* H.B.K. Bail., *Elaeis melanococca* Gaertn.

Dendé, corozo, coquito, caiauhé

NAME AND ORIGIN

Natives of the Amazonian rainforest call it "cai-auhé", whilst in Central America its name is "corozo". Its geographical area extends from Costa Rica, Panama to all South American tropical countries, growing mostly in swampy soils and riparian regions, under the dense shade of tall forest trees (Meunier 1975).

In comparison with other neotropical palm species, *E. oleifera* is of very low stature and can thus be easily recognized in its natural habitat. With its low height of 2–3 m it can be harvested easily.

MORPHOLOGY

Stems are short, sometimes inclined, but do not produce suckers. They are often creeping at the beginning with many roots emerging from the trunk. The size of this palm is no taller than a man.

Leaves pinnate (150–200 cm) form a dense crown. Inflorescences emanate from leaf ax-

ils, with separation of staminate and pistillate inflorescences in large head-like flower clusters on the same palm. They come into full bearing after 10 years. A tendency exists to develop parthenocarpic fruits. Production may last 40–60 years.

Fruits have a medium weight of 5 g, with 2–3-cm diameter, with a yellow-orange colour; inside 1–3 seeds, which contain a white fat.

The oily mesocarp represents 28–53% of the whole fruit. The extracted reddish oil has a high level of unsaturated fatty acids and is rich in vitamin A. Natives use it in its pure state as cooking oil.

Commercial exploitation should be stimulated by selection of higher-yielding plants, especially in view of the strong competition of the introduced African oil palm (*Elaeis guineensis*). Future breeding must be directed to hybridization with this genetically closely related palm species, which has a rather tall stem (Hartley 1977).

This wild-growing "oil palm of the New World" has many advantageous qualities, namely, resistance to serious diseases, good adaptation to marginal ecologies and reduced growth height; finally its oil can be considered superior to that of the African oil palm, due to its lower percentage of unsaturated oil acids. Its mesocarp oil has the highest iodine value and carotene content of all other American palms.

For all these reasons the palm oil industry has stimulated the creation of hybrids between American and African *Elais* with the same chromosome number. The resulting offspring often show hybrid vigour with larger leaves and dominance of advantageous fruit factors, combined with resistance against bud rot disease and sudden wilt. The difficulty is that by further sexual propagation such a combination of positive factors becomes lost. Therefore practical techniques for asexual propagation of the best hybrid types has to be developed.

## 6. Heart of Palms *Palmitos*

Under this commercial denomination we treat here a considerable quantity of palm species which have recently acquired high utility.

There is a notable trend at present in the European fancy-food market for expensive tropical food. Among these exotic delicacies range bamboo-shoots and palmitos = palmhearts. The demand surpasses the production. For this reason we recommend that such tropical countries where the labour force is cheap and the ecological conditions are favourable, should establish plantations of quick growing palm species. Unfortunately (and this is also the reason for the high price of palmhearts) for obtaining the "palmito" the tree must be killed.

The so-called "palmheart" is a cylindrical bundle of very young leaves, often several meters long with 10–18 cm diameter. They have a pleasant palatibility and are mostly sold as canned conserves (Fig. V.7). There exist a dozen species in the American Tropics, most of them still exploited from wild stands. But the danger of extermination is increasing now with the rising demand from the processing industry. So the cultivation of palmheart-producing species would circumvent the awful destruction of rainforests in the search for such palm species (Reitz 1974).

The following tropical palms belong to this group:

| | |
|---|---|
| *Acrocomia sclerocarpa* | *Geonoma diversa* |
| *Bactris gasipaes* | |
| *Chamaedora* spp. | *Iriartella setigera* |
| *Euterpe edulis* | *Oenocarpus bacab* |
| *Euterpe oleracea* | *Prestoa* spp. |
| *Euterpe precatoria* | *Sabal palmetto* (Florida) |

Until now the most important palm for the growing market of this luxurious vegetable is the genus *Euterpe*, native to the rainforests of Brazil and Venezuela.

The uncontrolled exploitation of the best palm groves with *Euterpe* and *Iriartella* has led to a near extermination of such palms as grow near accessible roads, e.g. along the Panamerican highway and the many roads which have been cut in Brazilian forests. Due to this excessive cutting down of natural stands, commercial plantings in Brazil, Venezuela and Paraguay are now beginning. Since 1980 Brazil has exported huge quantities of palmito conserves: 114,400 in 1980 alone, but it seems that the demand for this luxury product is greater than the supply.

The often-discussed nutritional value of palmhearts is not as low as is generally believed. Of course, most canned palmito is moisture (91%), but there are also 2.2% proteins, 5.2% carbohydrates and especially a fair amount of minerals, like calcium 86 mg, phosphorus 79 mg, iron 0.8 mg and also ascorbic acid 17 mg, niacin 0.7 mg and thiamine 0.4 mg. [This was calculated by Leung (1961) for 100 g edible substance of the palm *Prestoea*].

**Fig. V.7.** *Euterpe*-palm, used for preparing palm-heart conserves

### *Euterpe edulis* Mart.

Assai, manicola, palmito

Its areal extends from tropical regions of Brazil to the subtropical province Misiones in Argentina. This palm is easily recognized by its long (20–30 m) elegant stems, thin with a basis diameter of only 15–20 cm. The leaves are 2–3 m long, inserted mainly on the top of the plant. The inflorescences emerge only once in the year and produce 50–80-cm-long panicles with yellow, unisexual flowers. The fruits are drupes, similar to olives, 1 cm thick with violet skin and little pulp.

### E. oleracea Mart.

Gissard, pinot, nanac

**Fig. V.8.** Seeds of *Jessenia bataua* in Venezuela (Photo Braun)

Is the most interesting palmito for future industrial plantations. It belongs exclusively to the tropics and has a favourable growth characteristic as it always grows in clumps of many basal suckers. This allows rapid vegetative, clonal multiplication of selected individuals.

The palm has a thin trunk, often slightly curved, with 15–20 m height, often 20–25 stems together. With its crown of pinnate leaves and pendulous segments, *E. oleracea* produces an ornamental effect.

The fruits are produced nearly the whole year round. From these small black violet berries (12–14 mm diam.) the Indians make a tasty juice of high nutritional value. In the Amazonas region it is cooked with cassava meal as a fortifying beverage.

### 7. *Jessenia bataua* (Mart.) Burr.

Synonym: *J. polycarpa* Karst., *Oenocarpus bataua* Mart.
Batana, seje, pataua

The genus *Jessenia* consists of one single species: *J. bataua*, often confused with some species of the genus *Oenocarpus*, which grow in similar places in the rainforests of South America.

Some taxonomists in palms, like Balick and Gershoff (1981), consider both genera as a closed complex with genetic introgression; and natural hybrids have often been observed among wild populations of *Jessenia* and *Oenocarpus*.

The natives have several names for it: seje, patauá; or they gave it even a Spanish name: milpesos, which means in this case, that selling the fruits gives good money.

Its distribution is mainly north of the Equator, (especially Venezuela, Colombia, Brazil) but certain local biotypes (e.g. var. *weberbaueri*) have been described also from Peruvian rainforests. It is common in the entire Amazonas valley, and occurs also on the Caribbean Islands and Panama. The species has a wide ecological adaptation, from swampy lowland up to 1000 m mountain regions (Fig. V.8).

MORPHOLOGY

Stems are 20–25 m high, with 30 cm diameter at the basis. Whilst still young, they are covered with horrendous-looking, long, black spines. The Indians use them in their blow guns.

Leaves have a typical erect position and are inserted on the trunk at an angle of 15°, therefore enclosing the stem. The leaves are very long (7–9 m).

Inflorescences covered with a long spadix, (more than 1 m) and 3-m-long rachis. The large fruit clusters are similar to those of the African oil palm. Adult plants produce two fruit bunches per annuum, each 30 kg heavy and loaded with 1000 fruits.

Fruits black-purple with 2–4 cm diameter and have a thin, oily mesocarp. From this a valuable oil, similar to olive oil, can be extracted, which has also many uses in folk medicine, and is sold at local markets in Colombia. The chemical composition is approximately 80% oleic acid with 20% linoleic palmitic acid. The seed kernels have a horny consistency with high fat content. The expeller contains a fair quantity of protein with good biological value. During the last world war, Brazil exported considerable quantities of Batauá oil, which was obtained from wild-collected fruits.

*Jessenia* palms have not been domesticated so far. For this purpose the naturally existing variations should be studied primarily, combined with laboratory analysis of the (varying) oil content and quality. Based on this, germplasm collections should be stimulated, with preference to those regions where Indians know of promising stands, with pulpy fruits and thick mesocarp. Any future Batauá oil industry should be built up from the beginning on selected, high-yielding and quick-growing populations. The bad custom to fell good *Jessenia* trees for rapid harvest should be absolutely avoided.

At present mostly used by natives of tropical America with an important role in local diet, as construction material and for elaborating tools, weapons and fibres. But the Batauá palm has never been cultivated.

**Fig. V.9.** *Mauritia flexuosa*, a majestic palm of tropical plains (Photo Fougué)

temporary communities as nomadic hunters (Fig. V.9).

MORPHOLOGY

The stems reach 28–30 m in height, sometimes with thick trunks, which measure at the base 200 cm in circumference. They contain a starch similar to sago.

Leaves 3 m long and so heavy that two persons are needed to bear them away. Their tubular leaf petioles are so big that natives use them – lashed together – as rafts for water transportation.

Inflorescences develop heavy, 2-m-long panicles. The fruits are spherical (5–6 cm diam.) with a peculiar exocarp pattern of numerous rhomboidal brilliant brown scales. The mesocarp is rather thin and is composed of starch and a yellow oil of high quality. The seeds contain 50% fat and are rather easy to crack. Their walls are composed of mannocellulose, which is soluble in hot acid solutions. They are applied industrially for making "ivory buttons". The natives collect the fruits when they fall ripe to the ground.

### 8. *Mauritia flexuosa* L.

Moriche, miriti, buriti, aguaje

The genus *Mauritia* is represented in the humid American tropics by several taxa. The most important for human use are *M. vinifera*, and *M. flexuosa*, *M. setigera* and *M. minor*. The presence of the two last-named is a reliable indicator of water excess in the soil, and therefore not suitable for agriculture.

The first explorers of the Orinoco and Amazonas region, Humboldt, Bonpland, Wallace and Spruce, were impressed by the majestic growth of these palms and compared their stands with "natural temples". Still today hundreds of thousands of them cover the swamps and river borders in the "Llanos", a region difficult of access and where Guahibo Indians still live, mostly in

### M. vinifera Mart.

As its name indicates, this palm is appreciated for making alcoholic beverages. With this aim, Indians bore holes into the stem and may collect daily 8 l of sweet sap from a single palm tree. Sometimes they cut down the whole trunk – an undesirable method – and open it in different parts. A sweet liquid exudes, which begins quick fermentation. Equally important are the fruits which contain a yellow oil (20%) which has already attracted commercial exploitation of wild stands of this palm in Eastern Peru and which is rather similar to *M. flexuosa*.

### 9. Oenocarpus spp.

This neotropical genus consists of a dozen species native to the rainforests of Brazil, Colombia and Venezuela. Several of them produce a valuable oil which resembles olive

**Fig. V.10.** An *Oenocarpus* species from Venezuela

oil. All are used in the wild state and none has been cultivated so far. Typical for *Oenocarpus* palms are the thin, elegant and slender stems (Fig. V.10).

### O. bacaba Mart.

Bacaba

It has high (up to 20 m) elastic trunks. The leaves (5–6 m long) are crispate, dropping when ripe. The panicles bear a great quantity of small fruits. The fruits are nearly black in colour and are covered with a whitish, waxy-powdery substance.

### O. distichus Mart.

Icaua

The trunk is slender and of elegant stature (10–12 m high) and covered with spaced rings. The leaves are arranged distichously, like a fan. The inflorescences are covered by two long spathes. They produce many small fruits with violet exocarp and a yellow pulp. The mesocarp contains 25% of a yellow clear oil of good quality.
This palm grows in general as isolated individual plants, which are characteristic of the Amazon estuaries in the state of Maranhao.

### O. minor Mart.

Even though of minor stature (5–7 m high), this palm is appreciated by natives for the juice which is produced from its fruits and also for the "palmitos" of their young leaves. Its habitat is the rainforests of the Brazilian states Para and Amazonas and also the delta of the river Orinoco in Venezuela.

### O. mapora Karst.

Several stems grow from the same basis and reach 12–15 m height. They mature already after 4 years, and are for this reason interesting for future plantations. The leaves are 4 m long and cover the inflorescence. The fruits are small with dark epicarp. This palm extends its area from the Llanos of Colombia, through Venezuela mostly along the Rio Orinoco.

## 10. *Orbignya martiana* Barb.-Rodr.

Babassu, bassu, corozo, coruba

Of the genus *Orbignya* almost 20 taxa have been described from the neotropics, which makes it rather difficult to differentiate real taxonomical species. Its most septentrional representative is *O. cohune* (Mart.) Dahlgr., which is abundant along the Atlantic lowlands from southern Mexico to Costa Rica. *O. speciosa* and *O. oleifera* from semi-arid NE-Brazil have also been described, but the most important is the "Babassu" palm (Cavalcante 1977). *O. martiana* has its opti-

**Fig. V.11.** *Orbignya martiana,* the most important oil palm of Brazil (Photo Fouqué)

mal habitat in the transition zone between the southern part of the humid Amazonas basin and the drier part of NE Brazil, covering a part of the states of Maranhao, Piaui and Goias, in total an area of 200,000 km². This palm has a high adaptability to alluvial soils and even degraded and abandoned land following human occupation. It survives bad conditions, even occasional burning of surrounding forests. *O. martiana* has a special advantageous germinating system for such conditions. Thanks to its slow cryptogaeal germination, its seeds retard the growth of the embryo. The apical meristem lives several years hidden beneath the surface of the soil (Fig. V.11).

MORPHOLOGY

The stems are tall, (15–20 m height) and 40 cm diameter at the basis. However, there exist also "dwarf" forms only 5 m tall, which produce fruits.

Leaves are produced in a crown of a dozen leaves. They may reach a length of 6–9 m. Inflorescences are borne between the axil leaf, and are divided in masculine and feminine organs.

Fruit-bearing bunches are nearly 1 m long and may contain 300–500 single fruits. They resemble small coconuts and are 10 cm long and 6 cm wide, with a rather thin starchy mesocarp. The oil-rich seed kernels are enclosed in a very hard, woody endocarp (3–10 mm thick), which is difficult to crack. The seed kernels (3–8) have 65% fatty substance. The purpose of the thick hard shell is the protection of the seeds against potential predators, but for industrial exploitation of the Babassu palm, this represents a serious problem. The oil components are the following: 45% lauric acid, 15–20% myristic acid, 6–9% palmitic acid, with a relatively low content of stearic and caprylic substances. After pressing the oil, a seed cake with 27% protein remains, which represents an excellent animal feed.

The oil of babassu is similar to coconut oil and also used in a similar way. It is colourless and well adapted for different uses, e.g. in the toilet industry, for detergents and especially in margarine fabrication because it does not easily become rancid.

Juvenile growth is slow; the tree begins to bear fruits after 7–8 years, but can then yield for over 70 years. High production has been reported: four racemes, each with 200–600 fruits, similar to small coconuts, give 100 kg per year per tree.

It seems almost incredible that the major part of the 250,000 t harvested in 1980 originate from wild stands. The inhabitants of the marginal northeastern states of Brazil make a living from gathering the fruits and selling them to local factories. To overcome the insecurity of such collections, certain domestication projects with selected babassu biotypes are under way, similar to plantations with the African oil palm.

It has been calculated that the harvest of babassu fruits provides subsistence for 450,000 native households, who combine fruit gathering with some other minor rural activities. The economic value of babassu oil for Brazil alone has been estimated at five times (!) the value of the coffee crop in Brazil. The oil of babassu palms occupies the third place on the inland market of oils for Brazil. Its industrial exploitation and seed kernel collection has experienced a spectacular increase over the last 50 years: 1920: 6600 t, 1928: 30,300 t, 1940: 71,000 t, 1960: 99,700 t, 1970: 180,900 t and finally 1980: over 250,000 t of kernels.

The problem still exists, however, of the extraordinary hardness of the endocarp and the crushing of it. One needs machines which exercise more than 1 t force to break the endocarp shell open. Several mechanical devices (even deep freezing) have been invented for this purpose. I believe that in the long run the problem can only be resolved genetically, by selecting biotypes with a thinner endocarp. Until now the work has been done mostly by native workers in Brazil, who crack the shell by hand or with an axe to split the fruits longitudinally into several parts until the seed kernels fall out. By this time-consuming method a worker may obtain 5–8 kg of babassu kernels per day.

The large amount of broken shells can be used as fuel, the charcoal being a very common heating material in the state of Maraônon.

Finally we have to mention some neotropical palm species, which have aquired a certain technical and industrial importance on the world market:

## 11. *Copernicia cerifera* (Arr.) Mart.

Wax palm

This palm lives in the semi-arid regions of NE Brazil. Its leaves are covered with a thin wax layer called carnauba wax. Each leaf gives 3–10 g. For harvesting the wax the leaves must be cut, put on cloths and threshed. This is a rather difficult and time-consuming work and therefore prices of carnauba wax, which cannot be produced synthetically, have increased considerably on the world market. It has a high melting point (86°) and excels by its brilliance. The yearly production in Brazil exceeds 10,000 t.

### *C. australis* Becc.

is indigeneous to Paraguay and called there carandaaý. The wax covers the leaves in very thin layers, in the quantity of only 2 g per leaf. Nevertheless some natives of the Chaco collect this wax and sell it as substitute for real carnauba.

### *C. prunifera* (Miller) Moore

This wax is considered as the most valuable vegetative wax with a high melting point and good hardness. It is produced in powdery flakes on the leaves, which fold up, when dry, and protect the wax from being lost.

### *Syagrus coronata* (Mart.) Becc.

Urucuri, cabesudo

Considered as a substitute for the real carnauba, its wax must be scraped from the leaves by hand. This palm has a wide distribution in the arid regions of eastern Brazil, from Pernambuco to Bahia, where it is called licuri. Its yearly production varies between 200–450 t mainly for export.

The plants are rather short (8–10 m) and have large, dark green leaves covered with wax.

### Ceroxylon andicola Humb. & Bonpl.

This wax palm was discovered by Humboldt and Bonpland during their historic expedition to the Cordillera Central when they crossed it near Cartago (Colombia). It grows in nearly 3000 m alt. west of the Quindio pass. This altitude is quite unusual for a tropical palm, which usually lives in warm lowlands. With 60 m height this *Ceroxylon* species is considered the tallest palm tree of the world. Because of its impressive appearance this palm has been declared national tree of Colombia. The wax is produced on the trunk and the underside of the leaves.

FINAL REMARKS

A special case are the ivory nut palms. They have a perisperm as hard as ivory, with a similar white structure. Owing to this fact, many artistic objects are made from them, mainly from the fruits of *Phytelephas macrocarpa* Ruiz & Pavon. Its natural distribution reaches from Panama to Peru.

Here we should also mention that the fibres for the so-called Panama hats are not produced by a real palm, but a stemless *Cyclanthacea*. The best quality comes from Honduras, with very light hats (only 50–70 g), which need several weeks' weaving, and are called jipijapa.

### References

Balick MJ (1979) Amazonian oil palms of promise, a survey. Econ Bot 33:11–28

Balick MJ (1981) *Jessenia bataua* and *Oenocarpus* species: Native Amazonian palms as new sources of edible oil. Am Oil Chem Soc 12:141–155

Balick M, Gershoff SN (1981) Nutritional evaluation of the *Jessenia bataua* palm. Source of high quality protein and oil from tropical America. Econ Bot 35:261–271

Barbosa-Rodrigues J (1903) *Sertum palmarum Brasiliensium*. Bruxelles

Blaak G (1980) Vegetative propagation of Bactris gasipaes. Turrialba 30:258–261

Braun A (1970) Palmas cultivadas en Venezuela. Acta Bot Venez 5:ff

Cavalcante PB (1977) Edible palm fruits of the Brazilian Amazon. Principles 91–107

Clement CR, Mora Urpi JE (1987) Pejibaye palm (Bactris gasipaes), multiuse potential for the lowland humid tropics. Econ Bot 41:302–311

Dahlgren BE (1959) Index of American palms. Publ Nr 863. Field Mus Nat History, Chicago

Glassman SF (1972) A revision of Dahlgren's Index of American palms. Cramer, Lehrte, Germany

Hartley CWS (1977) The oilpalm. Longman, London, 900 p

Hodge WH (1975) Oil-producing palms of the world. Principles 19:119–136

Johannessen CL (1967) Pejibaye palm: physical and chemical analysis of the fruit. Econ Bot 21:371–380

Leung WT (1961) Food composition tables for use in Latin America. US Gvt Printing Office Washington DC

Lötschert W (1985) Palmen. Ulmer Stuttgart, 152 pp

Markley KS (1971) The babassu-oil palm of Brazil. Econ Bot 25:267–304

Martius KP (1837) Historia naturalis palmarum. Pars VI, München

Meunier J (1975) Le palmier à huile américain *Elaeis melanococca*. Oleagineux 30:51–61

Mora Urpi J, Clement C (1981) Aspectos taxonomicos relativos alpejibaye (Bactris gasipaes). Rev Biol Trop 29:139–142

Patiño VM (1958) El cachipay o pijbay, en la cultura de los indigenas de la America intertropical. Edicion Especial Inst Indigenista Interamericano, Mejico 39:176–203, 293–331

Pesce C (1985) Oil palms and other oil seeds of the Amazonas. Transl by D Johnson of the original 1941. Reference publ Inc Algonac, Michigan

Reitz PR (1974) Flora ilustrada Catarinense. Palmeiras Itajai Sta Catarina

Schultes RE (1980) The Amazonia as a source of new economic plants. Econ Bot 33:259–266

Seibert RJ (1950) The importance of palms to Latin America. Pejibae a notable example. Ceiba 1.65–74

Standley PC & Steyermark JA (1958) Flora of Guatemala. Fieldiana Botany 24, I., p 478

Uhl NW, Dransfield J (1987) Genera Palmarum, based on the work of H. E. Moore. publ by Bailey Hortorium, Cornell Univ

# VI. Industrially Used Plants

## Fibres

Two neotropical genera are of fundamental importance for American fibre production: *Agave* and *Fourcraea*, which include 200–300 xerophytic species. Both genera are morphologically similar, with the exception of their flower anatomy, and so can be grouped together here (Dempsey 1975).

The benefits from sisal fibre production are felt predominantly in underdeveloped nations. Therefore all projects to improve this neotropical species should receive international support. But it seems – on the contrary – that export from these countries is diminishing.

Brazil is exporting at present 150,000 t, East Africa (Kenya-Tanzania) only 105,000 t and Madagascar 20,000 t.

The production from Latin America is: Brazil 243,000 t, Mexico 92,000 t.

## 1. *Agave* spp.

Plants form large rosettes with 200 leaves inserted in spiralic order. Leaves succulent, evergreen and have strong sclerenchyme fibres which accompany the ribbon fibres. Fibre content is low, 3% per leaf. The margins of leaves are with spines, in contrast to *A. fourcroydes*. Roots are shallow (40 cm deep) but of enormous lateral extension (5 m around the plant).

Inflorescences 5 m high, when flowering begins after 10–20 years.

Thanks to its extraordinary resistance to adverse dry climate and poor soil, *Agave* give yields where other crops would fail. Never-theless, to obtain commercial competitive yields, plantations must receive potassium and lime regularly and precipitation should not be below 700 mm/year.

*Agave* plants have been used by Central American Indians since time immemorial for diverse purposes, mainly for their fibres and for their sweet juice, which after fermentation was converted into narcotic drinks (Gentry 1982).

They had also a considerable local importance as food and beverage. The baked "hearts" of *Agave* plants were a much-appreciated food in the arid regions of the American Southwest. Additionally, the sap of *A. pacifica* and *A. tequilana*, after fermentation, provided the Indians with alcoholic beverages.

Under the names pulque, mescal, tequilla, ritual intoxicants are produced as an expression of national pride and popular liquor of Mexico.

Early domestication has obviously led to the selection of polyploid hybrids, with loss of sexual propagation. Whilst the basic chromosome number of the genus is $x = 30$, several domesticated species, like *A. cantala* $2n = 90$, *A. sisalana* $2n = 138$ and *A. four-croydes* $2n = 140$ have reached higher chromosomal levels.

Between the many native species, (perhaps 100 taxa) we mention the following, in alphabetical order:

### *A. angustifolia* Haw.
Dwarf aloe, babsi
$2n = 60$

This diploid species is native to Mexico. It produces better and more leaves than *A. sisaliana* and is therefore employed as partner for hybridizations.

### A. *cantala* Roxb.

Magey, cantala, Manila Hanf
$2n = 90$

The wild form grows on the western coast of Mexico. From there, Cantala plants were introduced by Spanish sailors to the Philippines, then under the rule of the Spanish crown. Until the present century, it was the most important hard fibre there, besides the so-called manila hemp (Ramia with soft long fibres). The yearly export reached 14,000 t. The quality of cantala fibre was considered superior to that of sisal hemp. The plants also tolerate wet climate. Their life cycle is shorter than that of sisal (6–10 years). They produce 250 leaves in total and contain 4% fibres.

### A. *fourcroydes* Lemaire

Hennequen, sacci, weißer Sisal
$2n = 140$

This high polyploid *Agave* may have been selected by the Mayas – or even earlier – on the Yucatan Peninsula. In spite of several attempts to introduce and acclimatize hennequen for cultivation on other continents, it still thrives best under the agrobiological conditions and dry climate of East Mexico. It produces a coarser fibre than other agaves. Its export was 140,000 t in 1960, at which time it was mainly used in the USA as binder yarn for mechanical cereal harvesters. Its fibre content is low, approximately 4%.

*A. fourcroydes* has a long life cycle (20–30 years) and produces 230 leaves and finally a huge inflorescence on a 5–7 m-tall stem. On finishing flowering, the plant dies, but very seldom produces fruits when autopollinized. Nevertheless, its gyneceum is intact and fertile and can be used for species crossings.

### A. *lecheguilla* Torrey

Istle fibre

Its chromosome number has not yet been determined. We suppose that it may be low, because the ripe plants produce seeds and no bulbils.

Native to the very dry region of North Mexico-Texas, where it prefers calcareous soils with only 250 mm yearly rainfall. The plants are relatively small. The fibres are very coarse and are esteemed for their high resistance to breaking and bruising. Its main use is in fabrication of brushes and scrubbers.

### A. *sisaliana* Perrine

Maguey, sisal hemp, sisal tuxtlecs
$2n = $ approximately 138

This is the most important of all *Agave* species. Its name has been derived from the harbour town of Sisal on the Yucutan peninsula.

**Fig. VI.1.** *Agave sisalana*, plantation in East Africa (Photo De Geus)

The plant itself is considered a spontaneous pentaploid hybrid which was found growing wild in Eastern Mexico. To protect this biotype from foreign exploitation, the Mexican Government declared a ban on the export of living sisal plants already in the last century. Nevertheless, *A. sisaliana* was cultivated in 1890 in Florida, USA. From there the German farmer Hindorf in 1893 aquired a thousand plantlets and transported them to East Africa, (now Tanzania) but only 62 plants survived the transport. They were planted by the German East Africa Company on the banks of the river Pangani (Fig. VI.1). Thanks to these extraordinary efforts of German planters, the present state of Tanzania is the world's leading producer of sisal hemp, estimated now in 300,000 t hard fibre (Lock 1962).

From the phyto genetical view point, it is noteworthy that 62 identical clones were the basic breeding material, an extremely narrow "gene pool". After several vegetative multiplications they rose to 500,000 plants in the year 1900, producing 7 t. Just before World War I, the production had risen to 20,000 t. Surely – from the view point of modern genetics – it would have been advisable and economically necessary to widen the extremely narrow breeding basis for such an important national commodity, but: who cares about this in Tanzania?

### A. tecta **Trelease**
Maguey gigante

This agave is the largest of all. It is mainly cultivated in the Eastern part of Guatemala and used there for the preparation of the narcotic beverage "pulque".

## 2. *Fourcraea* spp.

Another useful genus for fibre production. We mention the following species:

### F. cabuya **Trel.**

Natives call it cabuya, or pia floja and use it mainly for domestic purposes, like hammocks. It is cultivated in Colombia.

### F. gigantea **Vent.**
$2n = 60$

Also called "Mauritius hemp", following its early introduction to the Island Mauritius by Portuguese merchants, in the 18th century. In Mauritius a secondary centre of diversification arose, constituting escaped plants which propagated in wild stands.

The species originated in the lowest-lying parts of Amazonia, where it flowers profusely. In cultivation the seed setting is rather poor, so propagation is done with bulbils, formed on the inflorescence poles. The leaves are 100–200 cm long with smooth fibres, which are longer and finer than those of the usual sisal; the fibre content is, however, low, estimated at only 2–3%.

### F. quicheensis **Trel.**

This species is found extensively in Peru, Colombia and Guatemala. Its use is restricted to the needs of natives.

BREEDING

Due to their long vegetation cycle all *Agave* and *Fourcraea* species are a difficult breeding object, even though German agronomists in East Africa 50 years ago began hybridizations with quicker-growing accessions. In the meantime some recurrent crossings have given rather promising results. There are hopes of obtaining 100% higher yields from certain artificial hybrids, which later may be multiplied vegetatively.

MORACEAE

## 3. *Castilla elastica* **Cerv.**
Ule-rubber tree, palo de hule, caucho
$2n = 28$

This binome was created by Cervantes in the year 1794 in honor of Juan de Castilla (not Castilloa). The Maya Indians used the latex of this tree in pre-conquest time to make rubber balls for their games and also water-proof clothing.

The Castilla tree has a high (25 m) un-branched trunk and is monoecious. It is easily recognized in the tropical forest by its enormous, buttress roots exposed on the surface of the soil, often 20 m long.

The seeds are enclosed in fleshy fruits and are rather perishable.

The latex vessels consist of elongated single cells. Once cut, there is an amazingly rapid flow of latex. The tappers climb with ropes as high as possible to obtain a maximum yield, open the bark with a very sharp machete, and make deep cuts through the cambium into the wood. This tapping procedure is quite different from that used in *Hevea*.

Often the natives cut the whole tree down. Such over-exploiting of this beautiful tree – especially during times of war – has greatly endangered survival of *C. elastica*, once abundant from Mexico to Panama.

While Wickham was busy collecting seeds from *Hevea* in Brazil, at the same time, the British India Office sent the naturalist Cross to Panama to study the future exploitation of "Ule" rubber trees (Fig. VI.2).

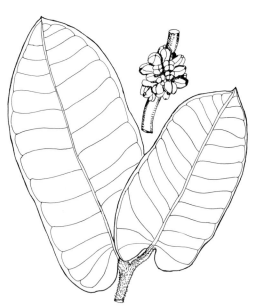

**Fig. VI.2.** *Castilla elastica*, ancient neotropical rubber source

BOMBACACEAE

## 4. *Ceiba pentandra* (L.) Gaertn.

Silk cotton tree, kapok, arbre à boutre, ceibo, pochote, Kapokbaum

$2 n = 72,80,88$

The radiation centre of the genus *Ceiba* is probably in Central America, even if these showy trees are extended at present throughout the whole of the neotropics. In pre-conquest times, the Indians of Honduras and Guatemala used to hold their religious gatherings under hallowed ceiba trees, sometimes of enormous size and many centuries old. Survivors of the Mayas in Yucatan still consider it as their most sacred tree.

MORPHOLOGY

High trees (30–50 m) with enormous trunks (often 2 m diam.) and long buttresses on the ground. The bark is covered with conical-spiny protuberances. The branches have a typical horizontal position. Branches are borne at right angles.

Leaves are palmate, subdivided in 5–10 leaflets. They are shed during the dry months, giving the tree a completely barren aspect. At this time the flowers appear, with delicate pink or white petals.

Fruits are large (12–18 cm), similar to cocoa fruits, and contain a hugh quantity of black seeds, which are embedded in a dense white fibre mass, like silk. These fibres arise from epiderm cells of the endocarp (quite different from cotton fibres, which sit on the seeds). The capsules contain 17% of such fibres. They are 12–35 mm long and covered with a thin layer of wax which inhibits spinning. The fibres are highly appreciated as fill material for life jackets due to their water-repellent characteristics. They are very light and resilient, and an ideal material for insulation or for filling cushions (Fig. VI.3).

The seeds are collected by natives for their high oil content (25%), they contain 30% linoleic acid.

The wood of the kapok tree is very light. I observed that Indians of Darien (Panama)

a

b

**Fig. VI.3. a** *Ceiba pentandra,* leaf, flowers and fruits. **b** *Ceiba* tree

used it to make dug-out canoes. Having felled such an immense tree in the forest, they had to work for several weeks in the spot, because of the lack of any transport facilities. Modern industry could use the wood for plywood and light furniture. On the world market fibres of other species of Bombaceae appear mainly from Southeast Asia, which are sold under the name "Kapok".

BIGNONIACEAE

## 5. *Crescentia cujete* **L.**

Calabash tree, calabassier, calabacero, totumo, cuia, otumo

$2n = 40$

For thousands of years in the neotropics, together with Cucurbitaceae (e.g. the pantropical *Lagenaria*), this was the basic household vessel for the Indians. Six different wild species are known, some from Central America, others from the Amazonas region. Most probably *Crescentia* was domesticated in distant places and quite independently, due to the ease of its propagation. Attractive mutations in the size and shape of the fruits have been selected and survive with indigenous populations, even in face of the present competition of cheap "plastic containers". We were astonished recently during our visit to the Kuna Indians, who live on the Islands of SE-Panama, to observe the diversity of calabash forms. For each need different samples and sizes exist. The natives grow them between their cocoa groves and collect them according to the demand (Fig. VI.4).

The trees are small (3–7 m) with widely extended side branches. The leaves are simple and attached in fascicles. Inflorescences with tubular flowers (5–7 cm long), borne cauliflorous on stems and large branches. Fecundation mostly by bats.

Fruits ovoid-spherical, often of large size (in general 20–30 cm diam.). Inside are many seeds, which are edible when unripe. For use as a vessel, the seeds must be removed. The shell of ripe fruits is thin, but hard, and has a durable cover (Fig. VI.5).

Ripe fruits of the calabash tree have a durable covering. Its seeds contain 8% fat and 8% protein and have been collected since time immemorial by the American natives. Once opened, they are used as practical containers for food and liquids, smaller ones serve as cups, spoons and scoops. Some ethnologists have confused the *Crescentia* calabash with the gourd calabash (*Lagenaria siceraria*), which has the same usage by the natives, but can be easily differentiated by its texture.

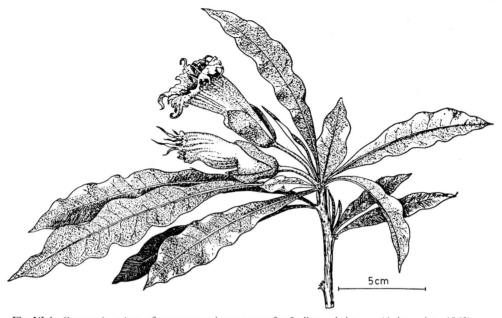

**Fig. VI.4.** *Crescentia cujete*, of paramount importance for Indian subsistence (Aristeguieta 1962)

**Fig. VI.5.** The calabash tree *(Crescentia cujete)* with fruits

### C. alata HBK

Known as "tecomate" in Middle America, jicaro, morro or guacal

The small trees are abundant on the Pacific coast from South California to Costa Rica. Fresh fruits are eaten, dry fruits have a use similar to *C. cujete*, as containers for many domestic purposes.

Oviedo y Valdez (1487–1557) in his *Historia General de las Indias*, mentions especially that the jicaro tree has leaves shaped like a cross, and concludes from this similarity that the Indians of this region must be addicted to Christianity. We can only hope that this mis-belief saved some of them from being tortured or crucified by the Spanish missionaries.

MALVACEAE

## 6. *Gossypium* spp.

Cotton fibre, algodon, coton, Baumwolle

The genus has pan-tropical distribution, and cultivars have been derived from wild forms in Asia and Africa, as well as in America. In accordance with the title of this book, we treat here only the neotropical *Gossypium* in detail.

A total of 36 taxa of wild-growing cottons exist around the world, some of them perennial shrubs, others even small trees. With the exception of *G. herbaceum* and *G. arboreum*, their fruit capsules are devoid of lint. In general they are diploids. These lintless diploid forms grow in the arid regions of the tropics and subtropics and have been classified in eight botanical sections. They are separated not only by morphological characters, but also by genetical barriers. The three Asian and African sections have the distinct genomes A, B and C. From the Australian continent nine species with the C genome have been described. In the Americas nearly a dozen taxa with the D genome exist.

Cotton is the world's most important and oldest fibre plant. That it has been gathered, used and planted on various continents for thousands of years has been proven by archaeological finds from different early civilizations. Remains of well-preserved cotton tissues have been recovered from the ruins of Mojendjo-Daro (West Pakistan, 2500 years B.P.), and in abundance from ancient Egypt. Further, several written reports exist from learned travellers of the pre-Christian era, like Theophrast or Plinius, about cotton plantations, spinning and weaving in India, Arabia and Nubia. These ancient people did not know our actual tetraploid cotton plants, but used the fibers of the diploid species *G. arboreum* and *G. herbaceum* and selected from them local races such as cv. *africanum*, *acerifolium* and *indicum* (Fig. VI.6).

The situation in the New World in ancient times was quite different. Many thousands of years ago a unique phylogenetical event occurred there: a combination, followed by polyploidy of the Old World A-genome with a New World D-genome. Until now, nobody has been able to explain how and where it happened; but that this genome fusion took place on American soil is without doubt. Therefore we consider the tetraploid cotton species: *G. hirsutum* and *G. vitifolium* (= barbadense) as neotropical crops.

Taxonomists have established two groups – which are genetically rather similar cultigens; calling the one *G. hirsutum*, "Upland cotton", with short staple fibres, and the other, *G. vitifolium*, "Sea Island cotton" with long fibres of silky aspect.

More than 50 years ago, Skovsted (1937) made the epochal discovery that the New World cotton cultivars are amphidiploids. His cytological work revealed that their genome is composed of two non-homologue chromosome sets of 13 units. He observed a pronounced difference in chromosome size. The American D-genome has small chromosomes, whilst the Afro-Asiatic A-genome is almost twice the size. Consequently, somatic idiograms of the allo-tetraploid *G. hirsutum* show a graduation in size from the shortest (D-genome) to the longest (A-genome) chromosomes.

This cytological observation was complemented by genetical experiments, i.e. artificial hybrids between different *Gossypium* species, bearing the A- and the D-genome, and doubling its chromosome number immediately with the help of colchicine. This decisive experiment was successfully accomplished in the same year independently by Beasly (1942) and Harland (1970).

Further cytological investigations were of great help to throw light on the phylogeny of the genus *Gossypium* (Gerstel 1953, Philipps 1966, Demol 1981). It was demonstrated that the different continental groups have distinct chromosome sizes:

A-genome of the Old World: moderately large chromosomes.

B-genome (Africa, Cape Verde) slightly larger A chromosomes.

C-genome (Australia) has very large chromosomes.

D-genome (Mexico-Peru) has small chromosomes.

E-genome (Arabia, Africa) has large chromosomes.

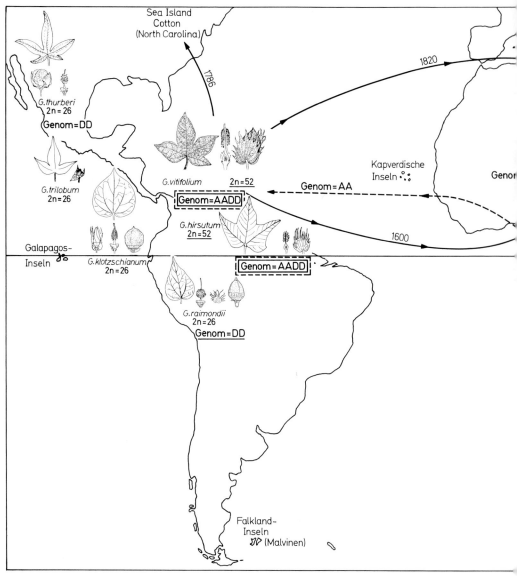

**Fig. VI.6.** *Gossypium,* world distribution of species; synthesis of the African A-Genom with Genom D on the American Continent

It was concluded that the B-genome is the prototype of the diploid species in the genus *Gossypium.*

The chromosomes of the A-genome are almost twice the size of those of the D-genome. This fact decisively helped to elucidate the cytomorphology of the amphiploid cultivated cottons and to trace their chromosomal components back to Africa (A-genome) and America (D-genome).

We may also conclude that the five amphiploid *Gossypium* species in the New World (*G. hirsutum, G. vitifolium, G. tomentosum, G. lanceolatum* and *G. mustelinum*) have a common A-genome ancestor. They show the same chromosome end arrangements, and in chromatographic assays a good correlation of protein-banding patterns exist. With certain exceptions in the "palmeri" group of *G. hirsutum,* the banding

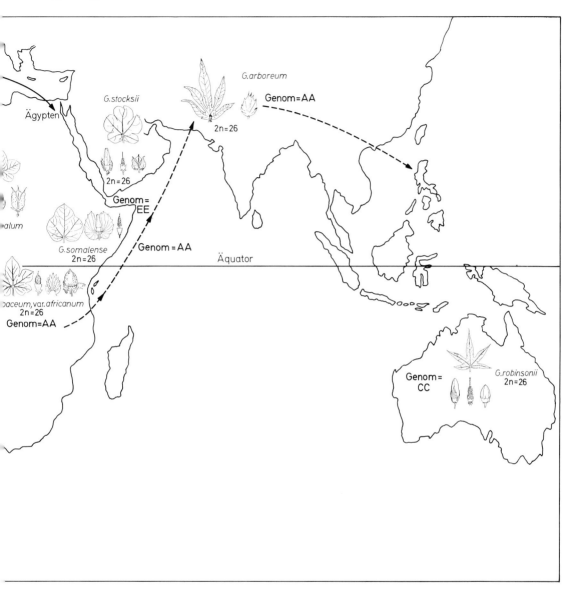

patterns of seeds from a number of accessions of the five amphyploid were identical (Endrizzi et al. 1985).

*G. hirsutum* L.

### *G. vitifolium* and *G. hirsutum* Lam.

$2n = 52$

In view of their relative similarity we treat them here, for reasons of space, together. In the case of doubt the two taxa can be distinguished by the presence or absence of "fringed hairs" surrounding the floral nectary; *G. vitifolium* does not possess them.

Name and History

Linné (1753) and his botanical contemporaries knew nothing about the existence of wild species of *Gossypium*, and recognized

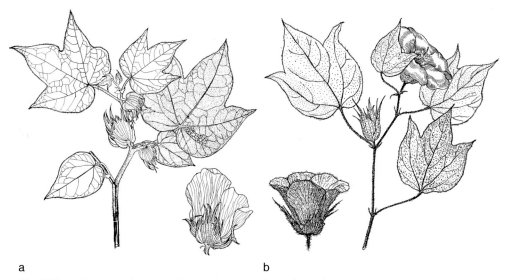

a                                                                          b

**Fig. VI.7. a** *Gossypium hirsutum.* **b** *Gossypium vitifolium,* the main cotton-producing species

only four cultivated taxa, two from Asia-Africa and two from America. As vernacular designation they used "cotton", which is a misspelling of the Arabian word "katun". The inhabitants of both Americas already thousands of years ago had innumerable applications for cotton fibres, besides spinning and weaving (Fig. VI.7).

According to Lanning (1967), cotton was cultivated in Southern Mexico by 3400 B.P. and in central Peru by 3600 B.P. In Central America it was *G. hirsutum,* and on the Pacific coast of South America it was *G. vitifolium* (generally called *G. barbadense* by archaeologists). The technique of dyeing and weaving improved in the course of thousands of years. Yet in the Chavin culture we notice spectacular developments, that reached their culmination in the Inka time. Although the Spaniards destroyed most of the marvelous clothes and artefacts, we can still admire some of the remains which are well preserved in local museums of Bolivia, Ecuador and Peru.

The archaeological excavations of Ica and Huaca Prieta indicate that the colour of the fibres was brown (not white), similar to the still existing primitive "algodon pardo", which grows in West-Peru and Ecuador. These remaining biotypes have small, round capsules with free seeds, with small bracts

and short teeth; the leaves are broad and very pubescent. Botanically they are *G. vitifolium.* The so-called wild cottons of Peru, which have been reported especially around the locality of Tumbes, at the border between Peru and Ecuador, are rather similar to the var. *darwinii* of *G. vitifolium,* found earlier on the Galapagos Islands.

Such primitive cottons have been compared with the *Gossypium* biotypes which are grown by Peruvian Indios near Iquitos (Amazonas) at the same geographical latitude, but separated by the almost impassable High Andes. These Amazonian plants are distinct from those of the Peruvian coast. Probably they have another wild ancestor (*G. hirsutum* genome?). In the humid tropic environment the cotton fibres are in general not woven, but mainly used for strings, for adornment, or cultic purposes. We observed, for example, when staying with the Waika Indians, that they twisted the yarn of cotton for fishing nets and for hammocks and into thick rolls, worn around the women's hips by way of adornment.

In general, the different Indian tribes who inhabit the rainforests of the Amazonas, Rio Negro and Orinoko consider – quite rightly – that in such a hot and humid climate clothes are unnecessary. Women go com-

**Fig. VI.8.** Wild-growing *Gossypium hirsutum* on an uninhabited island in the Caribbean sea, east of Venezuela

pletely naked, but for fiestas wear the cotton rolls around the hips.

According to our observations, at the Orinoco the fibres of *G. vitifolium* predominate. But this is no proof that *G. hirsutum* was not present in former times. In general *G. vitifolium* is better adapted to a humid, hot environment.

With respect to ancient *G. hirsutum* the opinion exists that its main range was on the Eastern coast of South America, the Caribbean Islands and Central America. Remains have been collected from the caves of Coaxtalan and the Tehuacan valley.

Probably the seeds had been ground, eaten or used for cooking oil. We assume that *G. hirsutum* was always associated with man, planted by him and maintained as "dooryard cotton" (Fig. VI.8).

Indian tribes cultivated such "chinari" shrubs on a small scale in both Central- and South America, long before the arrival of Europeans, as was reported, for example, by Magellan in 1519 from the Brazilian coast. There he observed the primitive spinning of *Gossypium* fibres, which were also used as fill material for sleeping mattresses.

Even if we do not consider *G. hirsutum* as an originally true wild plant, it can be found occasionally growing spontaneously. For example, we observed it often on the dry, hot coastal regions of Venezuela and on uninhabited Caribbean islands.

In this context we mention an interesting and recent contribution about "wild cotton in Northeast Brazil" by Pickersgill et al. (1975). They raise again the old question whether *G. mustelinum* Miers and Watt, which had been collected in 1838 in the state of Ceara as a "wild species", and more than 100 years later was described as tetraploid *G. caicoense* by Aranha (1969), may be a real ancestor or an important link in the evolutionary history of the tetraploid cottons. *G. mustelinum* has an AADD genome. The authors cannot exclude for sure that this biotype is a feral cotton, which during the last centuries has acquired certain wild characteristics which enabled it to survive in free nature, until recently, when goats and cattle have been introduced in Brazilian caatinga. Finally the authors conclude: "that it is a genuinely wild cotton, specifically distinct morphologically and perhaps genetically from both *G. hirsutum* and *G. barbadense* . . . ."

## MORPHOLOGY

Annual shrubs, 70–140 cm tall with well-developed tap roots which reach 2–4 m deep.

The stems are monopodial, with sympodial branches which produce flowers. Leaves cordate with broad triangular lobes; laminae as broad as they are long (7–15 cm across) with multicellular hairs and extrafloral nectaries on the veins of the undersurface.

Flowering extends over 3–4 weeks and begins at the lowest sympodium and finishes with the highest and youngest buds. Flower corollas are showy, large (5 cm diam.), white-yellow on opening and turning pink and red on the third day. Stamens are united to a tubular column, with style inside the staminal tube. Calyx reduced with several ring-shaped nectaries at base. Epycalyx with three to four long bracteoles, which protect the flower buds and persist until the fruit ripens.

Fruits, commonly called "bolls", are ovoid (4–6 cm long) capsules pitted with oil glands. They split when ripe along the carpel edges and expose the linted seeds. The epidermis of the seeds has two different sorts of hairs: the commercially important long and convoluted white lint and the short "fuzz" which is firmly attached to the testa and not suitable for spinning. Good lint hairs are 2–4 cm long. About 22% of the kernel is crude fibre.

The seeds are valuable forage because they contain 18–25% oil and 20% protein. Their content of gossypol, a pigment with toxic effects in animal food, is detrimental.

### Production and Breeding

"King Cotton", as this marvelous fibre was called before the advent of synthetics, is still strong on the world textile market, even increasing in acreage and output (Table 12)

**Table 12.** Cotton: world production in mio metric tons during the last decade (FAO)

| 1983 | 1984 | 1985 | 1986 | 1987 | 1988 (estimated) |
|------|------|------|------|------|------------------|
| 14,570 | 19,140 | 17,310 | 15,370 | 16,640 | 18,000 |

The leading countries for cotton fibre production are the USA with 3.4 mio t, China with 3.2 mio t and the USSR with 2.8 mio t. The plantation area, according to FAO estimates is given in Table 13.

**Table 13.** World area planted with different *Gossypium* species in mio ha

| | |
|---|---|
| India | 8.0 |
| USA | 5.6 |
| China | 5.3 |
| USSR | 3.2 |
| Brazil | 2.1 |
| Pakistan | 2.1 |

Different research groups in the world are working on the improvement of *Gossypium* varieties with high fibre strength, long fibres with a low degree of attachment of the fibres to the seed and other characteristics of better technological value.

As donor species the wild-growing diploids *G. anomalum*, *G. thurberi*, *G. raimondii* *G. stocksii*, *G. areysianum* and *G. longicalyx* are used.

### Pests and Diseases

Cotton – unlike other major crops – suffers from an extremely large quantity of pests. We cite in the following only some of the most important, where some possibilities of genetic resistance breeding exist.

*Anthonomis grandis* (= Boll weevil)
This insect causes enormous damage to cotton fields. Its origin is Central America, where it has lived together with *Gossypium* for thousands of years, as has been proven by an archaeological find from Oaxaca in Mexico. But no natural resistance has been found in this gene pool. The only hereditary resistance is the mutant *"frego bract"*. Plants with this recessive factor are inhibited in the full development of the bracteoles. Consequently flower buds are not covered. This is an obstacle for the weevil, which normally lays its eggs under bracteole protection on the young flowers. This genetic resistance is, however, only of limited value. Repeated insecticide spraying and planting of early varieties, which escape the main damage, are the most efficient solutions in field practice.

*Pectinophora gossypiella* (Pink bollworm):
The pink bollworm belongs to the most destructive insects in the world. The moth lays its eggs in the flower buds and from there thousands of small larvae emerge, which eat the developing seed. No genetical resistance exists against this pest.

*Lacadodes pyralis* (also called pink bollworm):
Against this insect certain immunity exists in the wild species *G. thurberi* and the dooryard cotton *G. maria-galante*, because they lack nectar glands, on which the parasite depends.

*Jassids* bugs destroy the leaves and inflict serious damage on cotton plants. Against them genetic resistance exists in those wild and primitive forms that have long hairs on the leaves. Hybridization with extremely hairy cotton from Cambodia introduced good resistance to African *Gossypium* varieties.

*Lygus* bugs destroy the flowers of cotton, but not of *G. tomentosum*. Crossings have been made to transfer its natural resistance to commercial varieties.

## Diseases

*Fusarium oxysporum*, f. *vasinfectum* (Fusarium wilt)
This disease spreads to a dangerous extent during rainy weather, but fortunately genetic resistance has already been found in *G. anomalum* and *G. arboreum* and intercrossed in high-yielding cultivars. The results were, for example, the new commercial varieties Auburn and Stonewilt.

*Verticillium albo-atrum* (Verticillium wilt)
Most subject to attack are plantations in irrigated fields. A certain tolerance exists in Egyptian *G. vitifolium* strains, but a reliable genetic resistance in commercial varieties is still missing. The cvs. Acala 4–42 and Acala SJ-1 from California are considered to be wilt-tolerant.

*Xanthomonas malvacearum* (Blight)
This bacterial blight must be combatted in the early stage, by disinfecting the seeds with mercurial products. Certain *G. hirsutum* races possess tolerance, which – curiously enough – is combined genetically with yellow pollen colour (Verma 1986). From the old landrace Upland cv. Allen the new cvs. Albar 49 and Albar 51 have been selected with good field resistance. Purseglove (1968) indicated that in the African species *G. arboreum* strains exist with complete immunity against *Xanthomonas*.

## 7. *Hevea brasiliensis* (H.B.K.) Muell.-Aarg.

Para-rubber, caucho, hévéa, seringueira, Kautschuk

$2n = 36$

### NAME, HISTORY, ORIGIN

The curious English name "rubber" is derived from the early use of *Hevea* latex to "rub out" marks made on paper. This denomination persists in spite of the fact that the actual use covers a legion of other applications. The Indian name "caá-uchú" means "weeping tree", because it sheds abundant drops of latex when the bark is wounded. It is also a significant name in view of the tragic exploitation and sufferings of the native "seringeiros", who collected the latex under subhuman conditions in the Brazilian rainforests during the last century. Para-rubber is an elastomere ( = poly-isoprene $C_5-H_8$). Such polymerized isoprenes are present in hundreds of plant species around the world, but there is only one vegetal which is absolutely superior in its polyisopren production: the tropical tree *Hevea*. At present 99% of world production of natural elastomeres comes from *Hevea* trees.

Local Indians used the seeds of *Hevea* as an emergency food (after eliminating the toxic substances, by cooking) and also for lighting. They extracted the high oil content, especially in the species *H. kunthiana* and *H. rigidifolia*. The latex was used only to a lesser extent. For making bottles and balls the Indios preferred the latex of the *Castilla* tree (Ducke 1946).

The history of the initial, occasional employment of *Hevea* by the natives of the South American rainforests until the highly sophisticated rubber industry of the 20th century is long and dramatic (Ule 1905).

European explorers of the 18th century, Humboldt, La Condamine and Aublet, took samples of elastic articles worked by the natives, (like shoes, bottles or balls) with them back to Europe. However, they found no application, until the English chemist Priestley discovered that such a piece of Indian latex would rub out what he had written with a pencil on paper. Things changed fundamentally when in the year 1839 Goodyear in North America and independently Hancock in Great Britain discovered that a mixture of *Hevea* latex and sulphur – after strong heat treatment – resulted in a quite new and elastic substance. The so-called "vulcanization" opened nearly unlimited perspectives for the industrial applications of rubber (Polhamus 1962).

Consequently this increased the demand for natural cautshuc from Brazil in an unexpected way. The year 1827 showed an export of 36 t, but 30 years later it had jumped to 2600 t. In the year 1865 the demand for latex exceeded the natural supply. The result was an irrational exploitation and even definite destruction of wild-growing *Hevea* trees. This finally created a feeling of insecurity for the rubber industry, and far-sighted authorities proposed the establishment of *Hevea* plantations in British colonies in Asia.

The director of Kew Botanical Gardens, J. Hooker, stimulated by earlier discoveries of the British plant explorer in Brazil, Richard Spruce, in 1873 and 1875 organized several attempts to bring living *Hevea* seeds from the Amazonas to England. But this failed for lack of adequate shipping facilities.

A British planter, H. Wickham, living at Santarem, tried it again. He was favoured by the fortuitous circumstance that at the same time as he collected seeds, a modern ocean liner *S. Amazonas* bound for Manaos, and with no return cargo to England, was available. Wickham loaded the perishable *Hevea* seeds (70,000) on this steamer, which proceeded immediately to Belem, the harbour at the mouth of the Amazonas river. There official customs clearing was accomplished, declaring that the ship was carrying "exceedingly delicate botanical specimens". With the help of the local Brazilian officials, the *Hevea* cargo was dispatched and soon arrived, in the first week of June 1876, in Liverpool. From there the valuable freight was sent by special train to the Botanical Gardens at Kew, where the seeds were planted with the utmost care in the greenhouses. However, only 2800 seedlings survived. When grown, the young rubber plants were sent in specially designed climatized chambers to the Botanical Garden at Ceylon and different other localities in Southeast Asia.

We have taken pains to relate the history in some detail and to repeat that the rubber seeds were dispatched with the cooperation of the local Brazilian authorities. Unfortunately, a myth exists – which is repeated again and again in newspapers and films – that Wickham "smuggled" the seeds out of Brazil. For the sake of the truth, we feel it necessary to emphasize again what was stated earlier by Purseglove (1969) and Schultes (1970), that no Brazilian law was broken, because such a prohibition of seed export did not exist at that time.

The few *Hevea* plants which survived in Singapore began fruiting in the year 1881, but were inadequately treated. Things changed only several years later 1888 when the botanist H.N. Ridley was appointed scientific director of the Garden. Purseglove (1969) considers him as "the father of the Malaysian rubber industry". Ridley lived for 100 years (until 1956), and witnessed the rise of a New World crop and played a decisive part in *Hevea* domestication. The well-planned *Hevea* plantations in tropical Asia also had a positive effect in tropical America. They put an end to the destruction of the natural *Hevea* tree reserves and the inhuman exploitation – and even slavery – of the Indians in the hands of Latin American Rubber Barons at Manaos and Iquitos.

Obviously the discovery, collection and spreading of *H. brasiliensis* from the rainforests of Brazil to Malaysia has many dramatic facets, but similarly, the history of field research by botanists and explorers in the tropical forests of South America in search of new species is also fascinating. Beginning with the exploration of French Guyana in the 18th century by Aublet, followed by Humboldt and Bonpland, who collected cautshuc plants in Venezuela, Spruce, who for 15 years (1877–1893) explored the rainforests of South America and laid the basis for the taxonomy of *Hevea*, Martius (1817–1820), who covered nearly the whole of Brazil in his expeditions, Huber, who from 1895–1907 was a director at the Museo Goeldi in Belem, investigating lacticiferous plants of Amazonas, Ule, who arrived in Brazil in 1883 and dedicated decades to field research of Amazonas and Guyana with intensive rubber-tree exploration; and finally Ducke (1876–1959), the infatigable explorer of the flora of Amazonia over five decades, describing in the course of them 762 species new to science, among them several of the genus *Hevea*. Compared with our present travel facilities and easy help against parasites and diseases when we explore the humid tropics, the defiance and perseverance of these naturalists deserves our full admiration.

The genus *Hevea* includes many different species, but only one has entered the economic circuit for its outstanding latex production: *H. brasiliensis*.

From the dozen or more taxa which have been described by taxonomists, only two others yield a cautchuc of acceptable quality; *H. benthamiana* and *H. guianensis*. Others are important bearers of genetic resistance, whilst yet others contribute certain valuable physiological factors, even if they are poor in latex output.

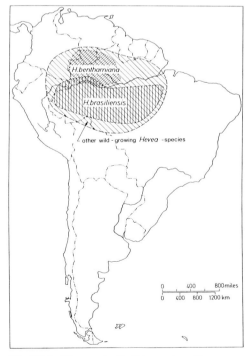

**Fig. VI.9.** The areal of wild-growing *Hevea* species in the South American rainforest

*Hevea* species grow wild in the rainforest of Brazil, Colombia, Guiana, Venezuela and Peru, with an especially high concentration of biotypes in the Upper Amazonas, Caquetá, Ucayali, Madre de Dios and Beni (Fig. VI.9). Especially in SE Peru enormous extended wild stands still exist. To discover single *Hevea* trees, one needs not only good eye sight, but also acustical help. Ripe fruits of *Hevea* open with a remarkable bang when they release their seeds. Experienced guides and indigenous seed collectors are able to differentiate local species by their distinct noises.

Some species have frondose growth and others are small shrub-like trees, like *H. camporum* Ducke, *H. pauciflora* Muell.-Aarg. and *H. rigidifolia* Muell.-Aarg., all natives to the Brazilian savannas and the semi-dry catinghas.

A quite different habitat suits *H. spruceana* (Benth.) Muell.-Aarg. and *H. benthamiana* Muell.-Aarg., which grow on swampy or sometime overflooded river banks. The first-named species grows abundantly in the lower Amazonas valley. Its typical enlarged trunk basis is often flooded, therefore this species may have developed a certain genetical immunity against root diseases. For this reason *H. spruceana* is a valuable rootstock for grafting of high-yielding clones of *H. brasiliensis*.

One very tall species is *H. guianensis*, found already in 1775 by the French naturalist Aublet and described as a "real giant" of the Guyanan rainforest, surpassing most of its neighbours in height (40 m). Also *H. nitida* Mart. ex Muell.-Aarg., native to the Upper Amazonas, Rio Solimoes and Caquetá, may reach extraordinary heights (Schultes 1953) (Fig. VI.10).

The best-known species, *H. brasiliensis*, has a very wide areal. It grows on the Upper Orinoco in Venezuela as well as on the lower course of the Amazonas River. The famous historical collection of Wickham[1] was gath-

---

[1] During my activities as Director of a Seed Project in Trinidad-Tobago (1970), I was lucky to come upon the last living tree on the Western hemisphere of Wickham's original collection. This is a magnificent tree in the Botanical Garden at Queen's Park in Port of Spain.

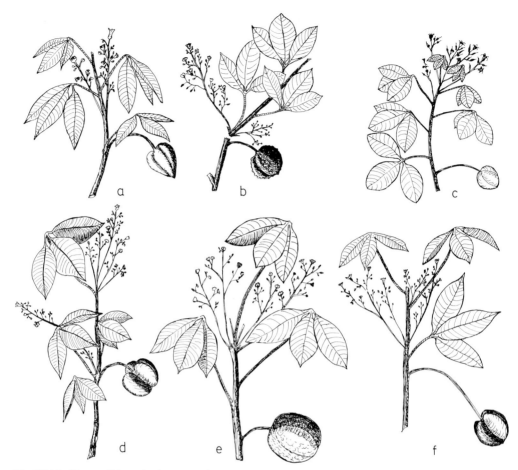

**Fig. VI.10.** *Hevea* wild species in comparison with the domesticated *H. brasiliensis*. **a** *H. microphylla*. **b** *H. guianensis*. **c** *H. camporum*. **d** *H. brasiliensis* **e** *H. spruceeana*. **f** *H. benthamiana*

ered between the rivers Madeira and Tapajoz, near Boim. It was not very homogenous, because most of the *Hevea* trees there are the product of hybridization and introgression (Baldwin 1947). Otherwise, the thousands of hectares with offspring of this initial collection could not have produced their well-known variability for further selections.

MORPHOLOGY

The different botanical species of *Hevea* have distinct aspects, but certain features in common. They are quick-growing trees with frondose canopies, and well-developed taproots with long lateral side roots (7–10 m long). During dry weather they shed their leaves.

The trunks are stout, 20–40 m high, and covered with a corky bark. The hard part of the bark is 1 cm thick, beneath it is the soft bark, which consists of sieve tubes and numerous latex vessels.

Leaves are spirally positioned, are trifoliate, glabrous, on long (10–20 cm) petioles. Young leaflets are purple, becoming dark green during hardening with 15 cm × 5 cm size.

Inflorescences are many-flowered panicles, unisexual, with male flowers appearing first and soon dropping. Female flowers sit at terminal ends of main and lateral branches. The entomophilous flowers are very small, without petals; calyx is bell-shaped with five triangular lobes. Their pollination occurs by small insects, midges and thrips. Fruits are large (3–5 cm diam.), three-lobed, with one seed per carpel. When ripe, the endocarp breaks open and releases the seeds explosively, throwing them often a distance of 10 m. Seeds have the typical spots (black, brown, grey) of a *Euphorbia* seed, similar to *Ricinus, Manihot* etc. Size 3 × 2 cm and weight 2–4 g. They have a very short germination time. Due to their high fat content, (40–50 % oil) Indians collect them for food, eliminating the toxic glucoside by cooking. Two things are noteworthy about *Hevea*: (1) It is the most recently selected major crop from the neotropics, which has achieved in only one century the highest status in world economy. (2) *Hevea*'s value resides in the bark and its subproducts, which is also an exceptional case in useful tropical plants, excluding *Cinchona* and some tannin-producing mangroves.

The commercial product of *Hevea* plantations is latex. This term refers basically to the physical state of the *Hevea* sap, which is an "emulsion" (i.e. a composition of two liquids which normally do not mix with one another). The substrate of the latex is the cell sap, which contains polyterpenes (= rubber globules) lutoids, proteins and starch grains floating in water (Roth 1981).

*Hevea* trees have articulated latex vessels, inserted in alternate, concentric rings around the cambial layer of the bark. They are situated externally to the cambium. Therefore, when the rubber tree is tapped by sharp incisions of a knife, the workers must take special care not to injure the cambium layer. Consequently, for the biological functions of a *Hevea* tree to be protected depends mainly on the ability of these operators. When the bark is cut, the latex flows abundantly, due to the interconnected system of secretory cells, the "latifers". The sap has high turgor pressure (10–14 atm), especially before sunrise. For this reason the tapping should begin early in the morning. Up to now, only little is known about the anatomy and physiology of the laticifer and secretory system.

Basically all genetic improvement of the latex emulsion secretion of a *Hevea* tree depends on this knowledge. The increase of polyterpene production, active metabolism and high photosynthetic activity of the leaf canopy depend on it.

IMPROVEMENT

The domestication and improvement of the rubber tree is one of the great pace-setting events in economic botany. In less than 100 years an absolute wild species of Amazonia, which even the Indians did not consider worth domesticating, was transformed into a high-yielding cultivar, dispersed and exploited for the benefit of mankind in the tropics of all continents. At the beginning of the *Hevea* domestication stands the Wickham collection. Its yield of latex was low: 225 kg/ha, at the end of the last century, some Asiatic plantations gave 300 kg/ha. By the year 1930 the average yield had been improved to 400 kg/ha, and some experi-

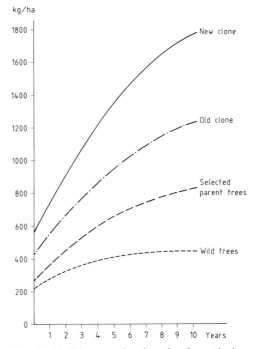

**Fig. VI.11.** Diagram showing the dramatic increase in yield of modern *Hevea* hybrids in East Asia (Ferwerda 1969)

mental plantations with selected clones reached 1000 kg/ha. In the meantime, new grafting and tapping systems (invented by Ridley, for example) raised the average yield to 1200–1600 kg/ha. In the year 1960 rubber yields in Malaya exceeded 2000 kg/ha, in 1970 the production of 1 ha reached 3000 kg. Since 1980 planting material is available with a potential output of 3500 kg/ ha.

The Malayan Rubber Research Institute, where at present the most advanced *Hevea* research and breeding work is being performed, predicts for the near future yields of even 5000 kg/ha, with the help of chemical stimulants for latex production (Fig. VI.11). Such a development of a comparatively young crop plant is unique in the world. We believe that no other one of man's crops has a breeding record equal to *Hevea*. At present it yields ten times more than only 100 years ago. Besides rubber there is a future large potential source of vegetable oil in *Hevea*. Its seedoil has a high content of linoleic acid (35%).

## DISEASES

*Microcyclus ulei* (South American leaf blight)

The paramount menace to *Hevea* monocultures is the fungus *Mycrocyclus* (formerly *Dothidella*) *ulei*. Its natural dispersal centre is in the Amazonas-Madeira region of tropical America, where most of the wild *Hevea* species live. It seems rather curious that this "leaf blight of rubber" causes little damage in the natural, wide-spaced, wild stands of *Hevea*, but becomes devasting in dense commercial monocultures of rubber trees. This was, for example, the biological reason why Ford in (1943, several decades ago), had to give up his cautshuc plantations in the Amazonas. Even today – with all modern phytopathological tools at hand – Brazilian enterprises have great difficulty in establishing prosperous rubber plantations, in the very homeland of *Hevea brasiliensis*!

It was incredibly fortunate for the British entrepreneurs in the last century that the Wickham collection did not transport, together with the *Hevea* seeds, also spores of the leaf blight to Asia. Today this devastating disease has still not appeared in Asia or Africa. Should it become disseminated there and prosper under the local ecological conditions, the world rubber industry would collapse. Therefore Imle (1978) warned: ... "there is no time to spare in developing... planting stocks with broadly based resistance to strains of the causal organism *Mycrocyclus ulei*."

From Brazil it has been reported that a breeding programme is under way to combine the high-yielding oriental clones with the blight resistance of indigenous selections. There are now several thousand blight-resistant progenies under field tests. In the meantime the application of protectant sprays with chemical products allows a certain control of *Mycrocyclus ulei*.

*Glomerella cingulata* ("die back") attacks especially trees growing on poor soils or in excessively dense stands. This disease is prevalent in the lower Amazonas valley. Good control results were obtained by applying N-P-K fertilizer to weak trees.

*Phytophthora palmivora ("pod rot, leaffall")*

Heavy leaf fall occurs during the three wettest months of the year and the damage is worst at the heights of the rainy season. Application of fungicides and planting of *Phythophthora*-tolerant top-budding clones may control this disease.

Genetical resistance to *Mycrocyclus* has been discovered in two wild species: *H. benthamiana* and *H. pauciflora*, which grow in the Peruvian-Bolivian border region of the rivers Madre de Dios and Madeira.

Rubber plantations in the eastern hemisphere suffer from their own diseases, like *Oidium heveae, Fomes lignosus* and *Corticium salmonicolor*.

## ECONOMY

Malaysia plants the world's largest area of *Hevea* with 5 mio hectars. World rubber production exceeds 6 mio t, including West Africa and East Asia. It can be considered as a positive sociological sign that nearly half of this is produced by smallholders.

Of the total rubber consumption, 70% is used by the automobile industry and its tyre requirements. Until the German discovery that polymerization of butadien results in high-quality synthetic rubber, *Hevea* cautshuc had a monopoly. After a certain period of preference for synthetic rubber, the rubber from *Hevea* latex has gained renewed importance. For example, the tyres of high speed aircraft must be produced from 100% natural rubber.

It is notable that whilst rubber plantations have increased in a spectacular manner in the tropics of the Old World, the original homeland of the genus *Hevea*, Brazil, did not succeed in establishing large plantations. Its production is only 1% of world output. Here is not the place to elucidate and judge this.

The once impressive exploitation of natural cautshuc from *Hevea* trees in the jungle of the Amazonas has become insignificant and even anti-economic. The natives had to install tapping circuits in the forests to cover 100 trees and to collect the latex daily in small cups put on the base of the rubber trees and treat it with coagulants. Stealing and even murder were not uncommon. Their living conditions were miserable, they were treated as slaves and exploited by rubber traders who made exorbitant profits. For all these reasons, the well-established Asiatic plantations can be considered as more humanitarian.

There are many other latex-producing plants in South America. We mention only the following in summary:

| Apocynaceae: | *Hancornia speciosa* |
| | *Landolphia hendelottii* |
| | *Funtumia elastica* |
| Asclepiadaceae: | *Cryptostegia* |
| Euphorbiaceaea: | *Cunuria* sp. |
| | *Joanessia* |
| | *Micranda* |

---

EUPHORBIACEAE

## 8. *Manihot glazovii* Muell.-Aarg.

Ceara-cautchuc, manicoba, Ceara-Kautschuk

$2n = 36$

This rubber tree of the *Manihot* genus is genetically so similar to the edible mandioka cassava, that the two species can be hybridized. Its advantages are quick growth (after 3 years the first yield of latex, which is of regular quality) and considerable drought resistance. For this reason German colonizers planted "Ceara Kautschuk" at the begin of this century in East Africa. In the course of war and expropriation, the 45,000 ha were lost.

Besides these major rubber-producing species, the neotropics offer a considerable quantity of other taxa with a minor latex content. Such species could achieve a certain, but restricted, importance in the case of severe damage to *Hevea* plantations (Ferwerda 1969).

The temporary scarcity of raw cautshuc during war times caused worldwide exploration for plants which contain elastomers. It was discovered that rubber-like substances are produced in hundreds of species in many botanical families.

---

SAPOTACEAE

In this family several species exist with an interesting content of "balata" (similar to guttapercha of Asiatic origin), which offers economic feasibility, especially in the genus *Manilkara*.

## 9. *Manilkara achras* (Mill.) Fosberg

Well known for its delicious fruits (called sapota, or sometimes also sapodilla) the tissues of this tree contain a milky latex. This is a mixture of 30–60% resins, 17% sugar and 20% coagulant. The trees are 20–25 m high.

To obtain latex balata the trunk is tapped in a similar way as the *Hevea* tree, but this can only be done every 2–3 years. The product is highly appreciated in the USA for the national custom of chewing gum ("chicle" chewing is an old Indian pastime; the word comes from the Aztecs).

### *Manilkara bidentata* A.DC.

Native to the Island of Trinidad, with trees of considerable height, which can be tapped three times a year. Its balata is a substitute for the authentic "chicle". The main use, after vulcanization, is for the manufacture of belts and transmission straps which were formerly produced from guttapercha.

---
COMPOSITAE
---

### 10. *Parthenium argentatum* A. Gray

Guayule

$2n = 18$

This slow-growing Compositae of the arid regions of Northern Mexico and Southern Texas is known there under its Indian name "Guayula".

Motivated by the shortage of tropical rubber during World War II, the Government of the USA launched guayule plantations in connection with its Emergency Rubber Project (ERP). Industrial tests have confirmed that guayule rubber has chemical and physical properties similar to *Hevea* rubber. The latex is produced in thin cell layers of roots and stems, mainly as a natural reaction to physiological stress. This factor should be kept in mind when *Parthenium* is planted under too favourable irrigation and soil conditions (McGinnies and Haase 1975). Polyploidy has no positive effect on resine or rubber porcentage according to Miller and Backhaus (1986), who investigated heteroploid stands of Guayule shrubs in northeastern Durango, Mexico.

MORPHOLOGY

Perennial shrubs of medium size, which can live more than 50 years. Leaves are small-elliptic, with dense grey pubescence on both sides of the lamina. Flowers are apomictic and the seeds of each individual represent genetically the mother plant. The elastomers are contained within the parenchyme plant,

but foremost in the stems and roots. Improved varieties may yield 20% rubber from dry weight.

Guayule was studied a century ago as a possible economic plant, but was abandoned because of its low yield (latex yield 480 kg/ha) and slow growth. Yields have been improved in the meantime by selection breeding. Technicians of the University of Arizona reported recently 860 kg/ha for test plots in California. Cultivation practice, fertilization and use of selective herbicides, which is obligatory in such a slow-growing crop, have also been improved. Direct seeding in the field is possible if the seeds are coated beforehand. Formerly it was necessary to rely on greenhouse production of cuttings. A certain stimulation for improving the guayule domestication is a recent contract between Firestone Tire and Rubber Company at Akton (Ohio) and the Indian Gila River Community at Phoenix (Texas) to establish on tribal land bigger plantations which produce industrial quantities of guayule rubber. In the Republic of Mexico plans exist for an extraction industry which should produce in the future 30,000 t annually of rubber from possibly wild-collected *Parthenium* plants in the Mexican Desert.

In the year 1979 the Carter Government put more than 30,000,000 at the disposal of guayule research and production. It seems doubtful if this huge sum gave adequate results, considering the hereditary obstacles to the lucrative exploitation of a plant which is still a long way from successful domestication (Nat. Acad. of Science 1977).

The chapter on industrially used plants of neotropical origin would be not complete if we omitted to mention briefly the waxes. Certain waxes cannot be synthesized on a commercial scale, therefore the industry depends on natural waxes, derived from tropical palms (*Copernicia* = carnauba wax, *Ceroxylon* = wax palm, *Syagrus*, licuri wax, carandayi wax) are important.

The Marantaceae *Calathea lutea* produces a protective wax for industrial purposes.

Other plants produce gums, resins and ethereal oils, for which there are no quality substitutes from other sources; but it would exceed our available space to describe them here.

# References

Aranha, C et al (1969) Uma nova especie para o genero *Gossypium*. Bragantia 28:273–290

Baldwin J (1947) *Hevea*, a first interpretation and cytogenetic survey of a controversial genus. J Hered 38:54–64

Beasley (1942) Meiotic chromosome behavior in species hybrids of *Gossypium*. Genetics 27:25–54

Demol J (1981) La seleccion cumulative. Application aux plantes autogames, exemple l'amelioration du cotonnier. Ann Gembloux 87:167–181

Dempsey J (1975) Fiber crops. University of Florida Press, Gainesville

Ducke A (1946) Novas contribucoes para o congecimiento das seringueiras da Amazonia brasileira. II. Bol Tec Inst Agron Norte 10:1–24

Endrizzi JE, Turcotte EL, Kohel RJ (1985) Genetics, cytology and evolution of *Gossypium*. Adv Genet 23:271–384

Ferwerda (1969) Rubber. In: Ferweda and Wit (eds) Outlines of perennial crop breeding in the tropics. Wageningen, Netherlands, p 511

Gentry HS (1982) Agaves of Continental North America. University of Arizona Press, Tucson, Ariz, p 670

Gerstel DU (1953) Chromosomal translocation in interspecific hybrids of the genus *Gossypium*. Evolution 7:234–244

Goldsworthy P, Fisher NM (1984) The physiology of tropical field crops, chapter Cotton. John Wiley, Chichester

Harland S (1970) Gene pools in the New World tetraploid cottons. In IBP handbook; Genetic resources in plants. pp 335–340

Kirby R (1963) Vegetable fibres. Leonard Hill, London

Imle E (1978) *Hevea* rubber – past and future. Econ Bot 32:264–277

Lanning E (1967) Peru before the Incas. Prentice Hall, Englewood Cliffs, New Jersey

Lock GW (1969) Sisal. Tropical Agricultural Series, London, 2nd edition

McGinnies WG, Haase EF (1975) Guayule, a rubber-producing shrub for arid regions. Arid Land Res Inf Paper nr 7. University of Arizona, Tucson, p 267

Miller JM, Backhaus R (1986) Rubber content in diploid guayule. Chromosomes rubber variation and implication for economic use. Econ Bot 40:366–374

Mueller-Aargau J (1874) Euphorbiaceae. Martius Flora Brasil 11:297–304

National Academy of Sciences (1977) Guayule, an alternative source of natural rubber. Natl Acad Sci, Washington

Philipps LL (1966) The cytology and phylogenetics of the diploid species of *Gossypium*. Am J Bot 53:328–335

Philipps LL (1974) Cotton (*Gossypium*) In: Handbook of genetics II. Plenum, New York

Pickersgill B, Barrett SH, Andrade-Lima D (1975) Wild cotton in Northeast Brazil. Biotropica 7:42–54

Polhamus L (1962) Rubber. Leonard Hill, London

Purseglove JW (1969) Tropical crops, Vol 3. Longmans, Green, London

Roth I (1981) Structural patterns of tropical barks. Bornträger, Berlin, p 600

Schultes RE (1953) Studies in the genus *Hevea* VII. Bot Mus Leaf Harv Univ 16:21–44

Schultes RE (1970) The history of taxonomic studies in *Hevea*. Bot Rev 36:197–276

Skovsted A (1937) Cytological studies in cotton. J Genet 34:97–134

Stephens SG, Mosley E (1973) Cotton remains from archeological sites in Central coastal Peru. Science 180:186–188

Ule E (1905) Kautschukgewinnung und Kautschukhandel am Amazonenstrome. Tropen-Pflanz, Beih 6:1–71

Verma JP (1986) Bacterial blight of cotton. CRC-Press Boca Raton, Florida

# VII. Aromatics, Narcotics, Stimulants, Spices

## General Remarks

This rather heterogeneous chapter includes native psycho-active plants and aromatics, as well as industrially highly developed stimulants and spices. We have refrained from discussing the many drugs and hallucinogens widely used by the natives of tropical America which have received outstanding treatment by Schultes and Hofmann (1980). In view of the urgently needed diversification of small-crop production in developing countries and their well-deserved aspiration to open new market opportunities, we recommend especially the spices, dyes and aromatics. They have the advantage of being long-storage and low-volume items, in addition to which most of them are work-intensive. The demand for spices on the world market is increasing, and the trade has continously expanded during the last decade. Producing dye plants and spices can act as an urgently needed stimulus for micro-economies of small-holder families in tropical regions (Purseglove et al. 1981). We begin with some examples.

## 1. *Bixa orellana* L.

Anatto tree, achiote, urucu, rocouyer, bija, onoto, uruku, achiote

2 n = 14

From the neotropical genus *Bixa* five species have been described, distributed in the rainforests of Brazil, Colombia, Ecuador and Peru. The binome was created in honour of the early explorador of Amazonia, the Spaniard Orellana (1541).
*B. orellana* is a cultivated shrub, which does not exist in the wild state, but several related species occur in the Amazonian basin, e.g. *B. excelsa* Gleason and Krukoff, a rather tall tree, which was considered by Ducke (1964) as the probable wild ancestor, further *B. platycarpa* Ruiz and Pavon, with nearly flat fruits without spines and *B. urucurana* Willd. from Ecuador, which is collected by Indians in the wild and used as body paint (Molau 1983).
*B. orellana* was an important trade object of the pre-Columbian inhabitants of tropical regions, who bartered it to the natives of cooler zones where onoto does not grow. According to Prance and Kallunki (1984), migrating Indians spread the use of onoto = achiote from South America through the Isthmus of Darién northward to Central America and Mexico (Fig. VII.1).
*Bixa* remains have been found in graves of ancient Peru. In Mexico it was used before the conquest as a medicinal plant, according to Hernandez (1651).

**Fig. VII.1.** *Bixa orellana*

## MORPHOLOGY

Small trees (3–5 m high), stems 10 cm diameter.
Leaves alternate ovate, simple (8–12 cm long, 8 cm broad), acuminate with smooth surface.
Inflorescence terminal panicles with showy flowers (4–6 cm diam.), pink or white, with numerous stamens and five petals.
Fruits ovoid capsules (3–5 cm long) red and densely covered with soft spines. Inside 15–20 g kernels which are covered with fleshy papillas and dark red soft arils from which the red colour of *Bixa* is extracted.
From a ripe plant 5 kg of red seed pulp may be harvested. This dissolves easily in hot water or in oil. *Bixa* is widely used by Indians as spice, as repellent against insects, and mainly as body paint in the tropical hot-moist climate, where clothing is unnecessary. Attractive body ornaments are designed with onoto.

## ECONOMIC IMPORTANCE

The red seed pulp was an important trade object for exportation to Europe in the 18th and even 19th century. In 1915, the port of Hamburg registered 40 t of onoto seeds. It seems that recently the *Bixa* products are in growing demand as a non-toxic and non-cancerogenous substance for colouring cheese, butter, lipsticks and ointments.

---
SOLANACEAE
---

## 2. *Capsicum* spp.

Chilli pepper, hot cayenne, Spanish pepper, poivre d'Inde, aji, uchu, paprika
2 n = 24 chromosomes

## NAME, ORIGIN AND HISTORY

These are only some few internationally used denominations for a related group of ancient Indian spices, but these are misinterpreted in several aspects. The English term "pepper" for this American condiment plant is wrong, because it is related to *Piper nigrum*, the "real pepper" from India. But there are more errors connected with *Capsicum*.

In the discovery of the New World by Columbus, the "peppers" played a notable role. His declared goal was to explore a direct sea route westward to the "pepper islands" and Cipango, where the expensive "black pepper" came from. When he landed on October 12, 1492 on a small island of the eastern Bahamas ("Samana Cay") which he called "San Salvador", Columbus was still under this illusion. When the Spanish fleet continued in the following month to Cuba and Haiti, they wrongly called it "West India" and were presented by the wrong "Indians" ironically with the wrong "pepper" (not *Piper* but *Capsicum*), which had red berries instead of black ones. When the natives of Haiti in January 1493, offered this "aji" to the Spaniards, they rejected it as "violentamente picante". However, as an intrepid merchant, Columbus realized immediately that this West Indian pepper belonged to one of the outstanding discoveries of his voyage. He took seeds of aji back to the Iberian Peninsula. From there plants of the "Spanish pepper" spread rapidly in the Mediterrean region, reaching the "true India" already in 1542. In the 16th century Arabian traders brought this new spice also to East Africa (Fig. VII.2).
The generic term *Capsicum* comes from the Latin "capsicus", meaning "of capsular form", and referring to the shape of the fruit.
The taxonomic situation of this genus was very uncertain until recently, because species names have been changed incessantly since the first botanical essays on this group, undertaken by Dierbach (1829) and Fingerhut (1832). Dunal (1852) established more than 50 distinct taxa, which proliferated until the end of the last century in nearly 100 botanical names. This confusion was reduced abruptly by Bailey (1923), who included all taxa in only *one* collective species. Terpo (1966), on the other hand, split the single species *Capsicum annuum* in 33 "concultas", creating numerous Latin names which were based on fruit form and size. His nomenclatorial system has to be rejected for the simple reason that these fruit characteristics had

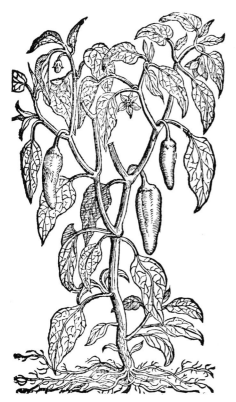

**Fig. VII.2.** *Capsicum annuum,* the oldest reproduction of a "pepper plant" in Europe (Krydboek 1581, Anvers) probably a drawing from Clusius

larger than those of the actual wild species. This suggests that the first steps in domestication and fruit size selection by primitive Indian planters were of very early date (Andrews 1984).

Different *Capsicum* species play a paramount role in the daily life of the Indios.

Most interesting: the more pungent a variety is, the more it is appreciated. One of the reasons is the strong anti-bactericide effect of the alkaloid capsicin. The fruits are very useful for their high vitamin content and for the preservation of meat and food in the tropics. Prepared "aji-sauces" are added as flavouring to nearly all meals, soups and maniok bread. Indios of the Altiplano take it, because it improves the blood circulation at great altitude. In the rainforests of Amazonas-Orinoco "aji" is fed to young Indios during the initiation rites as a curious proof of virility. The natives of Chiloé to Mexico use *Capsicum* as a stimulating condiment in great quantities and have a hundred different names for it. In all, *Capsicum* is the most important spice plant in the New World.

In his *Historia de las Indias* Padre Acosta (1590) describes very impressively the ubiquitous *Capsicum* species in the newly discovered territories: "... Hallase este pimienta de la Indias universalmente en todas ellas, en las Islas, en Nueva España, en el Peru y en todo lo demas descubierto, de modo como el maiz es el grano mas general para el pan, asi es el aji la especie mas comun para salsas y guisados ..."

Acosta mentions further that these "ajies" do not produce well in the cold climate of the Peruvian mountains, but only in the warmer valleys with irrigation. There he observed fruits of different colours: green, yellow and red. Some had a very pungent taste, others produced a special taste – like musk – in the mouth. Indians eat them green or red-dried and fine-ground. From this detailed description by Acosta we can deduce that *Capsicum* existed already in various domesticated selections, some used for condiment, others for food preservation, and also as vegetable.

Quite different from the European taste, which prefers a slight pungency, or even "sweet peppers", the American natives de-

been selected by man and not by nature. We question also Davenport (1970) and others, who proposed a monophyletic origin of all domesticates from a single ancestor: *C. frutescens*. Such a suggestion would be in contradiction to archaeological and morphological evidence (Pickersgill 1983; Eshbaugh et al. 1983).

Archaeological finds of *Capsicum* date back to 7000 B.P. in Central America (caves of Tamaulipas and Tehuacan in Mexico) and in South America from the graves of Huaca Prieta, dated at 2000 years B.P. In the early Chavin culture a stela has been found which depicts flowers and fruits of *Capsicum*. A considerable quantity of pre-Columbian painted pottery exists depicting fruits and leaves of *Capsicum* species. It is an interesting fact that in some early archaeological remains the "aji" fruits are considerably

veloped a near-addiction for the "hottest" *Capsicum* species and cultivars.

The biochemical reason why "peppers" are so highly appreciated and valued as a special flavouring by millions of inhabitants of this globe is the presence of the alkaloid capsaicin ($C_{18}H_{27}NO_3$), the decylenicacid derivate of vanillylamine. Nothwithstanding its extremely low presence of 0.1% in dried fruits, capsaicin is so pungent that the human tongue can detect it at molecular level. Of course this condiment was not developed in *Capsicum* species for the benefit of *Homo sapiens*. In free nature capsaicin gives the pepper plant excellent protection against predators, snails, insects and herbivorous animals. Strangely enough, certain tropical birds eat the pungent fruits and spread their seeds without negative effects to their intestinal tract.

Capsaicin is synthesized in the placentas of the ripening fruits and diffused from there to other tissues and the fruit pericarp. It has outstanding bactericidal properties. Since time immemorial the American natives have used wild *Capsicum* to preserve their fish and meat against putrefaction and later to season their food and beverages. Occasionally the "aji" were used also for ritual purposes, notably for virility tests. This objectionable custom of feeding the young men with food with high doses of pungent peppers still persists, as I have personally observed in the Upper Orinoco region, living together with nomadic Indios of the Makarikare tribe.

In addition to the capsaicin arousing the appetite and stimulating the salivary gland secretion, the pepper fruits are also a source of vitamins A and C. Recently the presence of vitamin P has been discovered, which regulates the blood pressure and circulation. The fruits contain also the red carotinoids (capsanthin, capsorubin, violaxanthin and alpha-carotin) and the antibioticum capsicidin. The vitamin C content in fresh chillies is high (300 mg to 100 g).

The pungency is very simply inherited and depends on one single dominant gene. The occasional appearance of "sweet-tasting" fruits is in fact a "loss mutant", which would not for long survive in free nature, because this alkaloid is a natural protection against premature extermination. From this point of view, the factor "sweet" is a defect in the metabolism of *Capsicum* plants to synthesize capsicin normally. Indians never selected "sweet varieties", because they had

**Fig. VII.3.** Different *Capsicum* species and cultivars in modern use (Photo Bayer)

no use for them. "Sweet paprikas" are modern creations according to European taste and survive only under the protection of man (Fig. VII.3).

Before considering the modern breeding aims in *Capsicum*, we have to ask for what purpose and in which direction the ancient Indians selected. The weight of yield could not have been the primary consideration, because one single "aji" plant produces sufficient pungent fruits for one family for a whole year. Even actual wild forms provide a satisfactory quantity, although their fruit size is rather small. We suppose that the primordial breeding aim during Indian domestication was to increase fruit size and pungency. The Indian motto was: "bigger and hotter". So they selected from the "piquinin" (pequeño) *C. annuum* larger cultivars like guajillo, with cylindrical and very piquant fruits, or the xatic corto with triangular fruits. Finally, they achieved the astonishingly large bell-shaped fruits in the species *C. chinense, C. frutescens* and *C. annuum,* as for example habanera roja or tabasco.

Recent researchers, like Eshbaugh, Heiser, Hunziker, Smith and Pickersgill-Heiser, have finally reached a bio-systematical agreement, which we take as basis for our description.

The genus *Capsicum* includes 30 wild-growing species, five of which have been domesticated by distinct Indian tribes in different and very remote places of the American continent. The best known wild-growing forms are the following:

*C. buforum* Hunz. 1969 in the vicinity of Sao Paulo, Brazil,
*C. campylopodium* Sendt. Brazil,
*C. cardenasii* Heis. and Smith, Dept. La Paz, Bolivia,
*C. chacoense* Hunz. 1950, Northwest Argentina, Paraguay,
*C. ciliatum* (HBK) Kuntze 1891, not pungent, Ecuador, Colombia,
*C. coccineum* Rusby,
*C. cornutum* Hieron., Brazil,
*C. dimorphum* Miers, Ecuador, Colombia,
*C. dusenii* Bitter,
*C. eximium* Hunz. 1950, Bolivia, South, Northwest Argentina,

*C. flexuosum* Paraguay, East Argentina,
*C. galapagensis* Heis. and Smith 1958, exists only on the Galapagos Islands
*C. hookerianum* Ecuador,
*C. mirabile* Mart. ex Sendt.
*C. parvifolium* Sendt., Brazil,
*C. praetermissum* Heis. and Smith 1958, Brazil, Belo Horizonte,
*C. schotteanum* Sendt., Argentina-Brazil. prov. Misiones to Porto Alegre.,
*C. scolnikianum* Hunz. 1958, Northwest Peru,
*C. tovari* Esb. & Smith, Central Peru, Highland,
*C. villosum* Sendt., Brazil.

Among these wild species, two have a special position: *C. ciliatum*, found in the last century, was originally placed in the genus *Witheringia*, but Hunziker included it in 1969 in the genus *Capsicum*. Its fruits are non-pungent. *C. chacoense* has a distinguished place in the system, because it is cytomorphologically distinct from the other taxa. McLeod et al. (1982) believe that a *C. chacoense*-like form was the ancestor for the white and the purple group of *Capsicum* and was an important phylogenetic forerunner in the ancient diversification area.

Some of the above-mentioned "wild peppers" are occasionally collected and used by natives in emergency or need as condiments. This refers especially to *C. chacoense, C. cardenasii, C. eximium.*

It should be mentioned that some lesser-known taxa, like *C. dusenii, C. ciliatum, C. schottianum* and *C. villosum* have pulp-filled berries and are not pungent.

With respect to the number of taxa and diversity of genomes, there is a certain concentration in the low-mountain region of South Brazil, with a dozen species. Eight have been described from Bolivia/Northwest Argentina, seven from Peru/Ecuador and four from Mexico.

This scheme differs substantially from publications of the Soviet school of agrobiologists, claiming a Central American (Mexican) gene center and place of origin of cultivated *Capsicum* species (see Hasenbush 1958).

Pickersgill (1983) concludes that ... "Evidence presently available therefore supports

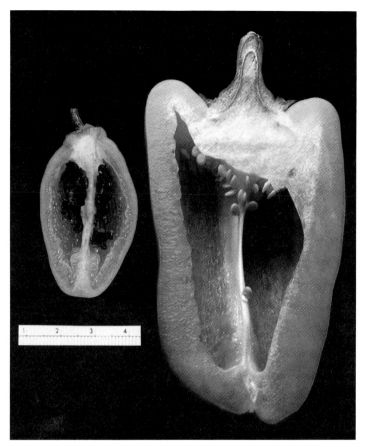

**Fig. VII.4.** Fruit sections of two *Capsicum* species: **a** a primitive, black-seeded *C. pubescens* from Bolivia; **b** a highly selected cultivar of *C. annuum* from the USA

suggestions of three independent domestications..." Matters were complicated by human influence, which spread the taxa from South to North and vice versa on the American continent. The Indians who migrated through the Isthmus of Panama contributed to this displacement of primitive cultivars. Pickersgill (1983) resumed the different opinions about origin, migration and domestication of man-used *Capsicum* in the following form, claiming independent domestications for:

*C. annuum*: origin in Mexico, still today in Nicaragua, Guatemala and Mexico the "common peppers", with the native names: aji, chilli, piquin, guajillo.
*C. chinense*: origin in Amazonas lowlands,

migration to West Indian islands and throughout the Isthmus of Panama, called panka.
*C. pubescens*: origin Bolivia-Peruvian Andes, migrated as far as Colombia, with the vernacular name: lokoto, rokoto.
*C. baccatum*: origin in Paraguay/East Bolivia, migrated to South Brazil, called ulupica.

These species have been selected and cultivated by different Indian tribes as far apart as Mexico, Lower Amazonas, Caribbean Islands, Ecuador-Peru, Bolivia and Paraguay, in a manner of polytopic domestication. There are wild forms with extremely small fruits: "forma aviculare". Of the cultivated pepper, *C. annuum* with innumerable commercial varieties at present distributed over the whole world, is of Central American origin. Seeds of this useful "aji" were dispersed very early, even by pre-agricultural Indians, who migrated southward through the Isthmus of Panama. It seems doubtful that some archaeological finds from Peru belong to *C. annuum* (Pickersgill 1969).

The cultivars of *C. chinense, C. frutescens* and *C. pendulum* are common in South America. *C. pubescens*, on the contrary, has a restricted origin and diffusion in the higher valleys of the Andes. Later it may have been introduced to the mountain regions of Costa Rica, Guatemala and Southern Mexico.

Some experts in Solanaceae are inclined to separate *Capsicum* into two phylogenetically distinct lineages. Their genetic and physiological differences inhibit crossings; further, they have different corollas.

A. Purple-flowered Lineage:

*C. pubescens* R. and P. of Bolivia, North Argentina, domesticated in the South American Andes
*C. cardenasii* Smith & Heiser, wild in Bolivian mountains,
*C. eximimium* Hunziker, wild in Bolivia,
*C. tovari* Esb. & Smith, Central Peru.

B. White-flowered Lineage:

*C. annuum* L., domesticated originally in Middle America, dispersed as wild as far as the USA
*C. baccatum* L., domesticated in Bolivia
*C. chacoense* Hunz., wild in the Chaco
*C. chinense* Jacq., domesticated in Amazonia
*C. galapagensis* Heis & Smith, domesticated or endemic to these islands
*C. frutescens* L., probably domesticated in Costa Rica and Nicaragua.

Cytogenetics does not help very much for the elucidation of *Capsicum* phylogeny and differentiation of the taxa. All have the same diploid chromosome number $(2n = 24)$ with similar karyotypes: eight pairs of chromosomes are large and metacentric, two pairs have medium size and finally there are two small pairs of chromosomes. In the

group of domesticated peppers slight differences exist. Cultivated *C. annuum* has two pairs of acro-centric chromosomes, i.e. in each chromosome one arm is shorter than the other. Genomes with this configuration are abundant in Central Mexico (Sinaloa), which may indicate, according to Pickersgill, that the domestication began there, based on the wild-growing *C. annuum* (forma "*minimum*"), which is common there. The South American peppers: *C. baccatum*, *C. chinensis*, *C. frutescens* and *C. pubescens* possess only one pair of acro-centric chromosomes.

Chemo-taxonomical research has also tried to differentiate *Capsicum* species. Ballard et al. (1970) directed special attention to the flavonoids. It has been demonstrated that the two purple-flowered Bolivian species *C. cardenasii* and *C. pubescens* are equal, but chemically distinct from the other Bolivian taxa. From the white-flowered group, *C. baccatum*, wild forms exist (= var. *baccatum*) and cultivars (= var.*pendulum*). They have identical flavonoids. This result underlines earlier morphological conclusions that both biotypes may be considered as a single species which developed from the same progenitor.

The use of electrophoresis and isozyme research performed by McLeod et al. (1979–1983) has allowed separating two units within the "white-flowering group". One is the *Capsicum annuum* complex (wild and domesticated) which has identical behaviour, but was quite different in electrophoretic assays from the *C. baccatum* complex (wild and domesticated).

The experiments with the North Argentinian *C. chacoense* Hunz were explanatory. By its different enzyme behaviour, this wild species does not fit easily into any group. It is considered as phylogenetically distinct from other *Capsicum* and may be the primordial ancestor of the two major groups. It is, moreover, predominantly autogametic, which results in low polymorphism and low heterozygoty.

BREEDING AND GENETICAL IMPROVEMENT

Cultivated *Capsicum* species have $2n = 24$ chromosomes. It is astonishing that no casual polyploidy has ever been observed and that artificially provoked genome duplications have never yielded practical results in commercial varieties. This in view of the fact that Indians have long been able to make use of the "gigas effect" in *Capsicum*. The dramatic increase of fruit size in *C. annuum* amounts to $1:100\times$, from the small "bird-pepper" to the local varieties in Central America. This has been achieved by factor combination alone.

Recently it was shown that hybrid vigour can be produced in certain species crossings, e.g. between *C. annuum, C. pubescens* and *C. baccatum* ( = var. *pendulum*). Hybrids *C. pendulum* × *C. annuum* have 50% hybrid vigour in plant height, but small fruit size was dominant.

Whilst the combination *C. pubescens* × *C. pendulum* showed growth depression, the reciprocal crossing resulted in 47% more vigorous plants.

Genic and cytoplasmatic sources of male sterility have been detected by Shiffriss and Frankel (1971) and are now being exploited in paprika breeding.

DISEASE RESISTANCE BREEDING

As a result of the huge monocultures of *Capsicum* in North America and Europe, each year the diseases increase in a spectacular manner.

There are fungus, bacteria and especially several viruses, which decimate the yields. Insects preferentially attack the "sweet peppers", whilst the pungent varieties are protected by their capsaicin alkaloid.

Many primitive and wild *Capsicum* strains have contributed worldwide to the genetic improvement of modern pepper varieties. Special conservation centres have been established in Europe (Gatersleben, Leningrad) and North America (California, Davis) to save the gene pools of Indian peppers for the benefit of mankind.

Among the wild *Capsicum* several donors for resistance genes exist. *C. frutescens* and *C. chinense* possess genes for immunity against *Verticillium* wilt. *C. baccatum* does not suffer from damage by *Phytophthora capsici*. Resistance to *Fusarium* occurs in certain primitive Mexican landraces.

Against different bacterial diseases (e.g. bacterial leaf spot) resistance exists in *C. chacoense*. Grubben (1977) reported tolerance in different wild forms for *Pseudomonas* and *Xanthomonas*.

With respect to virus infections, which are very common in Solanaceae, there is resistance in *C. baccatum* against virus Y and the cucumber mosaic virus.

YIELD AND WORLD PRODUCTION

According to the Production Yearbook of the FAO, the *Capsicum* acreage arose between 1970 to 1980 from 685,000 ha to 957,000 ha. During the same period, world production (fresh material) increased from 50,270 t to 70,550 t.

Sweet paprika and the milder varieties account for around two-thirds and the pungent "chillies" for one-third of the total trade in this category.

In the meantime, the production and trade of *Capsicum* pepper has exceeded that of black pepper (*Piper nigrum*) of Asiatic origin and gained the position of the most appreciated condiment in the world.

As often occurs in the history of leading world crops, *Capsicum* has also trespassed widely beyond the natural limits of its neotropical origin. Asia and especially the Indian subcontinent are today the main producers of "chillies". The annual export from India and Thailand together surpasses 200,000 cwt. Europe also holds a leading position in paprika production, especially in the Balkan countries. In Bulgaria, for example, more than one-third of the arable land is dedicated to *Capsicum* cultivation, second only after tomato production.

The United States of North America – by far the largest buyer for both paprika and chillies – imported an average of 9000 t annually from 1976–1980. In the same period the EC countries imported annually around 20,000 t.

RUBIACEAE

## 4. *Cinchona* spp.

### 3. *Cephaelis ipecacuana* (Brot.) Rich

Ipacac, golden root, raicilla, Brechwurz

$2 n = 22$

This efficient medicinal plant covers a wide area in various tropical countries of South America (Eastern slopes of Bolivia to Colombia, the humid regions of Brazil and crossing the Dairén gap to Panama).

It contains, mainly in its roots, several alkaloids; the most important are cephaeline and emetine, which proved to be an effective drug against amoebe dysentery. Natives dig out and collect the roots, which are sold dried. Put together on threads, they are simply called "el raiz". Several falsifications of the real root exist, obtained from certain Euphorbiaceae and Violaceae.

*Cephaelis* plants are low herbs, with branched yellow roots. The leaves are elliptic, 6–10 cm long with a pale green underside. The flowers are white, tubular with five lobes (6 mm long). The fruits are spherical (7 mm diam.) with bluish skin.

Decades ago, Brazil alone exported 50–70 t of dried roots, but the growing importance of antibiotics has exerted a negative influence on the commerce with ipecacuana. Only recently one notices in the modern pharmacoepia a tendency in favour of the natural remedy, as against strong antibiotics.

During my stay in the rainforest of Panama, I learned that Indians use raicilla against dysentery and intestinal pains and as an expectorant.

The same may be said of *Rauvolfia tetraphylla,* a neotropical representative of this pantropical genus of 100 species. Indians of the rainforests know its tranquillizing effects. The plants contain an extraordinary quantity of alkaloids, e.g. some with strong hypotensive effect, like reserpine.

Finally, we cannot omit another Rubiaceae, once world-famous as a medical herb and stimulant of South American origin:

It is an entirely neotropical genus and has recently acquired renewed industrial interest, as a source of quinine. Acclaimed for hundreds of years for its healing effects against malaria and at present widely used as tonic and stimulant in beverages.

### *C. officinalis* L.

Yellow bark, cinchona, quinquina, cascarilla, Chinarindenbaum

$2 n = 34$

NAME AND HISTORY

The genus is represented by many species and natural hybrids in the montane rainforest of the Andes in South and Central America, growing between 500 and 3000 m altitude. We mention only the following taxa, which were formerly exploited for their quinine content: *C. cordifolia* Mut. *C. ledgeriana* Moench, *C. succiruba* Pav. *C. calysaya* L. and *C. pubescens* Vahl. The nomenclature is, however, still confused and complicated by the existence of natural crosses. Cardenas, a Bolivian botanist who explored the *Cinchona* species of Bolivia and Peru, mentioned in his book (1969) more than a dozen taxa for this region (Fig. VII.5).

The generic name is related to the family name of the Viceroy of Peru (1629) Conde de Chinchon, whose wife fell ill with malaria, and was saved from the disease by Indian "curanderos", who knew about the antifever qualities of *Cinchona* bark. The successful treatment of the illness of Cinchon's wife with an Indian remedy caused much publicity at the time, and Linné used the name of Cinchon for the botanical classification of the marvelous "fever tree". Some historians later expressed their doubts and declared the whole thing to be a religious legend; but Linné had already given the name to this "fever tree" of the Indians. It is true that the bark of several *Cinchona* trees contains a potent drug against feverish diseases, among them malaria. Modern pharmaceutical analysis has determined 30 different alkaloids in *Cinchona* species. The

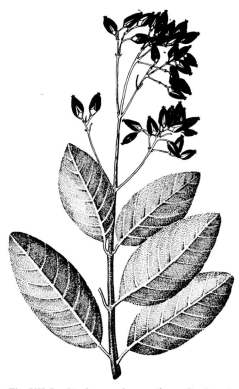

**Fig. VII.5.** *Cinchona calysaya* (from Cardenas)

most important is quinine, isolated already in pure state in the year 1820, with the formula $C_{12}H_{22}N_2O_2$. It acts as a protoplasm poison in the cells of *Plasmodium*.

MORPHOLOGY

Evergreen shrubs, or small trees with rough bark. Leaves simple, oblong-elliptic, with smooth surface.

Inflorescences in terminal panicles with small, fragrant flowers. These may have yellow or pink corollas, depending on the local variety. The flowers must be cross-pollinated to produce seeds. The resulting capsules are 1–2 cm long.

The bushes begin production after 4 years. The highest concentration of quinine is to be found in the bark of the roots and tree branches, but only after 8 years' growth does the bark achieve its maximum quinine content (Morton 1982).

For harvesting it is unfortunately necessary to uproot the *Cinchona* tree or bush. During the last century this custom was so destructive for the natural *Cinchona* stands in South America, that the colonial governments of that time decided to introduce the drug plant in their Asiatic colonies. Plantations began in Java in the year 1854, in India and Ceylon in 1861 and finally in Tanganyika in 1900. This policy saved the neotropical wild populations, as at this time the Indians kept the last locations of wild *Cinchona* secret. To appreciate the high degree of exploitation, it is sufficient to mention that at the end of the last century Colombia alone exported 3,000,000,000 kg wild *Cinchona* bark. Extermination of the species was at hand. Perhaps the salvation of the species was the discovery of synthetic anti-malaria drugs by the German pharmaceutical industry (e.g. atebrine and plasmochine).

This is, however, not the end of the *Cinchona* history. In view of the alarming "acquired resistance" of the malaria microbes to synthetic drugs, the natural *Cinchona* alkaloids have recovered importance. In 1938, world production of *Cinchona* bark rose to 14,000,000,000 kg, with 85% of the drug produced in plantations in Indonesia, where by genetic selections the alkaloid content has been increased threefold. It is claimed that *C. ledgeriana* cultivars contain 6–7% quinine.

It seems that the increasing demand of the pharmaceutical and beverage industries is at present higher than the natural production. For this reason genetical and phytochemical research should receive more support, with the aim of improving certain semi-domesticated *Cinchona* species.

RHAMNACEAE

## 5. *Colubrina reclinata* (Herit.) Brongn.

Snake bark, maubi, mambee

Small tree (3–5 m tall) with thin stem (6–10 cm diam.), covered with glossy brown bark, which can be easily detached. The bark contains a bitter aromatic (alkaloid?)

substance, which is the basis for a very popular stimulating beverage on the Caribbean Islands.

Native medicine successfully uses decoctions of *Colubrina* bark against stomach disorders. An exhaustive investigation of this neotropical species is recommended (Fig. VII.6).

**Fig. VII.7.** *Dipterix odorata*. Dried tonka beans

**Fig. VII.6.** *Colubrina reclinata*

---

LEGUMINOSAE

---

### 6. *Dipterix odorata* (Aubl.) Walld.

Tonka tree, cumaru, Tonkabohne

The genus *Dipteryx* is represented by eight species in tropical America. Most of them are beautiful solitaires, large (40 m high) evergreen trees with voluminous (100 cm diam.) trunks and valuable hard timber. Natives use the fruits and seeds, which contain a considerable quantity of coumarin, as spices. The species has 3% coumarin and 25% solid fat (Pound 1938).

Indigenous to South America, this tree grows abundantly along the banks of the great rivers, like the Orinoco, Rio Negro and Putumayo.

The compact trees are covered with a smooth greyish bark. The leaves are pinnate

(15 cm long) with three to six leaflets on a winged flattened rachis. The inflorescences are many-flowered and inserted in terminal panicles. The flowers are showy, 15 mm long, violet-coloured. The fruits are large (6–10 cm), indehiscent and form only one seed, surrounded by a pulpy mesocarp. The commercial product is the cured beans (Fig. VII.7). An average yield of 1 kg dried seeds is common, but with selected prolific tonka trees one can harvest ten times more. The product is used in perfumery and tobacco flavouring. Most tonka beans come from wild stands in Venezuela and Brazil. During my activities on the Island of Trinidad, I observed vast plantations of *Dipterix odorata*, now abandoned because of the low market price, resulting from the competition of synthetic coumarin.

---

ERYTHROXILACEAE

---

### 7. *Erythroxylon* spp.

Coca

$2n = 24$

This pantropical genus is represented in the New World with more than 200 taxa. Its leaves are used for chewing or in folk medicine, not only by natives of South America, but also to a lesser degree in Africa and Asia. Only on the American continent, how-

**Fig. VII.8.** A clandestine *Coca* plantation in Peru

ever, has the consumption of coca leaves as stimulant been developed in such a widespread and dangerous habit.

It is a depressing fact that *E. coca* climbed during the last decade to position No. 1 in world turnover among all "Useful Plants of Neotropical Origin". With an estimated 10,000,000,000 $ annual profit (80–90% for Colombia, 10% for Peru-Bolivia), *Coca* products now surpass the income of all American crops taken together, e.g. maize, mandioca, potato, cacao, tobacco, coffee, palm oil, banana, citrus fruits, peanuts, sunflower, etc. And what is worse: The world dealers of cocain have raised a former humble Indio herb to be the symbol of illicit plant commerce, corruption and criminality on American territory.

Unfortunately "Coca ipadu" possesses a wide ecological adaptability and grows well even in the hot plains, where *E. coca* had never before been cultivated. Covered by the jungle of tropical Ecuador, Colombia and Brazil, during the last years big secret plantations of "coca ipadu" have arisen. Even if its alkaloid content is only the half of the mountain coca, this drawback is compensated by the easy clandestineness of vast illegal cultivation (Fig. VII.8).

In the neotropics the following species and cultivars are represented.

### *E. coca* Lamarck

It extends from Northern Argentina, Bolivia, Peru towards Ecuador in several culti-

vated varieties, but is unknown in the wild. It must have been domesticated in very early times by Indian tribes, because by the time the first Europeans arrived, it existed only in cultivation.

According to Cardenas (1969), *E. coca* has taxonomic affinities with certain wild species of the eastern slopes (the Yungas) of the Bolivian Andes, like *E. anguifugum* Mart., or *E. Bangii* Rusby.

The var. *ipadu* Plowman (originally a Guarani-Tupi word) differs in several morphological aspects, has a consistently lower cocaine concentration than *E. coca*; it is mostly cultivated in the western valleys of the Amazonas region, where it is employed as a narcotic by Brazilian Indians (Plowman 1982).

### *E. novogratense* (Morris) Hieronymus

This cultivar is commercially known as Colombian coca. It grows well in lower, drier and warmer regions, mainly in Venezuela and Colombia. As it is more tolerant to different climatic factors than the first-named species, it was planted with irrigation in pre-Columbian times along the xerophytic coast of Peru, as archaeological finds indicate (Towle 1961). Plowman (1984) transcribed the taxon *E. truxillensis* Rusby into a variety which is cultivated in the dry western slopes of the Peruvian Andes, near Trujillo (up to 2800 m alt.). It was probably this variety which was slowly spread by migrating Indian tribes through the Colombian lowlands northwestward, crossing the Isthmus of Panama and reaching the region which is now Costa Rica, where scant archaeological evidence of the Cocle culture exists. There is no proof, however, that the coca plant reached Mexico. We believe that a narcotic plant widely adaptable to diverse climates and environments may have been enjoyed even in praehistoric times by Central American Indians.

During my recent stay in Panama – living together for a while with Choco and Kuna tribes and researching in the border mountains of Darién, where wild *Erythroxylon* exists – I found no evidence of coca chewing in Panama; I did, however, find an ancient text, in which Columbus himself reported, when he landed on the Atlantic coast of Panama, in the present-day province of Veragua the following: "Mientras estaban aqui, el cacique y sus principales no cesaban de meterse en la boca una hierba seca. De mascarla a veces tambien tomaban cierto polvo, que llevaban junto con la hierba seca, locual parecia cosa fea..."

This is undoubtedly an appropriate description of the Indian way of enjoying cocain. Later, Oviedo (1519) reported similar customs from the Pacific coast of Nicaragua. Finally, the Spanish conquerers severely suppressed the custom of chewing coca leaves and eradicated the *Erythroxylon* plantations in Central America. Of course, they could not destroy the wild-growing species in the rainforests of Panama and Costa Rica, described by botanists as *E. multiflorum* Lund and *E. panamensis* Turcz.

The first-named is a 5–6-m-tall tree, with elliptic oblong leaves, white flowers, arranged in dense clusters, which produce fleshy fruits (1 cm long) with a persisting calyx.

*E. panamensis* grows in the submontane rainforest of Darién. According to our recent inquiries, the Choco Embara Indios living there do not chew the leaves, but know the bush, which seemingly has low alkaloid content. The bush is 3 m high and has oblong, leathery leaves. The flowers have greenish-white petals (4 mm long) and produce red ellipsoid fruits, (10 mm long) with fleshy mesocarp and white seeds.

## MORPHOLOGY

The two cultivars of *Erythroxylon* are similar in their habit. They are medium-size shrubs (1–3 m high), with ramificated branches.

The leaves are long ovate-elliptic (4–10 cm long and 2–5 cm broad), simple, acuminate with a pronounced nervature and always with stipulae.

The spongy parenchyme contains several alkaloids, not only cocaine, but also hygrina, nicotine and several astringent substances. *E. novogratense* possesses also cynnamolcocaine and some aromatic oils (methylsali-

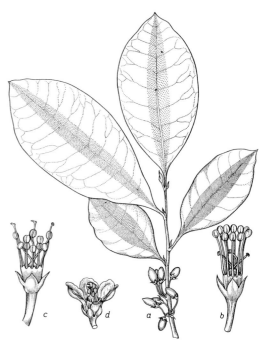

**Fig. VII.9.** *Erythroxylon coca* flower (Mansfeld 1986)

cylate), which explains the use of this species for flavouring beverages.

Flowers have an actinomorph corolla, with five short white petals and ten stamina; very often heterostyle. The calyx is conical (Fig. VII.9).

Fruits small (6–10 mm long) red berries of ellipsoidal form, containing 1–3 small seeds.

The two cultivars do not hybridize.

The chemistry of *Erythroxylon* was investigated more than 100 years ago by the German chemist Gaedecke, who in 1855 isolated "cocain" from the bark, giving it the formula: $C_{17}H_{21}NO_4$. He discovered that the drug has an immediate effect on the central nervous system and acts for some people as a sedative and for others as an excitant. In human medicine it is used as an efficient analgaesic. Coca leaves contain approximately 1% cocain. Samples from Huanuco in Peru taken from plants produced in higher elevations (1600–1800 m alt.) had stronger effects, which may be explained by the presence of aromatic oils. Natives use it as a tonic for nerves and stomach, or as an alcoholic elixir. Historically it is recorded that Pope Leo XIII regularly took coca extracts in his table wine. In any case, the use of coca subproducts is habit-forming.

## Production and Agricultural Methods

*Erythroxylon* is mostly cultivated on terraces, on the eastern slopes of the Andes, at 300–1000 m altitude. In the "Yungas" east of the Bolivian Capital La Paz, coca plantations were long the most lucrative crop industry, until the Government imposed severe restrictions on pressure from the USA. According to my observations in Bolivia and Peru, these plantations start from seeds, sown in nurseries and transplanted afterwards, or sown directly in small holes half a meter apart. For identical reproduction of high-yielding bushes, the planters prepare cuttings with root stimulants. The harvest begins after 3 years, and consists of repeated hand-picking of the leaves, often four times per year. To overcome the physiological damage and continuous loss of leaves, coca planters apply different fertilizers to maintain yearly yields of 1500 kg of dried leaves. The harvested leaves are bundled and packed tightly together by wooden hand presses and the packages covered with banana leaves for further transport.

The coca shrubs remain productive for several decades, but in the highly eroded soils of the Yungas the yield may go down to 250 kg/ha. The natives consider 800 kg/ha a good average yield, especially in clandestine plantations.

Production in Asiatic countries, where *Erythroxylon* was introduced some decades ago, is considerably higher. In Java, for example, biotypes with 2% cocain content exist.

Mechanical eradication of coca plantations has proved extremely difficult, when planted in inaccessible jungle. Plant pathologists have therefore proposed a certain form of biological warface. *Erythroxylon* is attacked by a new severe disease, called "estella", most probably a virus transmitted by sucking insects. This disease causes a deformation of the leaves, which take a filiform shape. The leaves, of course, lose their economic value, and the coca bushes finally wither away.

## History, Use and Abuse

Chewing leaves is an age-old custom of American Indians. It was not restricted to *Erythroxylon* alone, but included also for example *Nicotiana* or *Datura*. Proof of this is not only occasional remains of coca leaves in grave sites (= iscupurus) (from Ancon 1800 B.P., and Nazca 2000 B.P.), but still more indicative are the small ceramic containers (= coqueros) which had been used in the chewing ritual. These are receptacles for alkaline substances, lime, or ashes from *Chenopodium*, *Cecropia*, or many other plants. By combining these "Llipta" ashes with the alkaloid of the leaves, the stimulating effect is considerably increased. The archaeological evidence of coqueros and iscupurus dates back to 2500 B.P.; some show primitive faces with puffed-up cheeks, typical for coca-chewing persons. They have been found in Peru, Ecuador, and are abundant in Colombia and also in Central America. We take this as a certain indication that thousands of years ago Indians used coca widely, and even liked to depict this custom. In Colombia, in the Cauca valley, beautifully worked golden figurines related to coca use have been excavated. Very far from there – on the Santa Elena Peninsula of the Pacific coast of Ecuador – small lime containers and coqueros have been found dated back 2100 years to the Valdivian culture. From Peru many figurines are recorded with the typical cheek bulge of a coca chewer. We should mention here that this custom is highly repugnant, as a green, slimy saliva drops from the mouth of coca-addicted chewers.

During the Inka reign, the use of coca as a stimulant was reportedly a rare prerogative of the aristocratic classes. The later colonial system – interested mainly in exploiting gold and silver mines – abolished the privilege and distributed coca leaves to the mine workers, with the intention of dominating and tranquilizing the Indians and suppressing their hunger and thirst. Without doubt, chewing of coca leaves helps to overcome physical fatigue. In their Anden possessions, the Jesuits stimulated the use of coca among their Christianized Indios. Obviously, the consequent health damage, low intake of food and early death did not much disturb the ruling Spanish castes (Martin 1970).

To understand this complicated coca problem of the primitive Andine populations, we refer to Martin Cardenas, a Bolivian botanist, who in 1969 wrote in his book on *Plantas Economicas de Bolivia* the following:

"Only under the influence of coca can the unhappy inhabitants of the Andes support their existence, marked by the hopelessness of dominating a hostile nature... Up there in the bare solitary lands of the Cordillera you may encounter taciturn Indians, following their llamas, chewing coca leaves, with their mind submitted in the acceptance of a sad nature, mostly frozen, offering little living space".

Here the coca may stimulate the respiration, reduce and even paralyze the bodily feelings of fatigue, hunger and sickness. Under this influence, the natives of the Altiplano can walk more than 40 km daily, carry out hard agricultural labour in altitudes which surpass 3600 m, supported by the stimulating action of coca leaves, the only medium which allows them to survive the painful high mountain disease 'sorocoche'.

Of course, it is not my intention to make an apology for coca leaf chewing by American natives; I have also tried it when climbing high mountains, together with them. But I do not believe that educated and hygiene-prone persons could find a lasting pleasure in masticating coca leaves, with green saliva continuously dropping from their mouth.

Besides the danger of intoxicating the human body with alkaloids, it must be added that coca leaves contain a considerable fraction of vitamins (especially A) and minerals, a factor that should not be underestimated for inhabitants of inhospitable regions, nearly bare of vegetables. For millions of poor Indios, mestizos and caboclos in Bolivia, Brazil, Colombia, Ecuador and Peru, the coca plant is not only a mild stimulant but also an ancient tradition and house medicine for the lower classes. I consider it very difficult, if not even impossible, to eradicate this custom, by threats and punishment, as some military governments have tried recently. The authorities of these governments must find an attractive replacement for planting and chewing coca; not, of course,

another habit-forming plant like *Nicotiana* or smoking tobacco leaves.

To obtain 1 kg of cocaine, nearly 45 kg of dry leaves must be extracted. In spite of all well-intentioned efforts of governments and health organizations to reduce production in the main *Erythroxylon*-producing countries Bolivia and Peru, until now the results have been rather disappointing. Whilst the annual yield in Bolivia was 36,000 t in the year 1978, it increased steadily until 1984, when it reached 171,000 t. One should bear in mind that local consumption (mostly "chewed" by illiterate natives) in Bolivia is "only" 15,000 t. Ten times more goes into the international narcotics trade (Rivier 1981).

Unfortunately the decade-long advertisement of several brands of an artificial beverage, labelled with the name "coca", has created in the subconscious of primitive-minded humans the association with "something good". Even if the factories insist now that the former cocain aggregate has been exchanged for other stimulants, the damage persists. North America suffers consequently most from "coca" abuse. According to newspaper records, the USA alone has more than 5,000,000 cocain addicts (mostly of African descent). They spend yearly 9,000,000,000 . The dramatic increase of cocain consumption by the youth of the so-called affluent western society is not only a deadly menace to their health, but also causes a serious imbalance for the western powers in their world strategy.

---

AQUIFOLIACEAE

## 8. *Ilex paraguariensis* St. Hil.

Paraguay tea, Yerba mate, caá, Mate Tee
$2n = 40$

The genus *Ilex* is represented on all continents (and even in Oceania) with more than 500 species, which reach from 65° northern latitude to 35° south. Subgenus *Eu-Ilex* is the most extended, and includes also several neotropical species growing in Colombia, Ecuador, Brazil, Uruguay, Paraguay and

**Fig. VII.10.** *Ilex paraguariensis* badly damaged "mate" trees after harvest in Paraguay

North Argentina (Fig. VII.10). Some have a high content of caffein and are therefore used by different Indian tribes, but the most important is the meridional species *Ilex paraguariensis* (Giberti 1979). Further have been described:

### *I. argentina* Lillo

This species is botanically very similar to the above-mentioned taxon. It is not cultivated for its leaves, but is exploited for its timber in the Argentinian north provinces and in South Bolivia. It grows in the lower mountain regions between 900 and 1800 m alt.

### *I. dumosa* Reisseck

It is a small tree or shrub (1–3 m tall) with pubescent branches and grows preferentially in moist soils of Minas Gerais (Brazil) and East Paraguay, where it is called Kaa-Chiri.

### *I. guayusa* Loesner

Syn. *I. amara* Vell.

Grows at the foothills of the Andes in Ecuador and Colombia. The dry leaves are sold at native market places in Ecuador. The Jivaro Indians take daily tea infusions as a stimulant and herbal curative. Some ethnologists have reported that the Jivaros consume "guayusa" before the beginning of tribal battles.

### *I. brevicuspis* Reisseck

Its leaves are similar to *I. paraguariensis* and therefore used as substitute for the real mate tea. It grows in the border regions of Paraguay-Brazil-North Argentina.

MORPHOLOGY

Dioecic trees with a dense canopy, 18–20 high when growing in natural stands; in plantation the stems are maintained no higher than 4–5 m by continuous chopping with a machete to stimulate the side branches. Male plants are preferred in plantations because of their higher leaf production.

Leaves dark green, coriaceous (5–12 cm long, 2–5 cm wide), on short petioles; lamina with serrate margins.

Inflorescences in male plants dichasios with 3–11 small flowers inserted in the leaf axils. In female plants, fewer flowers (1–3) with short triangular bracts. Corolla inconspicuous, white with short stamens and subglobose (3 mm) ovaries. Fruits when ripe dark red or violet (5–7 mm diam.).

A reliable method is urgently required to determine the female individuals at an early stage and to eliminate them in the nursery and before transplanting in the Yerba plantations.

With respect to definitive domestication, the mate industry is still in its infancy. In Paraguay the objectionable custom still persists of going to the forests and collecting fruits of the wild trees and with these seeds setting up nurseries for marketable *Illex* plants. The offspring result in different and sometimes very low-producing populations. The renewal of Yerba mate plantations in Argentina should be based on seeds of controlled individual selection.

COMMERCIAL USE

The Yerba mate tea is of paramount importance for millions of inhabitants of the three Rio de la Plata nations. In the NE of Argentina alone 70,000 (perhaps 100,000) ha Yerba plantations exist. The same amount has been calculated for Brazil and Paraguay together. The production of dry leaves exceeds 160,000 t yearly in the Argentinian Republic. The beverage contains a considerable quantity of alkaloids, 1.5% caffeine, 6% tannin, and is consumed several times a day by Paraguayans and Argentinians as a national drink, mostly by the lower income classes. The usual expression yerba mate is somewhat confusing. Whilst "yerba" is the usual word for a dry plant, the term "mate" refers to the small vessel of a hollowed *Lagenaria* fruit. There is an age-old custom of

sucking the *Ilex* tea with a "bombilla" (a straw or even a silver tube) from the dearly loved "mate".

---
                                                                    SOLANACEAE
---

## 9. *Nicotiana* spp.

Tobacco, tobaco, Tabak

$2n = 48$

The genus *Nicotiana* has worldwide distribution and occurs on five continents. Until now 64 distinct taxa have been established, 45 of them of New World origin. They represent a vast polyploid series with chromosome numbers ranging from $2n = 18, 20, 24, 32, 36, 38, 40$ to $48$ (Goodspeed 1954, Gerstel 1966). In the evolution of *Nicotiana* cultivars, frequent hybridizations, followed by amphiploidy, have played an impressive role (Fig. VII.11).

NAME AND ORIGIN

The genus received its name in memory of J. Nicot, who travelled in South America at the end of the 16th century and took seeds back to France.

There are few plants in the world as closely connected with the industrialization of narcotics as tobacco and its alkaloids. Each year 1000 t of this poison are produced by the cigarette industry. An amount of 40–60 mg is already fatal for humans. A high percentage of lung cancer is undoubtedly caused by the increasing smoking habits in the western nations. Tobacco misuse counts among the most dangerous stupefaciants; but all attempts to suppress, or at least to reduce, the cancer-causing habit of inhaling nicotine are hampered for the simple reason that the sale of tobacco represents a major source of taxes and revenues for most governments. The international conference "Stop Smoking 1978" calculated the world expenditure on cigarettes at the astronomical sum of 10,000,000,000 $. The source of commerical tobacco are two annual herbs:

*N. rustica* L., the "Russian Majorka" known for its repulsive smell in many East-

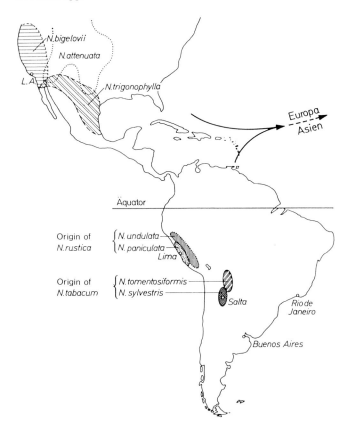

**Fig. VII.11.** Dispersal of wild *Nicotiana* species in America

ern Europe establishments, and *N. tabaccum* L. which produces the Western style of smoke.

American Indians used as many as ten different *Nicotiana* species for narcotic, medicinal, ritualistic and hedonistic purposes during tribal gatherings. The shamans and priests employed nicotine to enter into ecstatic trance. Aborigines had a great variety of uses for tobacco. It was smoked, chewed, sniffed and employed rectally in the form of enemas. According to Marin (1979), the last method was widely used by Peruvians and Mayas.

The question has been raised, how ancient is the use of tobacco in America? Furst (1976) stated that the two main *Nicotiana* species belong to the most ancient cultivars of the New World, perhaps even older than the earliest varieties of maize in Mexico. The earliest artistic evidence there is a bas-relief from a temple at Palenque (432 B.P.) in the state of Chiapas. A Mayan priest is depicted blowing smoke through a tubular artefact. The species *N. tabaccum* does not exist in free nature, nor does *N. rustica*. Both are, genetically seen, allo-polyploids with hybridogen origin (Gray et al. 1974; Sheen 1972). Their genome of 48 chromosomes is composed in the case of *N. tabaccum* of 24 units from *N. sylvestris* Speg. & Combes and 24 chromosomes of *N. otophora* Griseb. (or *N. tomentosiformis* from the same botanical section). The Majorca tobacco, *Nicotiana rustica*, is the result of a hybridization between *N. paniculata* L. and *N. undulata* Ruiz & Pavon, both native to the Western slopes of the Andes in Peru. We assume that the two cultivars originated in the areas of distribution of their wild ancestors. These diploid wild tobaccos live in geographical proximity in the border region of three South American countries, so that spontaneous crossings are probably the source of

a                                          b

**Fig. VII.12.** Wild-growing *Nicotiana* species. **a** Wild stand of *Nicotiana sylvestris,* discovered by my son Erik Brücher in a mountain valley of prov. Salta (North Argentina). **b** Wild-growing *Nicotiana otophora* in the vicinity of Camiri in south Bolivia

the two distinct tetraploid cultivars; we cannot, however, exclude human interference in their synthesis (Akehurst 1981) (Fig. VII.12).

Both amphyploid tobacco species spread very early, perhaps thousands of years before the arrival of Columbus, from their South American radiation centre, with the help of Indians migrating South and North. Especially the more resistant *N. rustica* reached the farthest limits of native agriculture, i.e., 42° degrees southern latitude at Chiloé and 50° north at the Saint Lawrence River.

Goodspeed (1954 and 1961) performed memorable experiments at the Californian University with the aim of reconstructing *N. tabaccum* from its putative wild ancestors. Of course we must bear in mind that the present existing wild tobaccos are not identical with their forerunners some 1000 years ago. Goodspeed crossed *N. sylvestris* with

*N. otophora* and artificially doubled the resulting hybrid genome. The offspring of this combination were, as the author says: ... "a great deal like modern *N. tabaccum*, in other words, a synthesized tobacco plant".

In search of the ancestors, Goodspeed (1961) undertook a collecting trip for wild tobaccos to South America and expressed his belief that the "meeting-mating" of the putative parent species occurred in the Argentinian province Salta, more exactly in the Sierra de Tartagal, where they are still growing together. Why they should not have hybridized there some 1000 years ago?

Clausen does not wholly agree. His opinion is that another species of the Tomentosa group, namely *N. tomentosiformis*, crossed with *N. silvestris*. The first-cited species belongs to West Bolivia, where the actual areas of these two species do not overlap, but this may have been different in earlier epoches. Recent investigations of Reid (1974) on the

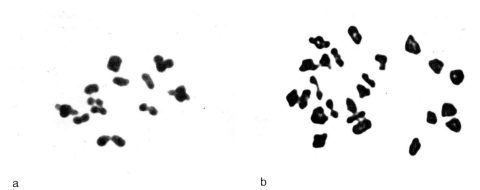

a                                                                b

**Fig. VII.13** Differences in chromosome size of *N. otophora* (**a**) and *N. tabaccum* (**b**) (Gerstel and Burns)

distribution of diterpenes in the genus *Nicotiana* favour the opinion that *N. tomentosiformis* was indeed involved in the species synthesis (Fig. VII.13).

## MORPHOLOGY

### *N. tabaccum*

Stems in mature plants are strong, erect, 100–600 cm high and 3–6 cm thick, terminating in a conspicuous inflorescence.
Leaves are ovate-elliptic (50–70 cm long and 12–20 cm broad), sessile on the stem. They may reach a maximum size of 18 dm$^2$ in the so-called Mammoth varieties. The lamina is covered with short simple glandular hairs. Its ripeness (for manual harvest) is indicated by a yellowish colour. Inflorescence rich-bearing with 100–150 flowers in one raceme. Corollas showy, 3–5 cm long, pink or white, with a long slender style and five stamens which are attached to the base of the corolla tube. The fruits are almost covered by the enlarged calyx (14–20 mm long). They contain an enormous quantity of minute seeds (estimation per plant: 1 million seeds), 1000 kernels weigh only 50–80 g. In spite of their small size, they have sufficient reserve substances which allow them to maintain their viability for 20 years, under adequate storage conditions. One single plant produces sufficient seeds for several hectares.

### *Nicotiana rustica* L.

Stems stout, 50–100 cm high. Leaves are arranged spirally around the stem. Their lamina is short-ovate (10–30 cm long, 9–26 cm broad) with dense hairy indument. Flowers are very different from *N. tabaccum*; the corolla is greenish-white with very short lobes. The style and the five stamens do not protude from the rand of the corolla.
*N. rustica* has a much higher nicotine content than *N. tabaccum* and perhaps for this reason was widespread in pre-Columbian times in the whole of America. North American Indians (the Senecas of the present New York region) were growing this species, when the first European settlers arrived, the latter changing during one century the local custom of smoking *N. rustica* for the "Virginia types" of *N. tabaccum*, probably introduced from Brazil (Wilbert 1972).
*N. rustica* is comparatively resistant to cold and diseases and in mountain regions (up to 2000 m alt.) develops a strong nicotine content (10%) and is therefore grown for nicotine extraction for chemical purposes, mostly as insecticides.
The typical physiological effect of nicotine in the body of warm-blooded species is caused by two alkaloids, $C_{10}H_{14}N_2$ and $C_9H_{12}N_2$, and additionally basin, anatabin and myosmin. During the smoking process considerable quantities of polyphenol also enter the lungs, which is well known for its cancerogenous effects. The alkaloid nicotine

is produced in the roots and transported to the leaves. Its content in the tobacco plant can be increased by nitrogen fertilizer.

## DISEASES, PESTS AND IMPROVEMENT BY BREEDING

Tobacco cultivation suffers from a considerable quantity of destructive leaf diseases (Marin 1979; Burk and Heggestad 1966):

a) *Phytophtora parasitica* (called black shank in North America) is considered one of the most dangerous diseases, which is favoured by humid climate. Thanks to genetic resistance from the wild species *N. longiflora* and *N. plumbaginifolia*, successful breeding work is under way.

b) *Sclerotinia sclerotiorum* attacks young tobacco plants in nurseries, but later also stems and leaves. The symptoms are soft brown spots. The destruction is worse when the ripe leaves are bundled in processing factories for bulk curing, with 100% humidity and 24–28°C temperature. *Sclerotinia* can cause a total loss within short time.

c) *Erysiphe cichoracearum*. This mildew disease causes damage in cool humid mountain climate. Heavy losses have been reported from African plantations in the hill region of Kenya and Tanzania. Hybridization with the resistant wild tobacco *N. glutinosa* transferred resistance genes to the cultivated tobacco.

d) *Thielaviopsis* (= black root) is destructive in African tobacco plantations.

e) *Corticium* and *Pytium* cause world wide the "damping off" of young seedlings, especially when sown too densely in nurseries.

f) *Peronospora tabacina* produces the terrible blue mould. This disease was originally restricted to Australia, but by negligence invaded Europe in the year 1958. Within only 3 years nearly all tobacco regions had been infected. Certain wild species possess immunity against the disease, e.g., *N. longiflora*, *N. plumbaginifolia* and to a high degree the Australian *N. debneyi*. With its help Bolsunov (1963) created the new variety Florida, which showed resistance and heterosis effect in tobacco fields in the Balkans.

## Bacteria

*Pseudomonas tabacci*, commonly named wildfire, spreads very rapidly under humid weather conditions in young tobacco plantations, causing heavy losses. For this reason it may be considered as decisive progress that geneticists have recently discovered a Colombian landrace with resistance, besides the formerly known immunity in the wild species *N. longiflora*, based on one single gene.

## Viruses

The tobacco mosaic virus (TMV) is one of the most studied viruses at experimental stations. Already in 1938 Holmes discovered immunity against this omnipresent virus in *N. glutinosa* and created the now famous var. Samson.

Leaf-curl virus is still dangerous, because it is transmitted by the difficult to eradicate *Bemisia* fly.

"Krom-nek" is a widespread virus disease in South Africa, which is transmitted by trips. Its symptoms are severe torsions of the stem. The worst virus for tobacco, tomato, and potato crops is Virus $Y^n$. Its origin is South America, probably North Argentina. Brücher (1969) studied its migration and possible introduction to Europe with British wild potato collections. This carelessness has caused multimillion losses over the whole world. The symptoms are vein necrosis and leaf drop, and finally, premature death of plants, which have been infected with the necrotic strain of Virus $Y^n$.

## Insects

We refer to *Thrips tabacci* and *Epithrix hirtipensis*, which mainly damage the leaves, and provoke a serious loss of quality.

## Nematodes

A considerable quantity of soil-borne nematodes cause great damage, first of all of the genus *Meloidogyne*. Against these root-knot-producing nematodes genetic resistance exists in the wild tobaccos *N. sylvestris* and *N. otophora* from North Argentina-South Bolivia.

YIELDS, ECONOMY, WORLD-PRODUCTION

The average yield of dried leaves amounts to 1000–1500 kg/ha, but under optimal planting conditions 1 ha may produce 4000 kg of leaves. Tobacco cultivation occupies in general the best soils of crop-producing countries. This fact may be open to criticism in the case of third world nations which suffer from undernourishment like Pakistan, Bangladesh, Tanzania and Zimbabwe. On the other hand, planting and harvesting of *N. tabaccum* demands much manual work and gives fair revenues for smallholders. World production is shown in Table 14.

**Table 14.** World production of tobacco between 1980 and 1985

|             | 1980  | 1983  | 1984  | 1985  |
|-------------|-------|-------|-------|-------|
| Area (mio ha) | 4.040 | 4.279 | 4.099 | 4.109 |
| Yield (mio t) | 5.563 | 6.000 | 6.480 | 6.590 |

At present, Continental China is the biggest producer of two narcotic plants *Nicotiana* and *Papaver*. This, with India, China and Pakistan, means that Asia produces approximately 2 mio t of dry tobacco leaves yearly. North America contributes with over 920,000 t, South America 570,000 t, Africa 260,000 t and Europe 88,000 t.

Most of the yield goes to the cigarette industry with an annual output of 4,560,000,000,000 units. The yearly cost for advertisement and other media to promote tobacco smoking swallows astronomical sums; in Germany alone 200,000,000 $. It is extremely curious and gives cause to reflection that *N. tabaccum* is the only crop in the world which needs constant stimulation to promote its consumption, while the really nutritive commodities sell well thanks to their inherent value.

## 10. *Paullinia* spp.

Guarana, paullinie, yoco

This genus is represented in the neotropics with several well-defined species (in Panama alone with a dozen taxa), but only few are considered useful for humans. Natives of Amazonia selected two species: *Paullinia cupana* HBK var. *sorbilis* and *P. yoco* Schultes and Killip. The first-mentioned is an important stimulant, today even industrialized in Brazil. Other species like *P. pinnata* L., are used for fish poisoning.

MORPHOLOGY

Climbing lianas, vines 10–15 m long and 5–8 mm thick. Leaves alternate, with five foliols, 20–35 cm long, the margin dentate. Inflorescenses in leaf axils, 10–20 cm long with many flowers, which are sexually differentiated. Five white petals (5 mm long). Fruits red, elliptic capsules (2–3 cm long and 1 cm large) with three black seed kernels (12–20 mm) enclosed by a red aril. One plant may produce 2 kg of dry seeds.

These seeds are the useful product of *Paullinia*, due to their very high alkaloid content (3–6% guaranatine) (Fig. VII.14). Indians use several species besides the most important *P. cupana*, such as *P. macrophylla* and *P. pterophylla*, and primarily the wild-growing *P. yoco*. This latter grows in the rainforests of the Putumayo-Caqueta river system. The Siona tribe take the whole liana, cut the stem in small pieces and rasp it. After squeezing, they obtain an alkaloid-rich juice of milk-chocolate colour. Indians make this a daily beverage as stimulant, as febrifuge, and mainly against hunger.

*P. cupana* exists in several varieties, some selected as small bushes for easier collection. Its wide areal reaches from Venezuela (Alto Orinoco, Rio Negro), Estado do Amazonas of Brazil to Colombia.

In Brazil there exists a popular tonic drink – often adulterated – called Guarana. For the needs of the rapidly expanding industry of this beverage *Paullinia* plantations now exist. The hard seeds are ground to a powder

**Fig. VII.14.** *Paullinia cupana,* the liane which produces "guarana"

which is mixed with mandioca cassava flour and kneaded to a paste. This is sold on indigenous markets as "pasta guaraná". The bitter-tasting tonic is made by scraping some of this paste and dissolving the powder in water.

| | MYRTACEAE |
|---|---|

### 11. *Pimenta dioica* (L.) Merrill

Allspice, toutépice, pimienta, malagueto, Nelken-pfeffer

$2n = 22$

Several species of the botanical family of Myrtaceae contain fragrant substances and etheric oils. In the neotropics we have the genus *Pimenta* which to some extent replaces three Asiatic spices in one: nutmeg, clove, and cinnamon. For this reason it is called all-spice.

The leaves of *Pimenta* contain a volatile oil, which is used in pharmacy, dentistry, food conservation; and in the Caribbean islands for production of bay rum, a potent beverage of the negro inhabitants.

The aetheric oil consists of 60–80% eugenol, besides cineol, phellandren and caryphyllen (Fig. VII.15).

**Fig. VII.15.** *Pimenta officinalis* (Aristeguieta 1962)

MORPHOLOGY

Small, evergreen trees (6–11 m tall), with slender trunks and silvery shiny bark.

Leaves elliptic-oblong (10 × 5 cm) glabrous dark green above and light green underneath with many oil glands.

Inflorescences many-flowered cymes which sit in the axils of upper leaves on small branches. The trees are dioecious, even if the inflorescences appear to be of hermaphrodite structure. Female individuals may sometimes produce pollen, but it is sterile. Flowers are small with numerous stamens and four white, 4-mm-long petals.

Fruits are 5-mm-thick globose berries with a thin sweet-tasting mesocarp. The berries are picked when green-ripe and stripped without great care from the branches.

There is a similar tree, native to the Caribbean region, *Pimenta racemosa* (Mill.) Moore., with the vernacular name bay-tree. It is still growing wild on the smaller islands of the Windward Islands. From its leaves is also distilled bay rum, with multiple inner and outer applications, when mixed with alcohol.

For improving the two species it would be recommendable to select high-yielding trees (70 kg dried berries) and multiply them vegetatively. Such future clones would substitute the semi-wild trees used at present, which have obviously low yields.

In commerce, Germany is the main buyer, but the current exports do not exceed 3000 t. Fifty years ago the export of Jamaica alone was 6000 t annually.

Stevia rebaudiana BERT.

---

COMPOSITAE

## 12. *Stevia rebaudiana* Bertoni

Kaahée, "the sweet herb of Paraguay"

2 n = 22

**Fig. VII.16.** *Stevia rebaudiana,* the Paraguayan sweetener

The genus *Stevia* is restricted to the American continent, with most of its species distributed north of the Isthmus of Panama; in Mexico alone it is represented with 70 species. Strangely enough, the only species of worldwide interest grows at the southern end of the distribution area.

The total estimates for the whole of South America include 120 taxa (Tanako 1980). We tested a certain amount of them with respect to their supposed steviosid content. None of the fresh or dried samples could compare with *St. rebaudiana* in its sweetener effect. We found a slightly sweet taste in *St. caracan* HBK, *St. lemmonii* A. Gray, *St. monardifolia* HBK and *St. punensis* Robins. The last-named plant seems to be used by natives in Peru as a sweetener (Fig. VII.16).

NAME AND HISTORY

*St. rebaudiana* is an endemic species of Eastern Paraguay, mainly in the Cerros de Amambay, bordering on Brazil. This remarkable plant is well known in the whole of Paraguay by its Guarani name (very difficult to pronounce and to write), Ka-A-Hé-E, and has been used there by natives since time immemorial as a potent sweetener, often together with the local tea, Yerba mate. The dried leaves are commonly sold on market places, sometimes adulterated with other plants. It is appreciated for its hypoglycemic effect. It is, however, definitely not true that Paraguayan women use it as contraceptive, as an Uruguayan author reported 20 years ago. It is documented that Guarani Indians used "Kaahee" before the Spanish troops arrived in Paraguay, according to documents of the National Archive in Ascuncion (Soejarto 1983).

MORPHOLOGY

The plants are annual or biennial herbs and grow to a height of 60–80 cm. The stem is slightly woody, covered with sparse glandulous hairs. The rootstock has many dormant shoots, which are suitable for vegetative propagation. The roots are cylindrical, with scarce branching, and spread horizontally rather than deep.

Leaves simple, ovoid to elliptic (22–30 mm long and 15 mm wide, with three visible nerves and slightly dentate margin. Insertion is opposite. Inflorescences according to our observations auto-sterile (isolated with bags no seed setting). Flowers small, 2–6 together in one involucre. Fruits (aquenes) with 16–18 hairs.

## USE AND FUTURE DOMESTICATION

*St. rebaudiana* plants collected in the wild contain approximately 7% steviosid, in dry material. Considering that in cultivation 1 ha yields 1000–1200 kg dry mass, one may expect 60–70 kg steviosid for commercialization. Pharmacological studies have shown that the sweetening element is not a carbohydrate but a glycoside. The formula is $C_{38}H_{60}O_{18}$, a sophoriside of oxyditerpenylic acid (diterpene glycoside). According to early tests by Rasenack (1908), steviosid is 300 times more powerful as sweetener than sucrose.

Besides the pronounced sweet taste, *Stevia* has a slight licorice aftertaste. Other *Stevia* species have certain ethnomedical use, perhaps induced by the presence of essence oils, bitter taste or aromatic odour. Pharmacological studies are under way in South America to test if the different uses in folk medicine have a scientific base. Physiologists discovered that steviol, a derivate of stevioside, has an action similar to that of gibberellin (Ruddat et al. 1963).

Paramount importance, however, is attributed to *Stevia rebaudiana* and its characteristic as a non-nutritive sweetener, which may replace saccharine and cyclamate. It is well known that in 1969 the North American Food and Drug Administration banned the use of cyclamate and similar sweeteners for human consumption, with special consideration for the millions of people who suffer from diabetes. This disease has increased to an alarming extent in the so-called affluent societies of Europe and North America, while primitive living populations are free of it. According to medical reports from Germany, 12% of the population suffer from different disorders of glucose tolerance, summarized as diabetes. Here *Stevia* can fill a gap as a nontoxic sweetener, suitable for diabetic patients. This is perhaps one of the reasons why at present in Japan several thousand hectares of *Stevia* plantations already exist, of course based on plants of Paraguayan origin.

STERCULIACEAE

## 13. *Theobroma cacao* L.

Cocoa, Cacaotier, Cacao, Kakaobaum
$2n = 20$, also 16 and 26

*Theobroma* with 22 species is an exclusively neotropical genus. It belongs to the botanical family of Sterculiaceae of pantropical distribution, which provides mankind with several other stimulation fruits, like the African cola nut (*Cola nitida*) and the "West Indian castaôna" (*Sterculia apetala* Jaq. Karst.).

## NAME AND HISTORY

The term cacao is generally used to design the tree and the crop, while cocoa means the seed kernel and the beverage which is prepared from the roasted and ground beans. Cacao is a Nahuatel-Maya word, originally spelled "cacahoatle" which also gave rise to the Mexican term chocoatl = chocolate.

Long before Linné gave the classical description (1753), the name "cacao" appeared in Europe to designate this outstanding neotropical fruit tree. Clusius (1605) referred in *Exoticorum libri decem* to "cacao-fructus". Hernandez (1630) described cultivated forms from Mexico with this name, accompanied by a good illustration of a criollo variety. The first taxonomic description was given by Tournefort in the year 1700, using drawings of Plumier, which depict clearly the curious flower, with its strangulated petals and erect staminodes (Fig. VII.17).

b

**Fig. VII.17.** *Theobroma cacao,* **a** tree covered with cauliflorous flowers and fruits (Photo de Geus); **b** flower of wild cacao (*Th. grandiflorum*)

a

Strangely enough, Linné objected to this name as indecent, exclaiming: "Cacao nomen barbarum, que rejecto".... He replaced it with a more poetical denomination: "food for the gods" = Theo-broma. Since 1753 this is the valid generic name. Nevertheless, Linné lacked detailed knowledge and consistent material from this tropical tree. Consequently, his short diagnosis was insufficient to separate the species. Only in 1806, Humboldt and Bonpland, using their own findings from Colombia, published the first satisfactory observations and illustrations of cacao trees, describing *T. bicolor* H&B in *Plantae Aequinoctiales.*

The first attempt at a monographical classification of *Theobroma* was given by Bernoulli (1869), after intensive field work in the tropics. This Swiss botanist lived for 20 years in Guatemala, and described between 1834 and 1878 18 binomials of *Theobroma,* 12 of them as new scientific species. His botanical survey made it clear that the only commercially important species is *T. cacao.* The others are wild species or even hybrids. In the meantime, the species *T. cacao* has been divided into two subspecies and three cultivated groups:

(a) "Criollo" of Central American descendency, and most probably domesticated there for 2000 years, whereby the astringent-tasting forms became eliminated. Criollo fruits have ten deep furrows, are markedly warted and pointed, with yellow or red skin. The pods have thin walls and contain large seeds, which are nearly round with pale violet or white cotyledons. They have excellent quality but low yield. (b) "Amazonian Forastero". The fruits have a smooth exterior, are not furrowed. Their skin is white or green. The pod walls are thick and sometimes woody. The seeds are flat, of astringent taste, with dark violet cotyledons, and high fat content. (c) "Trinitario", considered typical for the Island of Trinidad but introduced there from Venezuela, where it was the most popular variety. It is considered as an ancient crossing between a "Forastero" from Amazonia and "Criollo". This cultivar had a worldwide reputation for good cocoa quality (now lost). The surface of the pods is smooth to deeply sculptured. Probably this variety is still rather heterogenous.

There are no archaeological remains of cacao fruits in Amazonia. This is easily understandable in such an adverse humid climate. But this absence is not sufficient proof that prehistoric Indios did not collect, plant and use regularly the products of *Theobroma* trees.

When living in the Upper Orinoco valley, I observed that the nomadic tribe of Makarikari-Indians, whom I accompanied during their food-gathering wandering, had a special liking for the acid-sweet cacao pulp, sucking it as refreshment, but discarding the seed kernels. This custom does not mean that in other regions certain – more advanced – Indians would not have collected the seeds and even planted them around their huts.

Why should they not have detected – sitting around their open fireplaces – that roasted cacoa beans have a very pleasant aroma and a visibly stimulating effect (thanks to the high theobromine caffein content). Cacao seeds may later have been an attractive barter item on their wanderings to other tribes, living far North in the rainforests of the triangle between the rivers Napo, Caquete and Putumayo. These are the headwaters of the Amazonas-Orinoco river system, east of the frontier between Ecuador and Colombia (Leon 1984).

Thus *T. cacao* could have worked its way north and westwards from its originally reduced radiation centre on the Upper Orinoco-Amazonas to the Pacific coast or the Gulf of Uraba, inhabited by advanced Kuna Indians, renowned as bold seafarers. From there it is still a long distance to Central America and the tropical regions of Mexico, declared by many ethnobotanists and plant geneticists to be the genuine "domestication centre" of "cultivated cacao", while recognizing at the same time the lack of a suitable wild species in the same region.

This situation was already expressed by the authority on *Theobroma*, Dr. Cuatrecasas (1964): ... "in early times a natural population of *T. cacao* was spread throughout the central part of Amazonia-Guiana westwards and northwards to the south of Mexico..." This means that Central America is not at all a gene centre for cacao, but a secondary region of diversification and domestication under the guidance of the highly developed Oltec and Mayas (Rio Azul, Petén, Tikal civilizations), who dominated the Peninsula Yucatan-Guatemal-Honduras-coast (Lathrap 1973).

Schultes (1980) also voices a similar opinion when he writes: ... "it might have been pos-

sible for cacao to have travelled with Indians from its home on the eastern slopes of the Colombo-Ecuadorian Andes through some of the low passes, to be planted on the wet, hot tropical western slopes of the Andes... it would have no ecological barrier to travel westward into Central America and southern Mexico."

It was impressive to me how Indians used the hard wood of cacao trees to make fire. Because of the humidity, my matches were useless. The accompanying Indios resolved the problem of making fire in an unforgettable manner. After they had found an old dry cacao tree they cut some sticks, then they gathered soft wood and dry fungus. With this they began spinning the hard wood at high speed against the soft, until they obtained smoke. Blowing intensely, they achieved a glow, which was transmitted to the dry fungus. I never shall forget this impressive moment of "creating fire" in the rainforest, guided by "primitive" Indians. Some weeks later I returned to Caracas to continue my mutation work at the Atomic Reactor of the IVIC Institute, where the "burning elements" were somewhat different.

Columbus and his crew, on the fourth voyage to the *Indias*, noticed in the year 1502 the existence of this marvelous fruit, when their ship passed the Island of Guanja, off the coast of Honduras. But they did not bring cacao seeds to Europe. It was Cortes who in 1528 observed the popularity of an aromatic beverage, called xocaotl, during his conquest of Mexico, and sent cacao beans to Spain. It was, however, not the Aztecs – a race of highland Indians – who cultivated it. They received it from the Mayas, who domesticated *Theobroma* in the tropic humid lowlands of the coast and who maintained traditional trade relations with the Mexican highland. Cuatrecasas's (1964) opinion is: "Central American Indians undoubtedly developed the art of planting and selecting of cacao through several thousands of years, finally obtaining the high quality product which the Spaniards found at the time of the Conquest." They reported that the emperor Montezuma drank daily 400 cups of hot "xocolatl". It seems that only the elite class and the rulers of Maya and Aztec tribes were allowed to imbibe this precious beverage, which was mixed with maize flour, vanilla and pepper, obviously enjoying its high content of stimulating alkaloids. This mixture, however, was unpalatable to the Spanish conquerers and they could not understand how the rulers could imbibe big quantities of such a "loathsome drink".

Maya people perfected the agricultural methods for cacao and the know how to ferment and cure the seeds. In the Petén and Belize region there existed elaborated techniques and large cacao plantations with irrigation channels. They depicted cacao trees and fruits in stone stelae, belonging to the classic period of Maya culture (Acosta 1590), but all this was suddenly abandoned and destroyed, following the Spanish invasion.

Maya merchants used cacao beans as currency. The yearly tribute of lowland Indios to the Aztec ruler was paid in cacao. With 100 beans one could buy a slave, as the Spaniards reported with great astonishment. Even falsifications of this currency occurred – as always with money – by using empty bean shells filled with earth. Oviedo mentions that the rulers had to pass punitive laws against such falsifications. This cacao bean currency had wide application in Central America and seems to have survived as medium for exchange in Guatemala until the last century, even if it was not ideal for this purpose, because it contained too much fat and acetic acid.

The first commercial shipment of cacao beans arrived in Spain in 1585. The technique to produce chocolate from it was kept secret by the merchants jealous to monopolize this wealth from the New World. A half century later, other nations, like England and France, learned the preparation of this, then, luxurious beverage for the nobility. In the year 1652 it became fashionable in London and was so expensive that chocolate consumption was largely confined to wealthy people. One pound of seeds was equal to one pound sterling.

MORPHOLOGY

Small trees (8–10 m high) with scarce ramification. The growth of the tree occurs in periodical phases, which are manifested in different "etages" of lateral branches. The trunk is sympodial with plagiotrope ramification, sometimes twisted downward.

Leaves are elliptic-oblong, alternate, often of considerable size (40–60 cm long). The lamina is leathery and shows striking colour differences between the young leaves, which are red-brownish, and the older, turning dark green as they mature. Inflorescences are inserted directly on the trunk, or on very short woody branchlets, which produce flower cymes at their end.

Flowers have a fascinating structure. They sit on relatively long pedicels. The five petals are strangulated and separated by a narrow joint in two distinct parts. The lower part is carnose and rigid with the form of a hood, whilst the form of the upper half may vary in shape, according to the different species. Flowers are produced in enormous quantities (6000 per tree), but only few are pollinated, not by bees but by tiny insects, like midges, or even ants. The androeceum has a strange aspect. Five of the outer stamens are sterile, erect, carnose and transformed in petaloid staminodes, the rest of the inner circle of stamens producing fertile pollen.

The fruits are of different colour, yellow, green, red or dark purple, and, according to the variety, of different shape. They may be round, ovoid and long-elliptic (10–30 cm long) with thick pericarp. The fruit has five compartments, each containing 8–12 seeds which are imbedded in a white or pink pulp. The seeds are 2–3 cm long, filled with corrugated cotyledons, of different colour (white to deep purple). To ripen, they need nearly 6 months. They have very limited viability, germinate often inside the ripe pods and lose their germination ability after a few weeks. They are very sensitive to low temperatures and are affected by temperatures below 8°C; a fact which must be considered, for example, in airplane transport. Storing should be done at 100% humidity with temperature between 18° and 30°C.

Cacao is propagated in the swamp forests by bats and monkeys which suck the sweet and aromatic pulp surrounding the seed kernels. Wild-growing *Theobroma* species have been used since time immemorial by native fruit-gatherers of the rainforests, either to eat them fresh or to prepare drinks from the pulp.

The geographical distribution of the genus *Theobroma* on the western hemisphere is limited to the rainforests between 15° south and 18° north. Cuatrecasas (1964) gave exhaustive descriptions of the many (more than 20) wild species and concluded: "It seems that the richest region in species is around Panama and Colombia, where species with a very restricted area are found ..."

In the following we cite a few, which may be related to the domestication of the cultivated cacao tree:

### *T. angustifolia* DC

Often called cacao cimmaron. Its brownish, tomentose fruits contain many seeds which are highly esteemed by natives, who produce from it a beverage similar to chocolate. Its areal reaches from Mexico to the humid hot lowlands in Central America.

### *T. bernouilli* Pittier

Has been described from Panama and Costa Rica. Natives call it "cacao rana". The trees are nearly branchless and bear reddish-coloured fruits of smaller size than the usual cacao. The seeds are round, surrounded by a white odourless pulp and contain 27% oil which could be of potential importance as an aromatic substance, but until now has found no commercial interest (Fig. VII.18).

### *T. bicolor* Humb. and Bonpl.

Known by its indigenous name "pataste", this wild cacao grows in NE-Peru, where one of the first Spanish missionaries, Father Cobo, mentioned it. Also known from the Upper Amazonas and Rio Negro in Brazil and extended throughout Panama to Costa Rica.

"Pataste" grows mostly on open, even sunny, places quite in contrast to the cultivated cacao tree, which needs shade. The fruits have a cream-coloured pulp and a hard corrugated shell. They are not borne on the

**Fig. VII.18.** *Theobroma bernoullii* from the Colombian rainforest (Photo Cuatrescasas 1964)

main stem, but on side branches. The seeds have been used as a substitute for real cacao. According to my inquiries in Panama, the Choco Indians consider this fruit more important for their subsistence than the real cacao tree.

### *T. cirmolinae* Cuatr.

This wild cacao has been recently discovered by Cuatrecasas in the mountains of Colombia, where the big trees (20 m tall) grow on the Pacific slopes of the Andes. It is the cacao species with the highest elevation (1300 m alt.), and may have some tolerance to cold climate, which make this wild species interesting for further breeding work. Natives occasionally collect its fruits, which

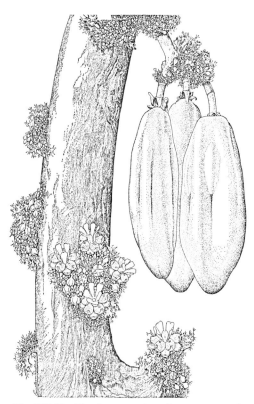

**Fig. VII.19.** *Theobroma cirmolinae,* a nearly branchless wild cacao tree (Cuatrecasas 1964)

hang down on long petioles (Fig. VII.19) quite differently from most other *Theobroma* species.

## *T. grandiflorum* **Willd.**

It is morphologically different from other species, because it is not cauliflor and bears flowers and fruits on side branches. The fruits have a velvet brown skin and hang down on long petioles. The seeds are round-flat and have a certain similarity with true cacao kernels, but have much more fat, with a consistency of "cacao butter". It is a native of the lower Amazonas (Para). Natives call it "cupu-assu" and plant this wild cacao for its delicious fruits.

## *T. microcarpum* **Mart.**

Its distribution is in the western Upper Amazonas region of Brazil and the Caqueta river of Colombia, where it grows as a significant element of the shadowy vegetation of rain forests. Mature fruits are small (6–9 cm long) woody and with many deep ribs. Natives have no use for it.

## *T. simiarum* **Don.**

Called "cacao de mico" in Costa Rica, where it is native. It has peculiar long-stretched fruits, like sausages. Its seeds are used by natives for a chocolate-like beverage.

## *T. speciosum* **Willd.**

Tall, slender trees with small fruits (8 × 6 cm). When ripe they are yellow and contain 20–30 seeds, which are surrounded by odourless white, sweet pulp; which is eaten by Indians. This species has a very wide areal which extends from Eastern Bolivia-Peru throughout the Amazonian Hylaea.

## *T. subincanum* **Mart.**

High trees (15–17 m), which are dispersed in the rainforests from Peru to Guyana. It produces many small fruits (6 × 4 cm) with relatively large seed kernels embedded in a sweet-tasting pulp, occasionally eaten by Indians, as we observed on the Upper Orinoco in Venezuela. This wild cacao has an enormous extended areal, which reaches from the foothills of the Andes in Peru, Ecuador and Colombia to the Orinoco, and covers the Amazonas river to its mouth in the state of Para (Addison and Tavares 1951).

### Diseases and Pests

Cacao cultivation suffers from an extraordinary quantity of cryptogamic diseases and an ever-increasing number of pests and parasites. Combatting them by spraying is a hazardous task, due to the high grade of precipitations in the humid tropics. Reduction of yield may reach 30–50%, under given circumstances (Urquart 1962, Entwistle 1972, Bastos 1981).

Fungus

*Phytophthora palmivora* causes the "black pod" disease, which not only destroys the ripe pods, but affects also stems and leaves, even the small seedlings. The fungus is extremely ubiquitous and spreads especially during long periods of high humidity in the atmosphere. This pathogen is considered by cacao planters of such importance that the presence of the diseases determines the profitability of an enterprise. Global losses were calculated at 50% (Thorold 1974).

Chemical treatment with copper fungicides depends on weather and gives only limited control. Breeding of resistance is based on wild cacao species. To the present, however, no introduction has shown sufficient immunity in view of the unfortunate fact that there exist several physiological races of the pathogen. After screening a large number of accessions in Costa Rica, only nine showed a promising degree of tolerance (Lawrence 1978).

Another severe fungus disease is "Witches broom" caused by *Crinipellis perniciosa* (= syn. *Marasmius*). The most obvious symptom – which gave rise to this name – is a hypertrophic growth of the shoots. There can be 200–300 "brooms" on one tree, causing early death of the plant; in easier cases heavy losses of flowers and pods occur. It seems that the disease is endemic to the Amazonas basin. From there it spread throughout the whole of Brazil, where nearly 80,000 ha have been infected since the beginning of this century. It was accidentally introduced to Trinidad, causing a severe outbreak in the year 1944. Until now the pathogen has not reached the big cacao plantations of West Africa, probably for adverse climatical reasons.

It has been recommended to cut the infected branches and burn them, but as the mycelium of *Crinipellis* grows within the plant, this is not the real solution. Fungicide application gave meagre results. In the Turrialba germplasm collection of 500 numbers, some showed considerable genetical resistance.

Further we must mention *Monilia roreri*, which attacks the pods and causes damage also in young plants. It seems that this fungus has spread from Colombia to neighbour countries in recent times.

*Ceratocystis fimbriata* produces a "wilt", also called "mal de machete", because it is transmitted by knives and machetes.

*Verticillium dahliae* did not cause considerable losses until recently in Brazil; but now heavy attacks have been reported in the states of Bahia and Espirito Santo. In the Uganda cacao plantations this disease caused a sudden dieback for many years.

Viruses

Several virus diseases cause damage; the most destructive of them is classified as "swollen shoot". Especially susceptible are the cultivars of "Amelonado" cacao. The symptoms of swollen shoot are vein-banding and angular chlorotic spots bordering the veins. The distinct viruses of the swollen shoot group have somewhat different pattern symptoms, but all can be transmitted by grafting and mostly by sucking insects, which belong to the Pseudo-coccidae (mealy bugs) and are often transported by ants.

This virus originated on the African continent. The first report came from Ghana 1936, where it existed before, but had not been properly recognized. The Government there began a campaign of cutting out infected trees in 1941. At the height of this eradication move – which from the beginning caused much suspicion and criticism – 1 million trees per month were being destroyed and by 1982 the incredible quantity of 185,000,000,000 cacao trees had been eradicated. The result was that new infections were occurring faster than the diseased trees could be removed. In the meantime the same disease has been reported from Ceylon in Asia and the Island of Trinidad in the Caribbean Sea.

It seems that the only feasible method is selection of genetic resistance. Between huge accessions of wild and primitive *Theobroma* some satisfactory numbers appeared, e.g. 85/799, for further resistance breeding (Legg 1982).

## Pests

Nearly 50 distinct insect species live on, or attack the cacao tree, making it impossible to enumerate them. Among the most dangerous we should mention:
A dozen species of *Coleoptera*, between them stem borers and bark-destroying Carambycidae. A dozen of Homoptera, like aphids, mealy bugs and jassids, Hemipterae, like capsids, *Empoasca*, thrips and chenilles, Heteropterae (Miridae), stink bugs and cacao capsids, like *Helopeltis*, Lepidopterae, several species and its larvae attack the ripening fruit and leaves.
According to Entwistle (1985), one can calculate that 1500 different insects and other parasites feed on *Theobroma*.

### YIELD AND WORLD COMMERCE

Cacao production and consumption show a steady increase since the first statistics available. In 1830 the whole production has been estimated at approximately 10,000 t. At the end of the last century, it had increased ten times to 100,000 t. World production of cacao exceeded 1,000,000 t in the year 1960 and in 1985 reached its maximum with 1,876,000 t (Wood Gar 1985).
At the beginning of this century, the American continent (the home of *Theobroma*) contributed 80% of world consumption and Africa only 16%. Since 1970, however, the situation has been reversed, Africa exporting 1,000,000 t, America 200,000 and Asia-Oceania 28,000 t. Actually the African Continent contributes over 60% of the world cacao (Ghana 30%, Nigeria 15% and other African regions 15%). It is worth mentioning that in West Africa cacao is grown almost exclusively on smallholdings – often less than 1 ha – supplying adequate living conditions to many families, sociologically a fortunate fact.
The world area planted with cacao trees is more than 4,700,000 ha with a continuous shifting from the original home region in tropical America to Asia and Africa. The rapid rise in cocoa prices from 1973–1977 stimulated more plantations. Unfortunately, this has caused a severe superproduction, followed by a notable erosion of prices paid

to the producer, but typically with no benefit for the average consumer. The International Cocoa Organisation (ICCO) has in vain made several attempts over the last years to halt the steady decline of cocoa prices, which occur at the cost of the small producer.
The industrial value of quality is determined by a sequence of heating and drying processes which depend on different fermentations. Therefore it is impossible to determine flavour and aroma of cacao beans in the fresh harvested pods. Only after defatting and fermenting would the professional cacao buyer judge the commercial value of a lot.
When living in Venezuela and visiting the once famous "cacotales" of the coast: Choroni, Chuao o Tukupita, which dictated cacao prices on the world market during the last century, we observed that they seriously declined through neglect. Quality was judged as low and yield was dropped to 200 g per tree/annum (!). The average per ha is 500–1000 kg, but selected plantation may yield 2500–3500 kg/ha.

---

ORCHIDACEAE

## 14. *Vanilla* spp.

$2n = 32$

This is the only orchid genus among the 140 genera which has acquired economic importance in human alimentation (leaving aside the worldwide trade with orchid flowers).
The genus *Vanilla* occurs in tropical latitudes of the Old and New World with nearly 100 taxa described, but only *Vanilla* species of the neotropics are of importance as aromatic and flavouring products, which in older times were in great demand and highly coveted on the world market. This position changed fundamentally, however, with the discovery 100 years ago of cheap synthetic "vanillin" with a similar odoriferous effect. Several species of the genus *Vanilla* are wild-growing from Mexico to Paraguay (Correll 1953). The most important is *V. planifolia* Andrews, with a wide areal in Central and

**Fig. VII.20.** *Vanilla planifolia,* wild-growing in Venezuela

South America. It is a terrestrial orchid with pronounced climbing tendency, mostly domesticated and exploited in small plantations. We have observed *V. pompona* Schiede in wild state in Venezuela, the Islands of Trinidad and Tobago and in Paraguay. Growing wild in the forests of Venezuela are *V. fragrans* and *V. pittieri* (Fig. VII.20).

### *V. planifolia* Andr.

Vanilla, vanille, echte Vanille

ORIGIN AND HISTORY

The original habitat of *Vanilla* are the tropical forests, especially in Central America (Guatemala, South Mexico) and the Antilles. From there the Aztec Indians introduced it as a domesticate, which was multiplied only by cuttings, vegetatively. The vanilla essence was used for flavouring the traditional Maya and Aztec beverage of cocoa. The Spanish explorers of the 16th century met this ancient custom and tried to introduce this aromatic plant to Europe and their African dependencies. At the beginning the result was a failure. Only in the year 1841, when Albius, a former slave on the island of Reunion, developed practical methods for artificial fecundation, did the situation change. The so-called "hand pollination" was the beginning of flourishing *Vanilla* plantations in Mauritius, Seychelles and Madagascar, which are still the foremost producers.

MORPHOLOGY

Stem flexuose, several metres long, climbing, monopodial, with many internodes (5–15 cm apart), with single roots.
Leaves flat, fleshy (10–25 cm long and 2–6 cm broad) long elliptic, with numerous parallel veins, inserted on short petioles.
Inflorescences: In contrast to many other orchids, which are well known for their exuberant showy flowers, *Vanilla* has rather unconspicuous dull, pale green corollas, with waxy aspect. The stigma is separated from the stamen by a flap-like rostellum, an arrangement which inhibits self-pollination.
*Fruits* are long (10–25 cm) capsules, which contain thousands of minute black, globose seeds (0.4 mm diam.), which are released when the capsules become ripe.
Fecundation takes place by insects, especially bees, and occasionally also by humming birds. This was not known when the first *Vanilla* plants had reached Europe, causing disappointment that they did not produce fruits as in America. Only in 1836 was it discovered that the sterility was caused by absence of natural pollinators (e.g. bees of the genus *Melapona*).
Yields are low compared with the expensive manual work involved. The average is 500–800 kg/ha of dried pods each year, until after 7 years the production of *Vanilla* plants diminishes considerably. Modernizing the

**Fig. VII.21.** Inflorescence with developing vanille pods (Theodose 1973)

culture may be possible by hybridation with other species and selection of high-yielding clones. As meristem culture is now widely used in orchids, this improvement seems feasible.

The typical vanilla flavour develops only after an elaborate "curing" process, which needs good practical experience (Fig. VII.21). The traditional method consists in alternate sweating and drying of the freshly harvested fruits until they have acquired a dark chocolate brown colour and a flexible, oily smooth aspect, which is accompanied by slow fermentation. Then the "beans" are spread on mats in the sunlight, where they shrivel for 1 or 2 months. After that 50 pods are bundled and put in sealed boxes for exportation. The pleasant vanilla flavour depends not only on the basic vanillin, which represents only 2–3% of the whole capsule, but on other distinct substances, resins and oily components. Therefore the natural product is still of superior value as against the synthetic cheap "vanillin". Its chemical composition is: $C_8H_8O_3$ (Theodose 1973).

In spite of the relatively cheap vanilla synthetics, a considerable world trade with natural vanilla capsules still exists. Between 2000 and 3000 t are traded each year, yielding an average of 55 mio, to the benefit of several underdeveloped countries of the tropics. The United States of North America is the largest importer, with 1000 t per year, followed by France and Germany.

## References

Acosta J (1590) Historia natural y moral de las Indias. Madrid

Addison G, Tavares RM (1951) Observaçoes sobre especies do genero Theobroma. Bol Tecn Agr Norte 25:1–20

Akehurst BC (1981) Tobacco. II edit Longman, London

Andrews J (1984) Peppers, the domesticated capsicums, artistically illustrated. Unversity of Texas Press

Ballard R, McClure J, Eshbaug WH, Wilson KG (1970) A chemosystematic study of selected taxa of Capsicum. Amer J Bot 57:225–233

Bastos CN (1981) A new hyperparasitic fungus (*Cladopotryum amazonense*) with potential of control of fungal pathogens of cocoa. Trans Br Mycol Soc 77:273–280

Bolsunov I (1963) Tobacco. Proc III. World Tobacco Sci Congr. Salisbury 264–272

Bohm B, Ganders R, Plowman T (1982) Biosystematics and evolution of cultivated coca (Erythroxylaceae). Syst Bot 7:121–133

Brücher H (1969) Observations on origin and expansion of Yn-virus in South America. Angew Bot 43:241–249

Buchtien O (1910) Contribuciones a la flora de Bolivia, La Paz

Burk L, Heggestad H (1966) The genus *Nicotiana*: a source of resistance to diseases of cultivated tobacco. Econ Bot 20:76–88

Cardenas M (1969) Manual de plantas economicas de Bolivia. Imprenta Icthus

Clausen RE (1941) Polyploidy in Nicotiana. Amer Nat 75:291

Correll D (1953) Vanilla, its botany and economic importance. Econ Bot 7:291–358

Cuatrecasas J (1964) Cacao and its allies, a taxonomic revision of the genus *Theobroma*. Smithson Inst 35:379–607

Ducke A (1964) Plantas de cultura precolombiana na Amazonia Brasilaira. Bol Tecn Inst Agronomico do Norte, Belem, p 24

Entwistle PF (1972) Pests of cocoa. Longman Harlow, London, pp 779

Eshbaugh WH (1980) The taxonomy of the genus *Capsicum*. Phytologia 47:153–166

Eshbaugh WH, Guttman, McLeod (1983) The origin and evolution of domesticated *Capsicum* species. J Ethnobiol 3:49–54

FAO Yearbook (1985) Trade, food agricultural statistics. Rome

Fingerhut (1832) Monographia generis capsici. Düsseldorf

Furst (1976) Hallucinogens and culture. Chandler & Sharp, San Francisco

Gerstel U (1966) Evolutionary problems in some polyploid crop plants. Hereditas, Suppl 2:481–504

Giberty (1979) Especies argentinas de *Ilex* (Aquifoliaceae), Darwiniana 22:217–240

Goodspeed TH (1954) The genus *Nicotiana*. Chron Botanica, Waltham Mass

Goodspeed TH (1961) Systematics of the genus Nicotiana. In: Handbuch der Pflanzenzüchtung. Parey, Berlin V:115–131

Gray JC, Kung SG, Wildman SG, Sheen SJ (1974) Origin of *Nicotiana tabacum*, detected by polypeptide composition of Fraction I protein. Nature 252:226–227

Grubben GJH (1977) Tropical vegetables and their genetic resources. Int Board f Plant Gen Res. FAO, Rome 197 pp

Hasenbusch P (1958) *Capsicum*. Kulturpflanzen Flora USSR (in Russian) 20:394–506

Hernandez F (1651) Rerum Medicarum Novae Hispaniae Thesaurus, Roma

Hunziker AT (1958) A synopsis of the genus *Capsicum*. Rapp 18th Int Congress Botany, pp 73–74

Lathrap DW (1973) The antiquity and importance of long-distance trade relationships in the moist tropics of pre-Columbian South America. World Archeology 5:170–186

Lawrence JS (1978) Screening of cocoa cultivars for resistance to *Phytophthora palmivora* in the collection of CATIE Rev Theobroma 8:125–131

Legg JT (1982) The cocoa swollen shot research project in Ghana 1969–78. Overseas Develop Admin, London

Leon J (1984) The spread of Amazonian crops in Mesoamerica. In: Stone (ed) Pre-Columbian plant migration. pp 167–170

Mansfeld R (1986) Verzeichnis landwirtschaftlicher und gärtnerischer Kulturpflanzen. Springer Berlin Heidelberg New York, p 714

Marin CM (1979) Tobacco literature. A bibliography. Technic Bull North Carolina Agric Res Service Nr 258, p 36

Martin RT (1970) The role of coca in the history, religion and medicine of South American Indians. Econ Bot 24:422–438

McLeod MJ, Guttmann SI, Eshbaugh WH (1982) Early evolution of chili peppers *(Capsicum)*. Econ Bot 36:361–368

Molau U (1983) Flora of Ecuador, Nr 20

Morton J (1981) Atlas of medicinal plants of Middle America. Charles Thomas, Springfield, USA, pp 1000

Pickersgill B (1969) The archeological record of chile peppers (*Capsicum*) and the sequence of plant domestication in Peru. Am Antiquity 34:54–61

Pickersgill B (1983) Migration of chili peppers (*Capsicum*) in the Americas. In: Pre-Columbian migration of plants from Amazonia. Peabody Mus Harvard University Cambridge

Plowman T (1982) The identification of coca (*Erythroxylum*) species. Bot J Linn Soc 84:329–353

Plowman T (1984) The ethnobotany of coca (*Erythroxylum* spp.). Adv Econ Bot 1:61–111, New York, Botanical Garden

Pound FJ (1938) History and cultivation of the Tonka Bean *(Dipterix odorata)* with analyses of Trinidad, Venezuelan and Brazilian samples. Trop Agric (Trinidad) 15:28–32

Prance GT, Kallunki JA (eds) (1984) Ethnobotany in the Neotropics. vol I, New York Bot Garden, 156 pp

Purseglove JM, Brown EG, Green CL, Robbins SR (1981) Spices. Longman New York, pp 813

Reid W (1979) The diterpenes of *Nicotiana* species and *Nicotiana tabacum* cultivars. In: The biology etc of Solanaceae. Linn Soc Sympos 7, London

Rivier L (ed) (1981) Coca and cocaine. J Ethnopharmacol 3:111–379

Rusby HH (1933) Jungle memories. Mac Graw Hill, New York

Ruddat, Lang O, Mosettig (1963) Naturwiss 50:23

Schultes RE (1980) The Amazonia as a source of new economic plants. Econ Bot 33:256–266

Schultes RE, Hoffman A (1980) The botany and chemistry of hallucinogens. Charles Thomas, Springfield, USA

Schultes RE, Hoffmann A (1980) Pflanzen der Götter, die magischen Kräfte der Rausch und Giftgewächse. Hallwag Bern-Stuttgart

Sheen SJ (1972) Isozymic evidence bearing on the origin of *Nicotiana tabacum*. Evolution 26:143–154

Shifriss C, Frankel R (1971) New sources of cytoplasmic male sterility in cultivated peppers. J Hered 62:254–256

Soejarto D et al (1983) Ethnobotanical notes on *Stevia rebaudiana*. Bot Mus leaflets 29 nrl Harv Univ, Massachusetts

Stone D (ed) (1983) Pre-Columbian migration of plants from Amazonia to the Isthmian region and their cultivation. Peabody Mus Harv Univ

Tanako O (1980) Chemistry of *Stevia rebaudiana* Bert. Saengyak Hakhoe Chi 11:219–227

Terpo A (1966) Kritische Revision der wildwachsenden Arten und der kultivierten Sorten der Gattung *Capsicum*. Feddes Report 72:155–191

Theodose R (1973) Traditional methods of vanilla preparation and their improvement. Trop Sci 15:47–57

Thorold C (1974) Diseases of cocoa. Clarendon, Oxford

Towle MA (1961) The ethnobotany od precolumbian Peru. Viking Fund Publ in Anthrop 30, IX, 180 pp

Urquhart DH (1962) Cocoa. Trop Agric Ser. John Wiley, New York, p 368

Wilbert J (1972) Tobacco, a shamanistic ecstasy among the Waroa Indians in Venezuela. In: Furst 1976, pp 55–83

Wood Gar (1975) Cocoa. 3rd edn of Urquhart. Longman, London, p 292

Wood Gar, Lass RA (1985) Cocoa. 4th edn, Longman, London

# VIII. Some Neotropical Timber

## General Remarks

With this theme we enter into one of the most controversial recent debates between ecologists, nature conservationists and biologists on the one hand and the representatives of big business, timber industrialists and investment groups on the other. Without offering any solution for the dramatic situation in forest destruction, we can only quote from an FAO report which speaks clearly: each year 7,500,000 ha of closed tropical forest and 3,800,000 ha of open woods are being destroyed. If this continues at its present rate, in a mere 20 years all will be over. For the year 1995 the world consumption of wood is forecast at 2,300,000 m³ of industrially used trees. This is a 75% increase of the present demand (Catinto 1972, UNESCO/FAO Report no. 14,1978, Neil 1981, Bruenig 1972).

In view of these astronomical figures, all reforestation projects with quick-growing trees must appear as timid palliatives. Such "artificial forests", as proposed by good-intentioned pioneers in tropical forestation, or by big investment groups suffering from development syndromes, are based essentially on *Gmelina arborea*, *Paulownia tomentosa* (from Asia) and some American coniferous species and *Caoba* (= *Swietenea*) (Meggers et al. 1971).

For better understanding of these species we give a short botanical description, extended to some other species with American origin.

## 1. *Araucaria angustifolia* (Bertol.) Kuntze

Pino del Parana, Pino, curiy

Two species are native to America: the subtropical *A. angustifolia* in N Argentina and Brazil and the frost-resistant slow-growing *A. araucana* of South Chile. *A. angustifolia*, locally known under the name Pino de Brasil, produces 40-m-high stems with a basic diameter of 150 cm. Its timber is rather soft (specific weight 0.5) but has long

**Fig. VIII.1.** *Araucaria angustifolia* from Brazil

**Fig. VIII.2** Different "mangrove" trees from the Caribbean coast

(10 mm) fibres which give an excellent cellulose paste. For industrial purposes it is planted on a large scale in Brazil and Northeast Argentina, whilst its natural stands are severely decimated. Its fruits need nearly 2 years for ripening. They produce the well-known "piñas" (3–6 cm large, brilliant brown) which are very nutritive and were therefore collected by prehistoric forest-dwellers. Hybrids between the two species give luxuriantly growing trees (Hueck 1966) (Fig. VIII.1).

## 2. "*Mangrove*" spp.

The mangrove biota is a special ecosystem, formed between sea coast, estuaries and "tierra firme". It has an important function in sediment fixation and especially as protection for many crustaceans, fishes and migratory birds (Teas 1984).
Mangrove forests have recently aquired growing economic importance. Extraction of tannin is one of the most interesting uses of the "manglares" in America, next in importance is the production of charcoal and common firewood. Unfortunately, the pulping properties of mangrove trees are rather poor and paper produced from them is of low quality. However, the comparatively short life utility cycle of 20 years is economically interesting in view of the extermination of other wood (Chapman 1976).
The mangrove vegetation of the tropical coasts of America is composed of different species which belong to the botanical families of Avicenniaceae, Combretaceae, Rhizophoraceae and Verbenaceae. In spite of their very different systematic position, the species exhibit a similar physionomy and have the same tendency to vivipary and pneumatophores (Fig. VIII.2).

### *Avicennia germinans* L.

This species produces 30-m-high trees. Abundant in the estuaries of Guayana, where it uses "pneumatophores" to receive sufficient oxygen for its metabolism. Its local name is "black mangrove". The stems are used for ships' masts. Its wood is heavy, with specific weight of 1.1.

### *Languncularia racemosa* Gaertn.

A usual element of the mangrove swamps in Middle America, where it is called "mangle blanco". Plants are small shrubs. Leaves are coriaceous-oblong with astringent effect. The wood is of restricted value, mostly used as firewood, with a specific weight of 0.86.

### *Rhizophora mangle* L.

Called also "red mangrove", it has long stilt roots, grows on the sea shore but also on river banks and on swampy ground in many places in America (and also in Africa). The trees sometimes reach a height of 25 m, but in general are lower. *Rhizophora* mangle stems have strong durable wood and resist even attack by termites and fungal diseases. Their bark contains 20–30% of tannin. Leaves are leathery, dark green, elliptic to obovate. Flowers have yellow petals. Fruits have a cigar shape, 3 cm long, leathery, with protruding radicle. The fruits begin germination while still hanging on the tree, then they fall perpendicularly down into the mud, where quick juvenile growth occurs.

---
                                                    BOMBACACEAE

### 3. *Ochroma pyramidale* (Cav) Urb.

Balsa tree, pochote

Exceedingly fast-growing tropical tree species, in Ecuador, Colombia, Panama, native from sea level to 1700 m alt.

Its timber has low density and is very light, with specific weight: 0.18. The tree reaches 15–25 m height, often with considerable circumference.

Leaves 30 cm long, three to five-lobate.

Flowers 10–12 cm long, with five petals. Fruits large (20–25 cm) with great quantity of black-coated seeds covered with dense kapok-like fibres.

The wood of balsa trees offers many uses for the natives: balsa dug-out canoes, floating rafts for fishing, etc., as also in modern industry for insulation material and airplane construction, as it is the lightest of all known commercial timbers (Roth 1984) (Fig. VIII.3).

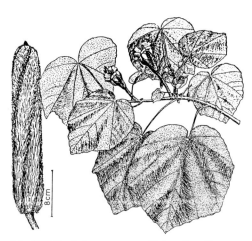

**Fig. VIII.3.** *Ochroma pyramidale,* the famous "balsa" tree (Aristeguieta 1962)

---
                                                    PINACEAE

### 4. *Pinus* spp.

The genus *Pinus* is represented in Central America with five species, but on a worldwide scale there exist more than 70 species. Pines are the most important timber trees of the Northern Hemisphere.

#### *P. caribaea*

It grows from the coasts of Mexico up to 700 m and is well adapted to a hot, moist climate. Its importance resides in its quick-growing lumber and high content of resins for industrial extraction.

This species achieved dubious fame in a giant re-forestation programme in Brazil. In the Jari Florestal of Amazonia the plan existed to plant 400,000 ha with this quick-growing conifer. It is not known if this Ludwig project is progressing, after a series of initial problems.

#### *Pinus radiata*

Its origin is California, where it resists rather dry environment conditions. Thanks to its quick growth and hybrid vigour, *P. radiata* is favoured for large plantations in South America, New Zealand and Australia. Cytological research has established the existence of chromosome fragments which may be related to its heterosis.

---
                                                    MELIACEAE

### 5. *Swietenia macrophylla* Jaq.
Synonym: *Cedrela mahagoni*
Mahogany, caoba, acajou, mahoni, Mahagoni-Baum

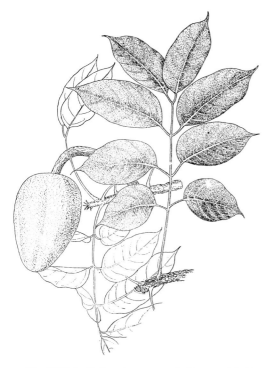

**Fig. VIII.4.** *Swietenia macrophylla,* worldwide known as mahogany or caoba (Aristeguieta 1962)

**Fig. VIII.5.** *Swietenia macrophylla,* an 8-year-old tree on the Island of Trinidad

The genus *Swietenia* is of pantropical distribution. Three species are of New World origin: *S. humilis,* in the septentrional region, from Southern Florida to Cuba, and from Jamaica to Guanacaste in Costa Rica. *S. mahagony,* mostly in Mexico and the Caribbean islands, easy to differentiate from the others by its small leaves.

*S. macrophylla,* a large-leaved species with a southern extension from Tampico (Mexico) to Venezuela, Brasil, Bolivia/Paraguay. The last-named taxa is well known in the whole world for its valuable timber, which has a reddish colour (dark red when freshly sawn) and beautiful texture (Lamb 1966) (Fig. VIII.4).

### NAME AND ORIGIN

The genus name was given in honour of the famous Dutch botanist M. Swieten. After the conquest of America, the Spaniards used

"caoba" on a large scale for ship-building and as wood for their cathedrals.

The American mahogany is sometimes confused with another Meliaceae, the African Khaya tree (*Khaya senegalensis*), which is similar to the American species and also gives valuable timber. The Dominican Republic took the mahogany flower as national emblem and it is said that the cathedral in Sto. Domingo was built in 1550 with selected mahogany timber (Fig. VIII.5).

### MORPHOLOGY

A quick-growing tree, which after 8–10 years reaches an average height of 9–10 m, with 10 cm diameter. An old tree can be 40 m tall. The specific weight of the wood is 0.5–0.55.

Flowers small cream-coloured. Leaves subdivided in three to four leaflets (30 cm long). Fruits are large pear-shaped very prolific in seed production. They contain a great quantity of "winged seeds" which are dispersed by dry wind. The seeds germinate immediately, without any resting period. Young plants are often seen on abandoned agricultural land so that it is easy to establish plantations and recuperate degraded tropical soils.

Future breeding work should be based on especially selected good mother trees with progeny produced by artificial and controlled pollination. This work should be speeded up in view of the quick extermination of good wild stands, which on the island of Trinidad have almost completely disappeared, as we observed in 1972. Venezuela cannot even supply its local demand, after having decades ago lost its position as caoba-exporting land.

coast on account of its dense *Chusquea* thickets, as I observed during my explorations there. *Guadua* is a tropical species, its areal reaching from Colombia to North Argentina. Bamboo species have a fascinating vitality, some species grow 10 cm in 24 h, in only a few months others produce culms 37 m high and 30 cm thick.

### *Guadua angustifolia* Kunth

Tacuara, tacuarazú, bamboo, Bambus

Like many other Bambuseae, *Guadua* has a typical interlocked root system, from which hundreds of stems grow up, sometimes making such a bamboo grove impenetrable. On the other hand, this system also prevents soil erosion and damage from casual overflooding.

---

BAMBUSOIDEAE

---

## 6. *Bambusoideae*

The numerous bamboo species belong taxonomically to the subfamily of Bambusoides. In view of their peculiar habit – woody culms and a complex rhizome system – they have been called "tree grasses". In reality some of the approx. 1000 species are small graceful grasses whilst others reach 20–35 m height with strong branching system and infrequent flowering, sometimes with intervals of 30–100 years. Not all the bamboo have tropical habitats, some species live in regions as far south as 47°, others reach altitudes in Asia of 4000 m and are covered with snow for several months (McClure 1973).

Its industrial importance in Asia is so great that half a million people are employed in its cultivation and manufacture. Therefore we can only recommend stimulating bamboo production in American countries.

In the "New World" the Bambusoides are rather poorly represented by the genera *Guadua* and *Chusquea* (Parodi 1941). The latter is inferior in size and reaches far south to Patagonia. For example, the Island of Chiloé is nearly impenetrable on its west

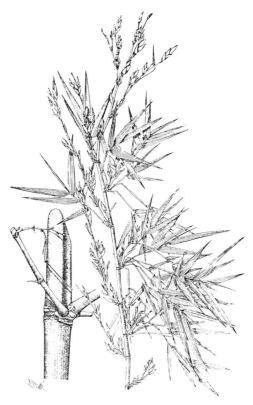

**Fig. VIII.6.** *Guadua angustifolia*, the bamboo of the River Parana in Paraguay and Argentina

The Tacuara stalks ( = cane) reach 25 m height and have thin walls (15 – 20 mm). The stems are hollow inside and have numerous nodes from which small side branches emerge. The internodes (30 – 40 cm long) are dark green at the beginning, but become, when ripening, yellow-brown with darker stripes. Between the nodes solid walls exist which inhibit internal communication. At the basis the internodes are enclosed by big sheaths, (30 – 40 cm) of deltoid form and with a small lamina on the top (Fig. VIII.6). Leaves lineal-oblong (10 – 15 cm long, 0.7 – 2 cm broad), persist for many months.

Inflorescences appear only once in the whole life of this species, possibly after 25 years. The most curious event is that all Tacuara begin flowering at the same time (Soderstrom and Calderon 1979).

After simultaneous flowering and synchronized seed ripening, the plants die away, completely changing the biocoenosis in the border region of great South American rivers. This seemingly catastrophic physiological behaviour, which still needs profound investigation, has been called a "suicidal bout" of reproduction (Janzen 1976).

*G. angustifolia* is used in the American tropics in the most diverse manners. For the natives, it supplies almost all they need for the construction of their rural dwellings. It is known that its canes have a specially high resistance to wood-eating insects. Modern industry uses it for water pipes and fancy furniture (Hidalgo 1974). Bamboo shoots of *Guadua* are not used as food, in contrast to Japan with its yearly production of 40,000 t (mostly from *Phyllostachys pubescens*). In Chinese cooking, different bamboo species are practically irreplaceable for many dishes. So it would appear reasonable to widen the use of bamboo also in tropical America (Ohrenberger and Goerrings 1986).

# References

Aristeguieta L (1962) Arboles ornamentales. UCV. Caracas

Bruenig E (1986) The tropical rainforest as ecosystem. Plant Res 24:15–30

Catinto R (1972) The present and the future of the tropical rainforest. Proc 7 World Forest Congr 2:2432–2441

Chapman VJ (1976) Mangrove vegetation. Cramer, Vaduz

Hidalgo O (1974) Bambu, su cultivo y aplicaciones. Estudios Tecnicos Colombianos Cali, Colombia, p 318

Hueck K (1966) Die Wälder Südamerikas. Fischer, Stuttgart

Janzen DH (1976) Why do bamboos wait so long to flower? In: Tropical trees. Linn Soc Symp Ser 2. Academic Press, London, p 135

Lamb F (1966) Mahagony of tropical America. Ann Arbor Inst, Michigan Press

McClure FA (1967) The bamboos, a fresh perspective. Harvard University Press, Cambridge, Mass

McClure FA (posthum 1973) Genera of bamboos native to the New World. Smithson Contrib Bot (Gramineae)

Meggers B, Ayensu S, Duckworth WD (eds) (1971) Tropical forest ecosystems in Africa and South America. Assoc Trop Biol

Neil PE (1981) Problems and opportunities in tropical rain forest management. Commonw Forestry Inst Oxford. Occ paper nr 16

Ohrnberger D, Goerrings J (1986) A preliminary study of the names and distribution of the Bambusoideae. Bibliography, p 152

Parodi LR (1941) Estudio preliminar sobre el genero Chusquea en la Argentina. Rev arg Agron 8:331–345

Roth I (1984) Stratification of tropical forests as seen in leaf structure. Tasks Veg Sci 6:522

Soderstrom TR, Calderon CE (1979) A commentary on the bamboos. Biotropica 11:161–172

Teas HJ (ed) (1984) Physiology and management of mangroves. Tasks Veg Sci 9:106

UNESCO (1978) Tropical forest ecosystems. A state of knowledge report by UNESCO FAO Paris, Nat Resourc Res 14

# IX. Tropical Pasture Plants

## General Remarks

The protein requirements of tropical peoples are covered, on a global basis, 70% by plants and 30% by animals. However, in some low-income regions, the provision with animal protein is much lower, perhaps only 10%. This deficient situation could be easily improved if the vast underexploited savannas of South America, which have been highly degraded by erroneous human management (burning, erosion, deforestation etc.) were recovered. The term "savanna", created 400 years ago by Oviedo, who described as "zavana" the enormous extensions of grasslands in South America, is now employed the world over (Cole 1983).

With respect to the degraded savannas (actually often mere badlands in South America) I must say that the reason for their poor yield is often the sheer ignorance of their owners. They could have learned how to manage them better, if they had followed the efficient practice, for example, of Australian cattle breeders with "artificial composed pastures". In this respect I acquired decade-long field experience from Argentina and Paraguay.

The omission of a chapter on "Tropical Pasture Plants" in my former book *Tropische Nutzpflanzen* (1977) earned me some criticism. The main reason for this admitted defect was that my editor had set a limit of 500 pages. In the meantime the knowledge of how to create artificial pastures by mixtures of perennial grasses and Leguminosae for tropical husbandry and meat production has increased spectacularly (Skerman 1977; Humphrey 1978). Nobody in tropical and subtropical America could have imagined such progress a decade ago, based as it is on the introduction of highly resistant cattle hybrids, raised on selected perennial forage mixtures. As I was personally involved in such selection work on my experimental fields in the Parana region of Paraguay, I feel competent to encourage Latin Americans to increase their livestock production, using such "pasturas perennes".

The successful breeding work of the Centro Internacional de Agricultura Tropical in Columbia deserves special mention. Recently selected strains and crossings of tropical forage grasses and legumes have been released, for example: *Arachis pintoi* CIAT 17434, *Centrosema* Nr. CIAT 5277, *Centrosema brasiliensis* CIAT 5234, *Stylosanthes capitata* CIAT 10280 ("Capica"), *Stylosanthes guianensis* CIAT 184 (cv. Pucallpa) with high tolerance against antracnosis, which in general causes much damage in this Leguminosae (CIAT 1985).

Other outstanding work in artificial pastures has been performed for decades in Australia by the Division for Agronomy of the C.S.I.R.O and in Queensland and by several private seed companies (Mott 1979; Walton 1983).

From the successful grasses for tropical America we mention *Andropogon gayanus*, even if it is of African origin. This highly productive perennial forage grass has, however, in only one decade made such advances in South and Central America (in Brazil alone more than 150,000 ha) that we may consider *A. gayanus* and its local selections as fully "Americanized". Estimates exist that 300,000 ha will shortly be sown in tropical America, often in mixture with autoctonous Leguminosae like *Stylosanthes capitata* or *Desmodium*, *Centrosema* and *Macroptilium*.

The immense forage production potential of the Latin American tropics is scarcely known in Europe. The savannas of Brazil, Colombia, Peru and Venezuela cover more than 300,000,000 ha. Including pastureland of Mexico, Nicaragua and Panama, this is perhaps a quarter of the tropical American land surface. Instead of continuing there with outdated so-called pasturas naturales, which in reality are only the "fire-resistant remnants" of earlier rich pasture compounds, skillful management should open the frontiers to this vast and undervalued hinterland.

Of course there are limitations: the soils often have a poor infrastructure, they are low in pH (oxisols with much aluminium) and poor in phosphorus and nitrogen (Sanchez and Tergas 1985). To overcome these serious deficiencies the establishment of legume-rich mixed pastures gradually improves the infrastructure and adds nitrogen to the soil. The legume-*Rhizobium* symbiosis is the most economic means of improving the N-content of tropical soils. Is it not surprising that in Latin America only 5% of tropical grasslands contain *Stylosanthes* species of American origin, whilst these Leguminosae have been sown extensively in NE Queensland and have revolutionized the livestock production there (Bogdan 1977, Bress 1980).

Artificial perennial pastures should be a balanced mixture of Gramineae with Leguminosae. In general shortly after sowing, the young plants of grasses grow more quickly and may become the predominant element due to their greater agressiveness. Of course this must be avoided. It is the task of the manager to assure that the original botanical composition of his grazing land does not deteriorate too quickly, knowing that also ruminants' activity and accidental fires have a negative influence on the longevity of intentionally composed perennial pastures (Pizarro et al. 1985).

# A. Forage Grasses

Some promising fodder grasses of neotropical origin are:

|                 |
| --------------- |
|      GRAMINAEAE |

## 1. *Axonopus affinis* Chase
Savannah grass, zacate, amargo

Of Central American origin, now widely sown in artificial pasture mixtures. It is a creeping grass which is rather resistant to overgrazing and poor soil. Introduced in Queensland, it began dominating other pastures there.

## 2. *Bothriochloa pertusa* (L.) Camus

A rather new forage grass, which spread during one decade in the coastal region of North Colombia. It is considered autoctonous to the Caribbean zone. A perennial species with strongly developed stolons. Plants 30–70 cm high with 10–20 cm-long glabrous leaves. Some cattle breeders like its high productivity and resistance to trampling, others consider this "colosuana" as invasory.

## 3. *Brachiaria mutica* (Forsk) Stapf

Whilst most of the *Brachiaria* species are African, this perennial grass may be native to Brazil, where it grows in huge extensions. As a valuable forage it stands in high esteem in many tropical countries.

## 4. *Cenchrus ciliaris* L.
Buffel grass

A prairie grass of worldwide importance covers uncounted thousands of hectares in the Paraguyan Chaco. Its North American origin is under discussion. Due to its optimal growth in subtropical-temperate savannas of both Americas we consider buffalo grass as native here. Other authors believe in an Indian or African origin. This perennial grass is very drought-resistant and does not suffer from strong heat and heavy grazing during the summer. When flowering, it produces typical "foxtail" inflorescences, the seeds enclosed in fine bristles. Abundant seed production favours its natural spreading. In the meantime, several commercial varieties have been selected, some tall (150 cm), others short tussock types.

## 5. *Euchlena mexicana* Schrader

Teosinte, mais silvestre

A near relative to the corn plant *(Zea mays)* and, like it, of Middle American origin. Used as annual forage-grass, due to its abundant phytomass and the high nutritional value of its leaves, similar to maize.

## 6. *Gynerium sagittatum* (Aubl.) Beauv.

Arrowleaf, vara de castilla, lata

A tall grass (3–5 m), which inhabits tropical regions, often river banks. Occasionally cultivated by natives, who use the long inflorescence stalks as arrows, and the leaves as emergency forage during dry seasons.

## 7. *Mellinis minutiflora* Beauv.

Due to its sticky secretion and volatile oil, it is called "molasses grass". Its origin is under discussion; we consider it to be a native grass of Brazil and Venezuela, where it covers thousands of hectares in the hill region. It grows well on acid, sandy soils.

## 8. *Panicum* spp.

From the approx. 60 taxa described from this pantropical genus, only few have New World origin. The following have aquired practical importance: *P. bulbosum* H.B.K. extended from Mexico to Argentina, highly resistant to drought and heat due to its reserve bulbs. *P. elephantipes* Nees, a successful forage grass in Brazil. *P. glutinosum* Swartz, cultivated from Mexico, Brazil to Paraguay and finally *P. virgatum* widely extended from North America to Central America as a valuable covergrass, which is also highly appreciated in Australia and India for animal breeding.

## 9. *Paspalum notatum* Flügge

Bahia grass, pasto horqueta

The last-mentioned species constitutes an important element of the natural grass vegetation of the humid pampa and subtropical steppe of South Brazil, Uruguay and North Argentina. The stems are erect and in the flowering stage reach 120–150 cm, in general with short rootstocks. The leaves are long, soft and very palatable for cattle. Similar in habit and economic value is the related taxon *P. dilatatum* Poir.

## 10. *Sorghum almum* Parodi

In spite of the well-known fact that the radiation centre of the genus *Sorghum* is in Africa, *S. almum* arose in Argentina, as a spontaneous hybrid between wild *S. halepense* and a cultivated *Sorghum*. As this perennial herb produces fertile seeds, Parodi was quite right to establish it as a new, autocthonous American species. Vigorous plants (2–3 cm tall) with short creeping rhizomes and broad leaves.

## 11. *Tripsacum latifolium* Nash

Guatemala grass

A vigorous perennial of the Maydeae family. The leaves have the same nutritional value as maize, with the advantage that the plants recover quickly after having been grazed down to the ground.

# B. Forage Legumes

## 12. *Arachis* spp.

Several wild-growing *Arachis* groundnuts species are promising forage plants (Grof 1985) and have been improved recently at Brazilian experimental stations. We mention the following:

### A. glabrata

The plants are perennial, with strong rhizomes, but weak in seed production. Therefore multiplication must be done vegetatively. Short cuttings are ploughed down, often in mixture with perennial grasses. The yields in such cases may reach 10–15 t of dry hay per ha per annum.

### A. pintoi

In the Southern States of North America, considered a promising protein-rich pasture. The crude protein content of its leaves is 14%. After several years, the soils have been considerably enriched in nitrogen.

### A. prostrata

This wild ground nut aquired in Australia a certain reputation as a perennial Leguminosae and good mixture with tropical grasses which does not suffer from pathogens. It also resists acid soils well.

## 13. *Aeschynomene americana* L.

The local names are "joint vetch", "pegape-ga" or "Yerba rosada". It is an erect herb (2–3 m tall) with pale orange or violet flowers ascending on other plants, with a woody base and strong pubescence of the stem. Native to tropical America, the species has now spread to Florida in the North and Paraguay and North Argentina in the Southern direction. There exist some other closely related species, for example *A. sensitiva* and *A. panniculata*, called also the "thornless mimosa", with 20% protein in the dry mass.

## 14. *Calopogonium coeruleum* Benth.

This promising perennial legume is still under experimental study in Brazil, where a local selection, called "calopo" is propagated.

**Fig. IX.1.** The pasture legume *Centrosema pubescens*

**Fig. IX.2.** Field for seed production of *Centrosema acutifolium*, cv. Vichada

### 15. *Centrosema acutifolium* L. and *C. pubescens* Benth.

Of the existing 30 South American taxa, these two species seem to be the most vigourous. Its origin is Southern Brazil, where the farmer call the first one "vichada" and the second "bejuqillo". Both are new, highly promising components of improved pastures (Schulte-Kraft 1986). Under good management they form dense mattress of 50 cm density, finally climbing on other herbs, highly appreciated by cattle and horses (Fig. IX.2).

### 16. *Desmodium* spp.
Several taxa with 2 n = 22 chromosomes.

This large genus comprises nearly 200 species, some of old world origin. We mention the following taxa, some of them already used in pasture mixtures:

### *D. barbatum* (L) Benth.

Known under the name "tick clover" or barbadino. A perennial dragging species with puberulous fruits.

### *D. ovalifolium* Wall.

Appreciated in Colombia for its vigorous growth and high nutritive value for cattle raising.

### *D. discolor* Vogel.

A perennial climber (2–3 cm tall) trifoliate with big leaves, leaflets up to 8 cm long. With a special deep root system. High protein content (17% crude protein in dry matter). Native in NE Argentina and Paraguay

### *D. tortuosum* (SW.) DC.

Known as beggar weed or mozoton. Very robust annual plants with 3-m-long vines. Fruits crooked.

### D. uncinatum (Jacq.) DC.

In several tropical countries propagated as selected varieties calles "silverleaf desmodium" or also "alfalfa of the tropics". In view of its excellent performance as a perennial pasture plant of high nutritive value.

## 17. *Dioclea guyanensis* H.B.K.

A perennial line with big violet flowers, forming broad pods with large (8 mm) seeds. Many palatable leaves. A new introduction as forage plant, which has shown excellent adaptation to acid soils.

## 18. *Lathyrus* spp.

Grass pea, bitter pea = almorta, Platterbse
$2n = 14$

This pantropical genus consists of more than 120 taxa. Outstanding practical importance was achieved by the eurasiatic species *L. sativus, L. latifolius* and *L. odoratus,* the former cultivated over many hundred thousand hectares in India for its protein-rich seed (see Brücher 1976).
Twenty species are indigenous to the American continent, with *L. latifolius* spreading recently in huge quantities in the irrigation zone of the Pre-Cordillere in Argentina and Chile, seemingly very well adapted to its arid climate. In view of its spontaneously increasing areal, we include it in our short description of autochthonous species:

### L. latifolius L.

This invasionary species may in the near future represent an important pasture plant for dry regions with their notorious lack of animal foodstuff. Its seeds contain 24–28% protein, 1–2% fat and the rest are carbohydrates. The handicap is its neurotoxical effect. This factor could be eliminated by mutation breeding, with results similar to those achieved by Indian scientists recently with

their "khesary dal" *(L. sativus)* plants. After intake of the protein-rich but toxic seeds, human beings suffer from a neuro-disease called in medicine lathyrism. The same substance (BOOA = aminopropionitrile) also disturbs the metabolism of warm-blooded animals (Bress 1980).

### L. macropus Hook. & Arn.

Extended in the Southern Cordillere, above 2500 m alt. Robust perennial herbs. Leaves with three pairs of leaflets, elliptical. The blue flowers together (8–18) in extended racemes. Fruits 5–6 cm long, when ripe, dehiscent. Not eaten by cattle, therefore still covering the Andine slopes.

### L. pubescens Hook. & Arn.

Now only rarely found in the Andes, because cattle devour it. From the perennial axonomorphical rootstock ascend winged stems, climbing on bushes emerge each year. Leaves uni-bi-jugate, with thick hair cover. Inflorescences with a dozen blue flowers. Almost extinct in the Cordillerean valleys, where it was formerly abundant, because domestic animals ravage it.
Other *Lathyrus* species of minor importance are *L. crassipes* and *multiceps* in the Pampa plains of Argentina.

---

Subfamily MIMOSOIDEAE

## 19. *Leucaena* spp.

Horse tamarind, leukena, guaje

Taxonomists have established ten different taxa of the genus *Leucaena* with its polymorphic centre in Middle America. The original habitat reached from South Texas through Mexico, Guatemala, Honduras to Panama, perhaps to Ecuador, but the actual area has been considerably enlarged in tropical America due to its use as green manure and forage plant, and now includes also Paraguay and North Argentina.

The genus *Leucaena* has its original centre in Mexico-Guatemala, with several species reaching as far North as Texas and South to Panama, Venezuela and Ecuador. Its actual areal has been enlarged worldwide. Indians used the pod as food and cattle breeders spread it as forage bush and green manure. Heavy frosts limit the expansion of *Leucaena* beyond the subtropics. Medium frost kills the above-ground parts, but the bush grows back from the basal trunk and forms multibranched shrubs. *Leucaena* roots reach very deep. From the central taproot many small lateral roots develop, which carry abundant nodules housing *Rhizobium* that fix atmospheric nitrogen. In dense *Leucaena* stands the production of nitrogen may reach 100–200 kg/ha (Brewbaker 1984).

Evergreen, short fast-growing shrubs of 4–5 m height. There exists also a "giant type" in cultivation, 20 m high, with large leaves and thick, almost branchless trunks. Leaves large bipinnate (10–14 cm) with tiny leaflets, which fold up under stress, for example drought, hot weather or darkness. Rich in provitamin A and carotene. Protein of good digestibility, similar to alfalfa.

Inflorescences multiflowered globose heads, like white fluffy balls, with diminutive flowers, self-pollinated.

Fruits are produced in drooping clusters. They are green-translucent when young, and extremely flat. On ripening, the pods become red and hard. They contain 12–28 seeds which are ejected when ripe and the pods split along the edges. Seeds have a waxy coat and a shiny brown colour. Quick germination after 1 week.

The genus has a serious drawback as forage plant. It contains a toxic substance (mimosine) which disturbs the metabolism.

With respect to diseases it must be mentioned that sometimes a bacterial blight on pods occurs, mainly in the humid climate of Panama, Colombia and Brazil. The fungus diseases *Camptomeris leucaenae* and *Fusarium semtectum* have been reported.

### L. diversifolia (Schlecht.) Benth.

This is a fast-growing small tree, which is tolerant to tropical acid soils. It is interesting for its low toxity (it contains only little mimosine) and good wood. The flowers are pink, the leaflets are thin. Grows up to 2000 m altitude and is rather tolerant to cold climate.

### L. esculenta (Moc. & Sesse) Benth.

Also with low mimosine content and good yield in wood, but forming tall trees (18 m) in the highlands of Mexico, where its fruits are eaten. Rather cold-resistant.

### L. glauca (L. Benth.)

$2 n = 104$

Its home region is Central America, but natural stands have been reported from several North Argentinian provinces where pods are eaten cooked, by natives. This forage bush is comparatively frost-tolerant.

### L. macrophylla Benth.

A drought-tolerant shrub of the Mexican lowlands. Excels by its rapid growth, large branches with large leaflets. Is comparatively tolerant to acid soils.

### L. pulverulenta (Schlecht.) Benth.

The most septentrional taxon, growing up to 35° latitude and comparatively cold-resistant. Forms rather tall trees with very dense wood. When crossed with other species, produces hybrid vigour, and the offspring maintain its low minosin level.

### L. leucocephala (Lam.) De Wit.

Has many abilities. First of all it is now used – world wide – as a forage plant, but it also enriches tropical soils, thanks to its symbiosis with nitrogen-fixing *Rhizobium* bacteria and its abundant leaf drop. *Leucaena* trees are also used as shade for valuable perennial crops and as windshields. The young leaves are eaten by humans as vegetable, in spite of their known mimosine content. Its para-

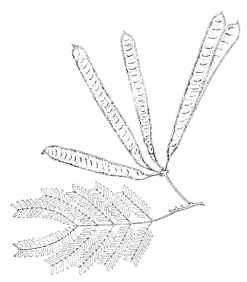

**Fig. IX.3.** *Leucaena leucocephala*

mount importance, however, lies in its high potential as a forage plant. Selected biotypes are sown as perennial pasture. With good management, such plantations may survive for two decades. It is always necessary to cut the small trees back to the ground, to produce new, soft shoots (Fig. IX.3).

## 20. *Macroptilium* spp.

A neotropical genus extending from Middle America to North Argentina. In Argentina alone seven perennial taxa exist (Burkart 1952) that are morphologically related to the genus *Phaseolus*.

Economically most interesting is *M. atropurpureus* (DC) Urb to which numerous varieties belong, some with underground fruiting and others with tuberous roots. Australian plant breeders developed from this outstanding South American fodder Leguminosae with the name Siratro. Strangely enough, the seeds of this forage plant, which grows wild on the borders of town streets in the capital of Paraguay, must be imported from Australia to Paraguay for lack of sufficient local seed material.

## 21. *Stylosanthes* spp.

This pantropical genus includes 30 species, with a dozen taxa on the American continent, from Mexico to Argentina. Some are annual with relatively low leaf production,

**Fig. IX.4.** *Stylosanthes capitata*

like *S. angustifolia* Vog. and *S. capitata* Vog. from Venezuela, sometimes used for crossings due to its immunity to fungus diseases (Fig. IX.4). *S. hamata* Taub. and *S. humilis* H.B.K. are native to the Caribbean region and Brazil. The latter is tolerant to high aluminium and manganese % in the soils (Staces and Edye 1984). For this reason, Australian breeders selected three commercial varieties which yield well under these soil conditions. The economically most attractive species is:

### *S. guianensis* (Aubl.) Sw.
Brazilian lucerne

It is native to North Brazil and Guyana. Erect herbs, with hairy stems (1 m long) and trifoliate leaves which are rich in protein (crude protein 12–18% in dry matter). The flowers are yellow. Appreciated for its drought-resistance and adaptability to grass mixtures with *Imperata cylindrica* and *Paspalum dilatatum*. Also called Alfalfa do Brasil or finestem stylo. Other South American species of minor importance are *S. scabra* ("shrubby stylo") and *S. viscosa* with pubescent fruits and viscose glandulas on the whole plant. Cytogenetic research in Australia has shown that the genus *Stylosanthes* represents a polyploid series from $2n = 20$ (*S. angustifolia* and *S. capitata*) to $2n = 40$ (*S. mucronatus, S. tuberculatus*) and finally $2n = 60$ chromosomes (*S. erectus*).

### 22. *Zornia* spp.
$2n = 20$

This American genus includes approximately 50 taxa, disseminated from the Carribean Islands, Panama, Brazil to North Argentina (e.g. *Z. glabra, Z. latifolia* and *Z. brasiliensis*).

### *Z. diphylla* (L.) Pers.
Barba del burro

An annual herb, with only two leaflets and yellow flowers (with big stipular bracts). Pods fragile when ripe. In good soils 50–70 cm tall. Considered as a valuable forage plant.

### References

Bogdan AU (1977) Tropical pastures and fodder plants. Junk, den Haag

Bress J (1980) Nutritional value of legume crops for humans and animals. In: Advance of Legum sci. Summerfield & Bunting. Kew Bot Garden, England, p 35–55

Brewbaker JL (1984) *Leucaena*, promising forage and tree crop for the tropics, 2n edn. National Acad Press, Washington DC

Brücher H (1977) Tropische Nutzpflanzen. Springer Berlin Heidelberg New York, p 500

Burkart A (1952) Las Leguminosas argentinas. ACME, B-Aires, p 570

CIAT (1985) Tropical pastures. Document no. 17, Cali-Colombia

Cole M (1983) The savannas, biogeography and geobotany. Dept Geogr Royal Bedford College Univ London

Grof B (1985) Forage attributes of perennial groundnut *Arachis pintoi* in a tropical savanna environment in Colombia. Proc Int 15th Grassland Congr, Kyoto, Japan

Humphrey LR (1978) Tropical pastures and fodder crops. Longman, London

Mejia M (1984) Nombres cientificos y vulgares de especies forajeras tropical. CIAT, Cali, Colombia

Mott G (1979) Handbook for the collection, preservation and evaluation of tropical forage germplasm resources. CIAT, Cali, Colombia

Pizarro EA, Toledo JM, Amezquita MC (1985) Adaptation of grasses and legumes to the humid tropics of America. Cali, CIAT, Proceedings

Sanchez P, Tergas L (eds) (1985) Pasture production in acid soils of the tropics. CIAT, Cali, Colombia, p 521

Schultze-Kraft R (1986) Natural distribution and germ plasm collection of the tropical pasture legume *Centrosena*. Ang Bot 60:407–419

Skerman PJ (1977) Tropical forage legumes. FAO, Plant Prod Prot, Rome, p 600

Stace HM, Edye LA (eds) (1984) The biology and agronomy of Stylosanthes. Townsville, Australia, p 656

Walton PD (1983) Production and management of cultivated forages. Restan Publ Co Restan VA USA

# X. Aromatic and Fleshy Fruits

## A. Fruit Trees

ANACARDIACEAE

### 1. *Anacardium* spp.

Due to the paramount importance of the globally known cashew nut, *A. occidentale*, it is often overlooked that in America eight more *Anacardium* taxa exist, some of them large trees, others medium-sized shrubs, and all useful for different purposes. The group of tall *Anacardium* trees includes the following:

### *A. giganteum* Hancock
Wild cashew, caju gigante, caju-assu

This large tree, 22–30 m high, is native to the Amazonian and Guyana rainforests. Similarly to the cashew tree, from its terminal inflorescences fruits develop with a fleshy peduncle, red-coloured and edible, which are collected by natives, who consider it even better than the common cashew nut.

### *A. excelsum* Skeels
Synonym: A. spruceanum
Espavé

In Darién, Choco Indians explained to me that this is the best tree for making dug out canoes. The trees I saw had 120 cm diameter at the basis, and may have reached 25–30 m in height. The bark is used for fish poisoning. The stems exude a gum, which is also utilized. Flowers are small, with 10-mm-long petals, 10 stamens, with one stamen much longer than the others. The fruits are in general not eaten. Before the destruction of the Panama forests, the species formed dense stands along the rivers.

### *A. rhinocarpus* D.C.
Mijao, Pauji

A really majestic tree of the Llanos and Venezuelan forests. Some are 40 m in height with a basic circumference of 7.5 m, as described by Pittier (1971). The leaves are simple, coriaceous. The flowers sit close together in terminal racemes. The fruits are biconvex, small, nearly the same as in *A. occidentale* only of smaller size and also edible; however, mostly used as pig fodder.

### *A. pumilum* St. Hil.

This is a low, bushy tree, native to the semi-arid central parts of Brazil (Minas Gerais). Its small fruits and seeds are collected and eaten by Indians.

### *A. humile* St. Hil.

This species has very short trunks, often twisted and sprawling on the ground. It can grow even on the salty soils in the torrid Brazilian interior.

### *A. occidentale* L.
Cashew tree, caju, merei, marañon, pomme acaju, Elefantenlaus
$2n = 42$

NAME AND ORIGIN

The Tupi Indians from East Brazil call the fruit "acaju", a name which still persists in

**Fig. X.1.** *Anacardium occidentale,* branch of the cashew nut tree with flowers and ripe fruits

our modern languages. The cashew tree is native to the semi-arid coasts of Venezuela and Brazil, where one can still find many biotypes in the wild state (Fig. X.1). The first description was made by Thevet, who already in 1558 gave illustrations of the cashew tree. The Portuguese navigators early recognized its future commercial importance and in the 16th century took seeds to India and the East African coast (Mozambique).

MORPHOLOGY

Tall spreading evergreen trees (10–12 m), with dense canopies. Leaves are simple, leathery, with prominent veins, 6–20 cm long and 5–10 cm broad, with short petioles, which have a swollen basis. Young leaves have a showy red colour. The inflorescences are long, many-flowered panicles, with small pentamerous white corollas. Inside the flowers, the style is longer than the filaments, which favours cross-pollination. Small insects are the preferred agents for fecundation. Nonetheless there occur deficiencies in fruit setting, only 5% of hermaphrodites give fruits. Such irregularities in

flower biology need further study, and at present make hand-pollination necessary. The fruits have a curious aspect. They are swollen fleshy pedicels ( = hypocarps), wrongly called apples, and a seed kernel on the top of the reddish or yellow "fruit". This is very aromatic, rich in vitamin C, 10–15 cm long and 4–8 cm broad. They are sweet tasting and may be eaten raw or used for juice and jam. But in general these "fruits" stay unused after the harvest of the seeds. The cashew nut is the most important product of the tree. It has a high fat content (40%) and is a good source of protein (15%). The nuts are 25–40 mm long and are covered with a hard shell. Unfortunately, the pericarp contains a caustic oil (cardol) which affects the skin. Cracking open can almost only be done by hand. Many attempts to replace the manual labour by mechanical processes have failed so far. The presence of the phenolic resine is still a nuisance for harvesting and shelling, which in Asia is done by cheap labourers, who consequently suffer from skin diseases. The same cardol resin has a good price on the world market because it is friction- and heat-resistant and represents a valuable by-product of the cashew industry.

ECONOMIC POTENTIAL AND IMPROVEMENT

Plantations in different tropical regions of
Africa, Asia and in its original homelands
Brazil and Venezuela are increasing. World
trade is estimated at 600,000 t, but a further
100,000 t are soon expected (Ohler 1979). In
general the yields are low, if local half-wild
plant material is used. Differences as high as
100% in local yields have been observed.
For this reason plus-variants should be selected and multiplied, if necessary also by
grafting and cloning. Actual average
yields are: 7000–9000 kg "cashew apples";
700–900 kg whole seed kernels, or 150–
300 kg of shelled nuts per ha. The nut is rich
in protein (21%) and fat (46%) and is a
good source of vitamin C (180 mg/100g).
Promising new varieties from Brazil are
Amarelo Gigante or Manteiga. On the Island of Trinidad, a selection exists with nuts
5 cm long; 78 of them have a weight of 1 kg.
Genetical improvement should be directed
at the following practical aims: higher proportion of functional bisexual flowers,
shorter flowering phase, better nut size and
kernels with higher protein content, possible
resistance against several diseases, e.g. *Corticium salmonicolor, Colletotrichum, Pythium* and *Fusarium*, which especially attack
the seedlings.

## 2. *Spondias purpurea* L.

Red-mombin, jocote,imbuzeiro, cajazeiro,
mombin rouge, Mombin-Pflaume

Probably of Central American origin, native
also in the Antilles. Small trees (3–7 m)
without spines on the stem. Rather resistant
to different soils and altitudes (up to
2000 m) in the tropical mountain region.
Leaves typically imparipinnate with 5–7
pairs of leaflets (12–22 cm long). Inflorescences are small clusters with unisexual and
bisexual flowers on the same raceme. Small
pentamere corollas. Fruits are ellipsoidal
drupes (3–5 cm long) with brilliant red epicarp. The pulp is sweet-sour with an agreeable taste. The seeds have a hard endocarp.
The fruits are a valuable source of vitamins
A and C (Fig. X.2).

**Fig. X.2.** *Spondias purpurea*

## *S. mombin* L.

Golden mombin, hog plum, jobo, cirulea del
monte, marapa

Distribution from Mexico to Paraguay, and
generally called by its Taina name: "jobo".
In Venezuela we observed very large trees in
tropical forests. Their bark exudes an astringent gum-like sap. During fruiting time,
thousands of yellow fruits cover the soil under such trees, avidly devoured by wild animals and hogs ("hog plum"). The stem has
spines, but the branches are free of them.
Leaves are similar to the foregoing species,
but larger (30–50 cm).
Inflorescences on long (20–40 cm) panicles.
Flowers small, yellow-whitish, with 5-mm-
long corollas. Fruits are 3–5-cm-long drupes with yellow epicarp, containing a soft
sweet-acid pulp, which is eaten raw or used
for many preserves, syrups and beverages.
The fruit is similar to European plums. Selections from the rich "gene pool" of wild-
growing biotypes in Brazil, Venezuela and
the Caribbean islands should be undertak-
en, especially in view of the existence of the

very attractive Asiatic species *S. cytherea,* the "Otaheite plum" from Tahiti.

Finally we mention two taxa of minor importance from the neotropics:

### S. tuberosa Aruda

Imbu, umbuzeiro

This is a small, often creeping tree from the arid Catinghas in the North east of Brazil and the sertaos. The 3-5-m-tall tree produces many superficial roots (not really "tubers") with watery tissue, 10–15 cm in diameter. Natives gather the watery roots in cases of thirst and drought. Its sweet-sour fruits are used for refreshing drinks.

### S. venulosa Mart.

Exists in different varieties which cover Venezuela, Guyana, North-east Brazil and the Atlantic coast of Rio de Janeiro. Exclusively wild-growing trees (10–18 m tall). Its fruits are collected by natives.

### 3. *Schinus molle* L.

Peruvian pepper tree, arbol de pimienta, aguaribay, molle, faux poivrier, Pfefferbaum

Slow-growing ornamental trees (10–25 m in height), with dense canopy of imparipinnate (10–14 pairs) leaves. Abundant production of coral-red odorific fruits (5 mm diam.), which contain an essential oil (also piperine). The berries are pulverized by natives to obtain cooling or weak alcoholic drinks.

The bark contains tannin (6–10%) and the stem exudes a valuable mastix. The "arbol de pimienta" occurs spontaneously from Mexico to Brazil, Peru and as far as Central Argentina.

---
ANNONACEAE
---

### 4. *Annona* spp.

The botanical family of Annonaceae is represented in the neotropics by two genera

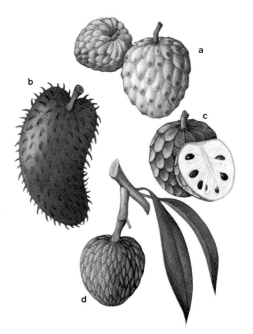

**Fig. X.3.** Different species of *Annona* fruits (Kranz 1981) **a** Annona mamillata; **b** Annona muricata; **c** Annona cherimola; **d** Annona reticulata

which are rather similar in morphology and practical use; *Annona* and *Rollinia*. On the American continent more than 70 *Annona* species exist, of which at least a dozen contribute aromatic fruits and have been domesticated. Several taxa exist only as wild plants, but its fruits are also regularly collected, for example: (Fig. X.3)

### A. glabra L.

We have seen it growing wild on banks of rivers in Panama. Its small fruits have a pleasant odour, yellow flesh and black kernels. It cannot be excluded that *A. glabra* may be an ancestral form of some of the domesticates.

In Brazil various wild species live in the "cerrados". For example *A. crassiflora* Mart. and *A. coriacea* Mart, distinguished by good drought resistance to seasonal lack of moisture. Finally we mention *A. purpurea* Moc.& Ses., called by the natives "torete", with a vast distribution area from Mexico to Panama to Venezuela.

## *A. cherimola* **Mill.**

Custard apple, cherimoya, chérimolier,
Rahmapfel

$2n = 28$

In contrast to the other edible *Annonas*, the cherimoya prefers the cooler climate of the tropical mountain regions (800–2000 m) of Colombia, Ecuador and Peru. Domesticated varieties have existed since Inka times, with an exquisite aroma and pulp consistence. We do not hesitate to consider cherimoya as the most attractive fruit of America.

The trees are of medium size (5–10 m high), with grey-pubescent branches. The leaves are oval-elliptical, dark green on the surface and typical brown-tomentose beneath. The flowers sit on the leaf axils and have a greenish white colour, similar to small magnolia flowers, and consist of three sepals and six petals (2 cm long). Flowers are auto-sterile. Fruits are fleshy syncarps, as a result of the congregation of many single carpels after fecundation. The original separation of the carpes can be easily observed on cutting such a multiple fruit of the *Annona* species longitudinally. The pulp is rich in sugar (12%) and aromatic substances, and encloses many shiny black seed kernels. Unfortunately, the cherimoya fruits have a soft exterior and ferment quickly. Therefore it is difficult to export them from South America to other continents. Selection of superior planting material is possible by grafting.

## *A. diversifolia* **Saff.**

Ilama, anona blanca

In contrast to cherimoya, which needs higher elevation for good production, the "ilama" can be grown under humid hot tropical climate. Its place of origin is the Mexican and Guatemalian lowlands along the Pacific coast. The fruits are larger (15 cm diam.) and even considered of better taste than other Annonaceae. Consequently this species should receive genetical improvement. The sweet pulp is cream-pink in colour and has an exquisite aroma. The species is self-fertilizing, an exception to other domesticated *Annona* species.

## *A. muricata* **L.**

Soursop, guabana, guanabana, jacá de pará corossol, Stachel-Anone

$2n = 14$

Its origin is most probably Brazil or the Antilles, where it was cultivated, but it was not found growing wild there. Before the Spanish conquest, the guanabana had been widely distributed in South America. The tree has an attractive habitus and is also used as an ornamental and shade tree in the tropics. It grows to 8 m in height. The leaves are ovate-oblong (6–16 cm long), dark green with visible nervature (6–12 pairs). Flowers are solitaire on 2-cm-long pedicels with long fleshy, (4–5 cm) petals, emanating a pleasant odour. Fruits are large (20–35 cm long) cordiform ovoid, green, covered with pointed outgrowths, similar to small thorns. Their weight is 2–3 kg. They are the biggest fruits of the whole genus. The pulp is white, juicy, acid-sweet and contains many black seeds. Most of the 12% sugar is glucose. The fruit is additionally a good source of niacin and riboflavin, also of vitamins B and C. The seeds have a toxic substance, which is used by natives as a repellent against body parasites. *A. muricata* is one of the most promising tropical fruits for commercial development, because its conserves do not lose their aromatic flavour and can be used in refreshing beverages, milkshakes, ice creams etc. The yields are low, one or two dozen fruits per tree. One of the reasons for this is deficient pollination. Commercial plantations need hand fertilization and protection against fruit flies and mealy bugs, which cause severe damage to the ripening fruits.

## *A. purpurea* **L.**

Custard apple, anona-colorada, bullock's heart, toreta, soncoya

In Central America, where it originated, it is one of the favourite fruits and is often offered for sale on the native markets. The custard apple grows well at middle elevations of the tropics. We have seen it in Panama growing wild. The trees reach 8–10 m in height.

Leaves are big (12–30 cm long), simple, elliptical-oblong and typically ondulate.

Flowers are solitary, yellow, inside red-violet. The fruits are globose large (14–16 cm), ovoid, and covered with a reddish tomentum. The carpels of the syncarp have prominent large rigid protuberances, typical for this Annonaceae. The pulp has a good consistency, is orange-coloured and tasty, and encloses rather large (3 cm) brown seed kernels. Fruits are often used in refreshing beverages.

### A. reticulata L.

Sweetsop, bullock's heart, corazon de buey, cachimantier, anona corazon, Rahmapfel

Small trees (3–8 m tall), of Central American origin. Well adapted to hot climate and quick growth; producing the first fruits already after 3 years.

Leaves (8–26 cm long) lanceolate-oblong, simple, with acute apex.

Flowers in clusters (3–12), with pleasant odour, rather small.

Fruits globose-heart-shaped (8–12 cm diam.) with a smooth surface, which give them a more attractive appearance than other Annonaceae. The outside is yellow-brown, the pulp inside cream-coloured with many brown seeds. The "sweetsop" deserves genetical improvement, due to its favourable fruitform.

### A. squamosa L.

Sugar-apple, Anón, Attier, Saramuyo, Pomme canella, Atá, Anona-blanca, Zimtapfel

$2n = 14$

Native of Central America and the Antilles, with high tolerance to alkaline soils and drought spells. The species is now dispersed in all tropical countries, especially in India, which is erroneously considered as their homeland.

The trees are of medium size (7–8 m) and are branched at the basis.

The leaves are lanceolate elliptic (5–15 cm long) with a brillant green surface.

The flowers are small, hang down in small clusters on the side branches and are greenyellow in colour. The floral biology inhibits self-pollination due to the fact that the pistils begin to wither as soon as the anthers open.

The fruits are small (5–10 cm diam.) with a juicy pulp of agreeable flavour, but the excessive quantity of small black seeds is a disadvantage. The fruits must be harvested before they are completely ripe; otherwise the carpels become separated and the fruits suffer immediate decomposition. An analysis of the fruits, made in Panama, gave the following results: 69% humidity, 2.5 g fibre, 1.3 g ash, 4.4 g calcium, 0.4 g nitrogen, 55 mg phosphorus, 1 mg iron, 42 mg ascorbic acid, 1.3 mg niacine (Esquivel).

Occasionally seedless mutations occur, a very important character for vegetative multiplication of high quality fruits.

Important for future commercialization are fertile hybrids between *A. squamosa* and *A. cherimola*. These hybrids, commercially denominated "Atermoya", combine the outstanding characters of the two species, i.e. the sweetness of the former, with the exquisite aroma of the cherimoya. These plants show hybrid vigour and produce large quantities of olive green fruits (8–15 cm long) with white pulp.

The genus *Rollinia* is similar to *Annona* and includes also some valuable fruit trees, for example:

### 5. *Rollinia deliciosa* Saff.

Biriba

The tree is 8 m high and needs a hot humid climate; common in North Brazil. The fruits are 8–12 cm in size and have a sweet whitish pulp, with good sugar and vitamin content (Fig. X.4).

### R. jimenezii Saff.

Anonillo

Grows wild from Mexico to Panama, and is occasionally collected by natives, although it has a quite acid taste. The fruits have prominent protuberances.

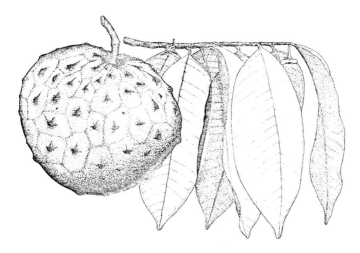

**Fig. X.4.** *Rollinia deliciosa*

### R. multiflora DC.

Cachimen

This species is appreciated in Venezuela more for its timber than for the edible fruits. The trees are 10 m high with straight stems of compact texture.

### R. emarginata Schlecht.

Arachichú

This is the most southern growing *Rollinia*, spread from Brazil to Paraguay and North Argentina. It has a bushy habit, 2–4 m high, and produces small fruits (2–3 cm diam.) yellow in colour when ripe, which are occasionally collected by the natives.

APOCYNACEAE

### 6. *Couma utilis* (Mart.) Müller

"Cow-tree", vacanosca, sorveira, guaimaro macho.

Showy trees (20–30 m tall) native to the Amazonas region, but spread also to Venezuela, Colombia and Peru, probably by wandering Indians who use the rich flowing latex as substitute for milk, and making chicle. Certain tribes of the Amazonas valley celebrate fiestas when the fruits are ripe. Stems are thick with a smooth bark. The leaves are long elliptical (8–20 cm) glossy dark green. The flowers are white, pentamer, with a 1–cm-long tube. The fruits are round berries (5–8 cm diam), yellow when ripe. The exocarp is thin, the pulp is white, sweet-aromatic, with numerous small seeds. Not only are the fruits valuable, but the resin is also exploited for chicle fabrication. When the bark is cut, there is a copious flow of latex, of agreeably sweet taste. Obviously its consumption is without side effects, because the natives of Venezuela drink it in quantity, like milk.

### C. macrocarpa Barb.-Rodr.

This species is similar to that described above, with a distribution north of the Darién gap. Natives use the latex and also the timber. It is remarkable that the latex of these two species has no noxious effects, considering that several species of the Apocyanceae are very toxic.

BOMBACACEAE

### 7. *Pachira* spp.

Several *Pachira* species are represented in Central and South America, and are appre-

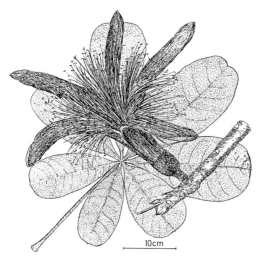

**Fig. X.5.** *Pachira insignis,* flower and leaf (Aristeguieta 1962)

ciated by natives for their nutritive fruits and seeds. *P. macrocarpa* Walp. grows in Mexico and Costa Rica. *P. insignis* Sav. is known from Trinidad-Tobago and Venezuela.

*P. aquatica* Aubl. is the most important species, and appears to have been domesticated by Indians in Central America. We found it in Panama but also in Paraguay. The edible seeds are sometimes called "chestnut of America". The trees are 5–15 m high and grow along waterways and on swampy rivers (Fig. X.5).

## 8. *Matisia cordata* H.B.K.

Chupa-chupa, chichi, zapote de monte

Extended from Panama southward to Colombia-Peru and West Brazil. Humboldt and Bonpland collected this fruit tree for the first time at the borders of the Magdalena river (1801) and were impressed by its tasty fruits. The trees are small (4–6 m) and ramificated. The leaves are large (25 cm) and mainly inserted at the end of the branches, which gives the tree a peculiar aspect. The flowers are white, the fruits are egg-shaped (8–12 cm diam.) with a brownish skin, covered with short hairs. The five big seeds are connected with thin strings. The pulp is yellow with a sweet taste and rich in vitamins A and C.

## 9. *Carica papaya* L.

Pawpaw, papaya, lechosa, mamon
2 n = 18

Historical dates about pre-Columbian cultivation of papaya are rather scarce. Due to the highly perishable substance of fruits and seeds, no remains have been found in caves or grave sites. According to Prance (1984) we have to discard an Amazonian origin or domestication; most probably the early selection of useful biotypes occurred on the western side of South America, where several wild *Caricas* live in the tropical valleys of the Anden foothills. About the early use of papaya in America we have the following dates: Oviedo y Vadez reported in 1526 the existence of papayas in the Quebore tribe (Nicaragua), who called it "olocoton" and appreciated its good taste and also used it as medicine, with a great variety in taste and form.

Pizarro met the plant in 1531 on the dry coasts of Ecuador in irrigated fields. Zamora found it on the Eastern flanks of the Ecuadorian Andes.

Several taxa with rather large and edible fruits which have been considered by some authors as probable ancestors of the cultivated papaya are most probably hybrids between wild forms and cultivar, e.g. "*chrysopetala*" Heilb. from Ecuador with large oblong fruits (12 cm) produced on 5-m-high trunks. The same applies to *C. pyriformis* Willd. from Peru and *C. pentagona* Heilb. from Ecuador. The latter exists in different cultivated clones (e.g. Babaco) with astonishingly well-formed fruits, used by natives for fresh consumption and preserves.

MORPHOLOGY

Papaya is a short-lived perennial with a tree-like trunk. In reality the plant is a large herb without lignified tissues, and not at all a woody tree. In tropical climate, papaya has a vigorous vegetative growth and a quick fruit setting, and can survive for several years (Badillo 1971).

**Fig. X.6.** *Carica papaya*

The trunk reaches 3–8 m in height and is covered with conspicuous scars from leaf petioles. It contains a milky juice, rich in "papain".

The leaves are orbicular (30–80 cm diam.) palmately lobed. They often form a terminal crown and are inserted on 1-m-long, thin petioles (Fig. X.6).

The flowers are mostly dioeciously distributed. The genus *Carica* is polygamous. Thus both unisexual and bisexual flowers occur. Two are hermaphroditic types, (denominated *elongata* and *pentandria*) and two are unisexual, called male = staminate and female = pistillate. Besides these normal types, also abnormal teratological forms occur, which we cannot consider here (see Storey 1967).

Female inflorescences are short and pauciflorous; the corola segments are yellowish, narrowly triangulate (3 cm long and 1.5 cm wide) and the ovary is unilocular.

The staminate flowers are green-yellow, inserted on long, pedunculate male inflorescences, which are sometimes 1 m long. The corollas of the staminate flowers sit on long tubes (2–4 cm) and are twisted. The anthers are 1–3 mm long, narrowly elliptic.

The fruits (= berries) have different colours: green-yellow, or red when ripe. The soft sweet pulp is also sometimes red with contrasting black seeds (5–7 mm). The size of the fruit depends on the variety. In Venezuela local cultivars exist which are famous for their impressive fruit size (60–100 cm long), weight 10 kg. On the average, the commercial varieties are medium-sized: 15–30 cm long. Not seldom hermaphrodite flowers appear on male inflorescences which develop fruits hanging down on long petioles.

In such cases, male flower organs have changed sex, caused by temperature or mechanical stimulants. Natives cut the stem with knives or put wires around it, to provoke such changes.

*C. papaya* does not exist in the real wild state. It has been selected by indigenous planters from wild species, perhaps even hybrids, and long domesticated in their gardens. Still today natives of Panama, Venezuela and Ecuador collect and use wild-growing papayas. Recently I was shown such primitive selections by Choco-Indians in the Darien rainforest in their cultivated land. My guide translated their local name into Spanish "tapaculo". The fruits were small egg-sized (6–9 cm long) outside smooth, green, with whitish, slimy pulp and brownish striped seeds. The taste was not too bad.

According to my guess it was *C. cauliflora* or *C. microcarpa*. The Choco Indians use this wild papaya in different ways. Besides eating the ripe fruits, they apply the seeds as vermifuge, whilst the leaves are used to wrap meat in it to make it tender (effect of papain!).

An unquestionable wild ancestor of *C. papaya* has not been found so far, due to the widespread tendency for open pollination between species which grow in the same habitat, at present also mixing with escapers from numerous native plantations; we believe that this question will remain unsolved. Perhaps the only way in the future would be an artificial re-creation of *C. papaya* using the closest wild relatives.

The geographical origin should be looked for in the region where most of the wild species of *Carica* occur. This is in the eastern valleys of the cooler tropical mountain region of Peru, Colombia and Ecuador. There, for example, the following species exist:

*C. candicans* Gray, *C. crassipetala* Bad. *C. cauliflora* Jacq., *C. omnilingua* Bad., *C. microcarpa* Jacq., *C. horovitziana* Bad., *C. sprucei* Bad., *C. sphaerocarpa*, *C. pulchra* Bad., *C. parviflora* A.D.C., *C. weberbaueri* Harms, *C. augusti* Harms, *C. quercifolia* , *C. glandulosa* Pavon., *C. candicans* A. Gray, *C. pubescens* Lenne & Koch, *C. stipulata* Bad., *C. goudotiana* Solms, *C. monoica* Desf. and *C. chrysopetala* Heilb. (perhaps a hybrid).

Rare exceptions to this species concentration at the Eastern slopes of the Andes are: *C. chilensis* Solms, which is restricted to Chilenian mountain valleys, where it resists also lower temperatures. It produces small, inedible fruits (3–4 cm) with glabrous teguments and few seeds. At the other geographic extreme, in Mexico and several Central American countries, occurs *C. cauliflora* Jacq., maybe originally from Colombia, where it is also widespread in montane forests.

In the following we give a short resumé of such wild papaya species as may have played a role in the phylogeny of the cultivar itself.

### *C. cauliflora* Jacq.

Its distribution reaches from Colombia-Venezuela northward to Panama and many Middle American countries as far as Mexico. It must definitely be considered as a wild species, spread mainly by fruit-eating animals and also by people, who collect the small tasty fruits.

Plants are dioecious, 5 m tall, with a glabrous stem. The leaves form a dense terminal crown. The flowers are 3 cm long, with yellow, five-lobed corollas. The fruits sit near the trunk on a 3–5-cm long pedicel. They are 5–8 cm long and attain an orange colour when ripening. The mesocarp is sweet with an aroma of fresh grapes.

### *C. microcarpa* Jacq.

A heterogenous taxon which has been divided into various subspecies, which inhabit

**Fig. X.7.** *Carica pentagona,* planted in North Chile in arid land with scant irrigation

Venezuela, Colombia and Ecuador. The ssp. *heterophylla* extends from Peru to Panama. It includes biotypes with simple leaves and three- and five-lobed lamina. The fruits are often produced in small racemes (size 1–4 cm). This wild papaya is occasionally planted by Indians in the mountains of Darién. The trunks are weak and need support from other shrubs.

### C. monoica Desf.

This is the only homogametic and monoecious species of the genus *Carica*. Hybridization experiments have demonstrated that *C. monoica* does not possess X-Y chromosomes. The two sexes exist in the same individual. Its habitat is the eastern slopes of the Andes (Bolivia, Peru, Ecuador, Colombia) with subtropical climate (Fig. X.8). Small shrubs with dense leaf clusters in a terminal leaf crown. Fruits are small to medium-sized (5–7 cm), ovoid, pendulous. When ripe they have a red-yellow pulp with good taste.

**Fig. X.8.** *Carica monoica* fruits (8–10 cm long)

## C. parviflora (A.DC.) Solms

These small shrubs have red or rose-coloured flowers. Also the small fruits, which on becoming ripe have a reddish colour. Natives of Ecuador and Peru collect them in the forests.

## C. pubescens Lenne & Koch

Synonym: C. candamarcensis Linden, C. cundimarcensis Kuntze, C. chiriquensis Woodson.

This species is still cited with the above-mentioned synonyms, a custom which should be discontinued. C. pubescens has a wide areal, which covers the yungas of Bolivia, and the wet forests of Colombia, Venezuela and Panama. The plants are 3–10 m high with a pubescent, stout trunk, which bears the scars of dropped leaves. The fruits are obovoid (7–12 cm long). When ripe they are orange-coloured and have a pleasant fragrance.

## C. stipulata Badillo

Its major distribution is in Ecuador, in the montane region of 1400–1600 m altitude. The trunk is tall and is covered with thorn-like stipules. Before flowering it sheds its leaves. The fruits are 8–10 cm long and 4 cm wide, with ten pronounced longitudinal furrows. Wild animals collect the fragrant fruits. The species produces spontaneous crossings with C. pubescens, which lives in a similar habitat.
C. stipulata possesses high virus resistance and may be used in bridge crossings with other wild papayas to transfer this valuable factor finally to Carica papaya.

## C. quercifolia Solms

This species has a meridional distribution from the Argentinian provinces Cordoba, Catamarca to the Bolivian border, entering Paraguay as far as South Brazil, where it is called "naracatia", or "mamon silvestre". With small fruits (3–4 cm), oblong yellow-orange, sometimes eaten or used for pre-

serves. Plants 5–7 m high. The leaves display a notable foliar polymorphism. The lamina is oblong and somewhat lobed.

For reasons of habitus, fruit setting and general morphology, we exclude the following taxa from the ancestry of C. papaya:

C. chilensis (inedible fruits), C. cauliflora (different fruit setting), C. glandulosa (morphology different), C. horovitzii (climbing stems), C. microcarpa (thin stems, small fruits), C. quercifolia (different leaf and fruit morphology), C. pyriformis (probably a hybrid).

### PRACTICAL USE AND ECONOMY

The value of this delicious fruit is based on its easy digestibility and tonic effect. The acidity is low (0.1%), the sugar content is 10% (50% is saccharose, 30% glucose and 20% fructose). The carotinoid content is 4 mg/100 g and the vitamin C 50 mg/100 g. The fruits are difficult to store and ship over long distances, therefore the Hawaiian papaya industry has developed mechanized harvesting methods to handle 50,000 kg per day and apply a rapid spray and waxing treatment as decay control.
Papaya plants can live 15-20 years if they are not infected by virus, but the general practice is to exploit papaya plantations only 3–5 years.
The average number of fruits per plant is 50, with different weight, according to the variety. The yield could reach 30–50 t/ha. The 50-cm-long fruits sometimes have 7 kg weight; for trade purposes, 700–1000 g are preferred.
World production is constantly increasing and may have passed now 2 mio t with expansion from the exclusively tropical to subtropical regions, 32° lat north and south of the Equator.
Recently, the papaya plant has gained major industrial interest for extraction of proteolitic ferments. Two enzymes are derived from Carica species; namely "papain" and "chymopapain", the latter with medical use and the former as a meat tenderizer. The product is tapped from the green fruits and stems. Several wild species surpass the cultivars in papain content and should therefore

**Fig. X.9.** *Carica papaya*, a heavily virus infested plantation in Trinidad

be given preference in future plantations for exclusive papain extraction.

The highest content has been found in the wild species *C. quercifolia* which grows in North Argentina and Paraguay.

Traditionally the latex is collected by making longitudinal incisions in fruits and trunks. The crude papain is obtained by different tapping procedures. The present world demand oscillates between 200 and 400 t.

DISEASES

A serious viral desease is "bunchy top", a leaf curl virus which can destroy the yield (Fig. X.9).

The upper leaves bunch together, become smaller and finally die; the apex and the trunks fall. As the transmission is caused by the omnipresent "white fly" (*Bemisia tabaci*) to combat it in tropical countries is nearly impossible.

Lesser damage is caused by the ordinary "leaf mosaic" virus, which is transmitted by various aphids and leafhoppers. In general most of the *Carica* plants are infected and show mottled or yellowish leaves. The consequence is often a drastic decrease in fruit production.

EBENACEAE

## 10. *Diospyros* spp.

This is a rare botanical family with few (5–6) genera distributed in all tropical hemispheres. The best known is the asiatic *D. kaki* L., now widely cultivated as a prolific fruit tree. In America four species are native:

*D. virginiana* L., *D. digyna* Jacq., *D. ebenaster* Retz. and *D. inconstans* Jacq.

### *D. ebenaster* Retz.

Synonym: D. brasiliensis Mart, D. diguna
Black zapote, barbacoa, black sapote, ebano

This 12-m-high fruit tree has its origin in the tropical forests of Central America (Guatemala and Mexico) and grows also in the Antilles. Its bark is dark coloured. The leaves are large (10–20 cm), oblong-elliptical and 3–5 cm broad, with a dark green shiny lamina.

It was adapted to cultivation in pre-colombian times and spread northwards to Florida. In the Southern part of USA the fruits are more appreciated than the local *D. virginiana*, both called "persimmon" or black sapote.

The flowers are unisexual and are borne in the leaf axils. Their corolla is formed by five white petals. The male flowers have 16–60 stamens. In the pistillate flowers the calyx becomes enlarged and persists on the ripe fruits. The fruit colour is dark green with a sweet brown pulp. The seeds (3–5) are flat and 1 cm long. The species still needs genetic improvement, especially selection of hermaphrodite plants.

Besides the neotropical species of *Diospyros* there exists in Asia the well-known kaki-fruit (*D. kaki*), which has been introduced to America with great success. Its fruits are considerably larger and have an attractive orange-yellow colour. In the Southern States of the USA, *D. virginiana* is found, and is called there persimmon, with plum-like, acid-sweet taste. In the Antilles *D. inconstans* Jacq. grows with glossy brown fruits, eaten by the natives.

**Fig. X.10.** *Mammea americana* (Kranz 1981)

similar to apricots. Inside are two to four oblong brown seeds (Fig. X.10).

Natives of Panama believe that eating much of this fruit raw kills parasites. Its probable antibiotic action should be investigated.

---
GUTTIFERAE

## 11 *Mammea americana* L.

Mammey-apple, mamey, apricotier des Antilles, Mammi Apfel

This delicous neotropical fruit is botanically closely related to the Asiatic species *Garcinia mangostena*, which is considered by experts in tropical fruits as the "loveliest fruit of the world". The mammey-fruit also has an exquisite taste.

The tree is native to the West Indies and the tropical part of Central America, some reaching even Florida in the USA. It is of stately growth (15–27 m) with an evergreen closed canopy. The stem and branches exude a yellow latex from which the natives prepare tinctures for medical purposes. It is said that parts of the plant have an antibiotic effect.

The leaves are obovate-elliptical (10–15 cm long), shiny dark green, with strongly developed secondary nerves. The lamina is covered with glands. The flowers sit close to the branches, and unisexual or hermaphrodites are found on the same tree. The corolla has four to six petals, exuding a pleasant aroma,

## 12. *Rheedia* spp.

This genus is represented in tropical America by at least six taxa. All are small trees which produce edible fruits. Some of them could easily be developed as useful garden trees.

### *R. acuminata* (R. & P. ) Planch. & Trian.

Madrono, Machari, naranjito, Satra

With a wide distribution, from South America to Central America, where it may have been naturalized by natives, generally known as "madrono". In Colombia it has been domesticated and grows in house gardens, the fruits being sold at local markets. The trees have a medium-sized (8–10 m tall) canopy of pyramidal form. The leaves (6–15 cm long and 5 cm broad) have a typical dark green brilliant glossy surface with pronounced lateral nerves. The flowers are monoecious with four yellow petals. The male flowers have many stamens, which surpass the petals.

The fruits have the form of an elongated berry (3–7 cm long) with a prominent apex. Its pericarp is yellow and slightly rough. The pulp is white, sweet-sour and contains three to four seed kernels.

### R. brasiliensis Planch et Triana

Pacuri

This is the most Southern-living species, extending from Brazil to Paraguay, where we have observed it on the riverbanks of the Parana River. There the natives collect the yellow fruits and call it "pacuri". The trees are small (3–5 m) and deserve further selection.

### R. edulis Planch et Triana

Limao do matto, jacomico

A wild-growing tree (8–20 m tall). Stems exude a yellow sap and have a red inner bark. Leaves 15 cm long, shiny green. Small flowers with an aromatic smell. Fruits ovoid (3 cm diam.), with warty surface, thick mesocarp and good-tasting white fleshy pulp. Inside, two big oblong seeds which are covered with a brownish fibrous layer. Grows in Mexico, Panama, Colombia and Peru, where natives gather the fruits wild and offer them at local markets.

### 13. *Platonia esculenta* Arr. da Gamara

Synonym: *P. insignis* Mart.
Bacuri do Parana

Probably of Brazilian origin, where this tree is sometimes also cultivated. Tall trees (20 m) with hard, glossy, rhombic leaves. The big flowers have 4–6 pink petals. The fruits are spherical (8 cm diam.) with a smooth yellow exocarp. The mesocarp has an agreeable aroma and contains four to five large seed kernels which taste like almonds. Finally, we mention only the fruit- and latex-producing genus *Clusia* and the tropical species *C. utilis* Blake and *C. odorata* Jacq.

LAURACEAE

### 14. *Persea americana* Mill.

Avocado-pear, ahuacatl, aguacate, avocatier, palta, Butterfrucht
$2n = 24$

NAME, ORIGIN AND DISTRIBUTION

In Central America several wild species of *Persea* grow. Quite rightly it has been concluded that the cultivated forms must have been developed in Mexico and Guatemala at a very early stage of man's history.

There still exist two different native names in America; the nahuatel denomination was "ahua-cacua-huitle", from which the other Central American names, like ahuacatl, alcuahte, aguacate, have been either derived or shortened. Ahuacatl is the common name in Aztek idiom. From this grotesque word corruptions developed like "abacata" (Portuguese), "alligator" pear (English), "Advokaten-Birne" (German).

In South America the fruit is called "palta", probably derived from the Indian tribe "Pal-

**Fig. X.11.** *Persea americana*, a richly bearing avocado tree

ta", the Ecuadorian inhabitants of prov. Palta. Under this denomination *P. americana* is known in Argentina, Chile, Bolivia and Peru.

The proof for early man's use of avocado fruits are seed remains which have been discovered in cave deposits of the Tehuacan valley (Puebla, Mexico). The oldest cotyledons came from Coxcatlan Cave, dated by C. E. Smith at 10,000 years B.C. Then follow many more finds from younger horizons, indicating that *Persea* fruits were a common food in the southern part of Mexico.

The first historical report on the existence of avocado trees in South America goes back to Martin Fernandez de Enciso, who travelled with one of the first Spanish expeditions to Colombia. He wrote in his *Suma de Geografia* (1519) that the Indians of Santa Marta have a fruit..."which contains inside something like butter and is of marvellous flavor" (Fig. X.11)

MORPHOLOGY

Due to the varietal diversification and polymorphism of clones, it is impossible to give a standard description,

In general, medium tall trees (10–20 cm) with a strong root system. The leaves are simple, ovalo-elliptical (16–20 cm long) with short petioles.

The leaf aspect varies according to the position on the branches and the age. Young leaves are pubescent, older ones have a shiny surface and short hairs underneath. According to the variety, leaves possess many glands which contain anis essence.

The inflorescenses are formed in panicles at the end of new branchlets and produce huge quantities of small flowers. The bisexual flowers have a short whitish perianth, 12 stamens and a short pistil. The flower biology is complicated. For a long time the enigma existed of how it is possible that with such an enormous quantity of flowers the fruit production in aguacate cultivars is so extremely low (perhaps only 5%). Finally it was discovered that there exist distinct times for pollination and female reception.

In certain varieties the pistil is receptive in the morning, but the anthers do not open

before the afternoon of the following day. This, of course, inhibits auto-fecundation. Other biotypes have pistils ready in the afternoon, but the anthers shed the pollen only in the early morning of the next day. The practical solution is to plant different biotypes of floral (anthesis) rhythms.

The fruits have different characteristics in shape, colour and kernel size. In general the excessively large kernel is a handicap for commercialization. Therefore "seedless" mutants should be promoted. The mesocarp is fleshy, rich in fatty substances and protein, but has low sugar content. It is surrounded by a strong epidermis with a sclerenchym layer, which separates easily in ripe fruits. Fruit size in certain varieties is small and spherical (8–10 cm), in others large and pear-shaped (16–22 cm), the latter having a weight of a half kg or more (cv. Russel, Trinidad). At Lake Ipacaray in Paraguay a local biotype exists with thin, light green exocarp devoid of anis glands, and with a rather small seed kernel. The tree gives good yields. This variety deserves interest for local breeding and improvement work.

Different *Persea* species have been described from Mexico to Chile. They grow in general in protected mountain valleys, but never in the extremely humid/hot lowlands.

We begin with the most meridional representant of the genus *Persea*. This is *P. lingue*, which in cold tolerance exceeds all other taxa. The fruits are too small to be used for human consumption.

In the coastal cordillere of Venezuela we found *P. caerulea* Ruiz & Pav. which differs in flower biology from the cultivated aguacate and has black-coloured fruits only 1–2 cm in size.

In comparison with the infrequent presence of *Persea* species in South America, Central America is rich in biotypes, considered as a real species by some authors, whilst others give them only varietal rank inside *P. americana* lat. due to the elaborate crossing system of this species.

### *P. nubigena* L.

A wild-growing tree, called aguacate silvestre by natives, widely distributed from the mountains of Mexico and Guatemala to the border of Costa Rica with Panama (in the Chiriqui mountains). It has small, insipid fruits.

According to Williams (1976), the Indians of Guatemala selected some cultigens from wild-growing *P. nubigena* which he calls *var. guatemalensis*. Their origin is the cool valleys of the montane forest. "It is the best avocado known to have appeared in pre-conquest America."

### *P. schiedeana* Nees

Synonym: *P. frigida* Bailey, *P. pittieri* Mez
Aguacatillo, Coyo, Yas.

A wild-growing *Persea* with a wide physiological range from the sea coast to nearly 2000 m elevation.

The trees are 10–25 m high, their branches are covered with dense, brown pubescense. The fruits differ enormously. some are small (5–8 cm) with little flesh and many fibres, others are quite large with yellow brown pulp, free of fibres and with a pleasant taste. These latter, better fruits are sold on local markets in Guatemala. The species is an important rootstock for grafting selected cultivars. Its present dispersal from Mexico-Guatemala to Costa Rica, Panama and Colombia may have been influenced by migrating Indians in early times.

ECONOMIC VALUE AND IMPROVEMENT

It seems strange that such a typical fruit of Indo-America has never aquired commercial success in those countries, but rather in the United States of North America. California (with 33,000 ha) and Florida (3,500 ha) have the largest *Persea* plantations in the world, and have developed their own varieties, like Hass, Pollock and Duke. Whilst there is an efficient horticultural and industrial infrastructure around the avocado fruit, the Latin American countries were not able to gain a leading position in the export of an authochtoneus species. We observed in several countries, like Venezuela or Paraguay, that a great part of the yield is lost before reaching the consumer, or is not even harvested. It is diffficult to understand why such a nutritious fruit has never become a popular food in Latin America.

World consumption is in steady progress and doubled from 1960 to 1980, estmated now at 1,300,000 t. The USA has more than 90% of world production. Yields in South America are low. For example, in Colombia they vary between 8 and 12 t/ha with an average of 10 t/ha. The total surface area in Colombia is 4900 ha with a production of 54,300 t.

Mexico produces 300,000 t, Brazil 135,000 t, Venezuela 58,000 and, for comparison, such a small country as Israel 28,000 t, exported to the European markets, where the "butter-like consistency" is highly appreciated.

The nutritional value depends, of course, on the different varieties. Californian selections gave the following values:

Water: 60–80%, protein: 2–7%, fat 10–30%, carbohydrates 3–16%.

European analysis gives: Water 68%, protein 2%, fat 25–30%, carbohydrates 3.4%.

The calorie figures are high: 240 kcal (= 1008 kJ) in 100 g fruit pulp. Several vitamins in considerable quantities (A,B,C,D,E) are present. The total food value is high. In comparison with ordinary meat, the same weight of palta fruits gives nearly double the energy. For this reason alone Third World countries, where people suffer from a chronical protein shortage, should plant *P. americana* over great areas. But as we observed e.g. in Panama – quite the contrary – poor people cannot afford to buy 1 kg of aguacate, because it costs one dollar and is rare on the market.

Aguacate oil commands a very high price (ten times that of peanut oil) thanks to its high (77%) percentage of oleic acids.

DISEASES AND PESTS

The most serious problem in tropical and subtropical soils, especially when they have a high degree of moisture, is the fungus: *Phytophthora cinnamoni* (root putrition) This disease exists now – by contamination – in all American countries where avocado fruits are produced. The fungus enters from the roots in the vascular system and obstructs the flow of sap. In most cases, after a certain time the whole tree dies. Actually the disease is so severe in California that 25 % of the trees are condemned to extermination.

North American plant breeders discovered tolerance and resistance in the wild species *Persea schiedeana* and *P. steyermarkii*. They could be used as patrons for grafting, but unfortunately they are intolerant to implants with *P. americana* cultivars. Recently also some *P. americana* introductions showed promising results, e.g. G-1033 and G-1038.

Some fruit diseases which belong to the genera *Cercospora* and *Colletotrichum* may be controlled by chemical fungicides. *Physalospora persea* causes brown scale on the fruits, diminishing their commercial value. Among damaging insects we mention especially the coleoptera *Oncideres poecila* (3 cm long), which causes considerable damage in palta plantations of North Argentina and Paraguay. With its strong mandibles it cuts even 8-cm-thick branches into short segments to place one egg in each, with the result that many branch segments cover the soil under the attacked trees.

ly collected as a basic food of the Choco Indians, who eat them raw or cooked, claiming that the fruits taste similar to avocado. Besides *G. superba*, the natives of Panama also use the fruits of *G. nana*, which is a small tree with fruits of inferior quality. The Choco Indians told me that they eat 1 kg per person daily as a main meal.

The two species grow spontaneously in the humid forests, but the natives plant them also in their house gardens. When the Spanish conquerors entered Panama, they were astonished at the abundance of *Gustavia* trees, which they called "membrillo", recalling this European fruit. Oviedo y Valdes, who directed some mining work in the Gulf of Uraba, confirms the wide use of this fruit by Indians, when he writes: ..."membrillo is thrown in a pot with or without meat as good and healthy food... bearing fruits throughout the larger part of the year."

---

MIMOSOIDEAE

## 16. *Hymenaea courbaril* L.

Coroboro, guanipol, jatobá

The species is of economic importance for various reasons. Its hard wood is highly appreciated for tool-making, with a specific weight of 1.1. Its bark is used by natives for canoes. From its trunk and roots there exudes an especially valued resin ( = "copal americano"), sometimes excavated by natives from the ground as "fossil" incense for their religious ceremonies and adornment at places where formerly the species grew. Finally the pulp of the fruits is eaten. *Hymenaea* is sometimes – erroneously – called "Algarrobo", an arabic word for the South American *Prosopis* species.

These stately trees, often 30 m high, grow from Mexico, Panama to Peru and also on the Caribbean Islands. Leaves with two asymmetric leaflets, 10 cm long. Flowers in terminal corymbs white with 2-cm-long petals and 3-cm-long stamens. Fruits oblong-oviform (5–12 cm), ligneous hard, indehiscent, inside a reddish mesocarp

---

LECYTHIDACEAE

## 15. *Gustavia superba* H.B.K.

Synonym: G. speciosa DC.
Paco, membrillo macho, tupu-membrillo (Kuna)

In the neotropics some 40 species of genus *Gustavia* exist, but the most important for human use is *G. superba*, which is indigenous to North Colombia, Panama and Costa Rica (Croat 1978).

These trees with slender stem reach 12–15 m in height. The leaves are concentrated at the top, showy with their long (30–70 cm) tongue-like, elliptic lamina. The flowers are stem-borne, large (10 cm), with pink-coloured petals and many stamens, which extend beyond the corolla. Fruits abundant spherical (10–16 cm diam.) with yellow flesh, similar to pumpkin, and 3–4 cm large seeds. In the Dairén region they are regular-

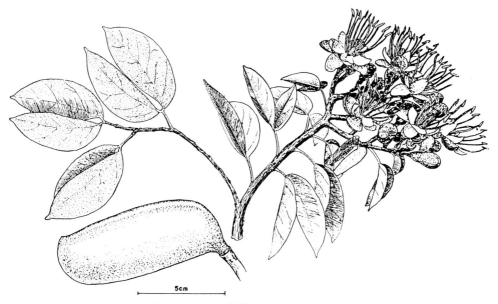

**Fig. X.12.** *Hymenaea courbaril* (Aristeguieta 1962)

which turns powdery when the seeds are ripe. This pulp is considered a very nutritive delicacy by the native people (Fig. X.12).

## 17. *Inga* spp.

The genus *Inga* embraces in America nearly 300 different taxa; several of them have been selected by Indians as food trees. Proof of this is the presence of *Inga* pods in Peruvian graves of Ancon and Paracas and the frequent use of *Inga* motifs on prehistoric ceramic. In the Inka civilization, the planting of *Inga* trees along the roads of the empire was obligatory, probably as shade-tree and as emergency food for messengers. It underlines the importance of *Inga* fruits that the last emperor Atahualpa offered the invading Pizarro, as a sign of friendship, a basket with "guama" pods (*I. feuillei*); but the Spanish conquerors killed him anyway. The big seeds are really not attractive for European taste, so it is difficult to imagine that this gift from an Indian monarch should have impressed the brutal Pizarro.

From the many *Inga* species (Leon 1966) we cite only the following in alphabetical order:

### *I. cinnamonea* Benth.

Wild-growing in the Amazonas region. The cylindrical fruits are gathered by Indians, who suck its sweet pulp.

### *I. edulis* Mart.

Inga sipo, guabo bejuco

Widely cultivated in the American tropics as shade tree and especially for its appreciated fruits. Its origin is the tropical forest of Guayana, Venezuela and Brazil. The trees reach 12 m in height. The dense leaves consist of five to six pairs of leathery leaflets, with smooth surface and pubescent underneath. Inflorescenses are borne in leaf axils. They consist of small corolas with long (3–5 cm) stamens. Fruits are cylindrical, long (60–100 cm), brown outside and inside with a spongy white pulp (arils around the large seeds). This is the edible part of *Inga*.

### *I. feuillei* DC.

Guamo, pakay

Its origin is the tropical part of Peru. Trees are 8–12 m high with dense foliage. The fruits are long-linear (30–60 cm) with many seeds, which are covered with sweet white arils. The embryos sometimes begin germination inside the pod.

### *I. nobilis* Willd.

Guamo

A small tree (4–7 m) with globular canopy and useful as shade tree for coffee plantations. Flowers white, fruits 6 cm long, edible.

### *I. spectabilis* (Vahl) Willd.

Guabo real

This species extends from the northern part of tropical South America through Panama

**Fig. X.13.** *Inga spectabilis* (After Forrero, Colombia)

to Costa Rica. The trees of medium size (10–12 m) have a dense foliage. The leaves are large, with broad leaflets (25 cm long). Inflorescences are terminal panicles with white flowers; the corolla is tubular, 2 cm long, with 4-cm-long stamens. The pods are large (50–100 cm long and 7 cm wide). The white pulp is highly appreciated by natives (Fig. X.13).

## 18. *Prosopis* spp.

Several *Prosopis* species have a wide diffusion through the whole American continent, from Colorado in the USA to Central Chile and Patagonia in Argentina. Nearly all produce abundant fruits, which have been used by Indians since time immemorial. Still today some "Algarrobos" represent an important wild aggregate to human food, especially in the under-developed regions of Chile, Paraguay and Northwest Argentina. Others are considered as aggressive weeds, difficult to eradicate, once they have invaded cleared fields.

From the botanical point of view, there exist a hugh diversity of species, local races and even hybrids; among them various noxious weeds, like *P. juliflora* and *P. glandulosa* (= "mesquite bush"), which invaded several million hectares of rangeland in North America. Similarly harmful is *P. ruscifolia* in the Chaco of Argentina, called "vinal", a bush with 20-30-cm-long spines, which also invades abandoned areas of former cultivation (Burkart 1976).

### *P. alba* Gris.

Algarobo

High trees (15 m tall), very common in the semi-arid hot regions of Argentina, Paraguay and Uruguay, where it is highly esteemed as timber, livestock food and for human consumption, its fruits being, for example, milled into flour. Fruits are compressed yellow pods (8–16 cm long) with many seeds inside. The fruits are produced in abundance. They have a sweet pulp and protein-rich seeds (30% protein, 7% oil);

**Fig. X.14.** *Prosopis chilensis,* from the semi-desert regions of Chile and Argentina, abundant in autumn with fruits under the algarobo trees

when ripe they fall to the ground, sometimes covering it, and are easily gathered. It is remarkable that the pods stay closed for a long time and do not split open immediately, losing their seeds as is usual for most of the Leguminosae. The mesocarp of the algarobo pods contains 26% glucose, 11–17% starch, 7–11% proteins, 14–20% pectins and several organic acids, which render them highly nutritious for humans. There are still some popular algarobo products in Argentina, such as pastas harinosas ("patay") and fermented beverages like "aloja" and "anapa". Natives of the Chaco lived in former times during several months of the year nearly exclusively from *Prosopis* products.

### *P. tamarugo* Phil.

Tamaruga

This species survives in several thousands of hectares in the Atacama desert of North Chile, even on soils which are covered by thin salt crusts. Efforts have been made by the Chilean authorities to extend the species in artificial tamaruga plantations in 40,000 ha.
Evergreen bushes or small trees (6–8 m tall), with a deep root system. The plants are relatively quick-growing when they receive water. Mostly used as forage for sheep and goats.

### *P. chilensis* (Mol.) Stuntz

Algarrobo blanco

Medium sized trees (3–10 m) with brown-red bark which loosens easily. Branches with 4–6 cm long spines. Leaves with many pairs of small leaflets. Inflorescences cylindrical racemes (6–12 cm long) yellow-green. Fruits compressed, falcate, 6–18 cm long, with a sweet edible mesocarp. Rather quick-growing. Extension from Peru to prov. Mendoza in Argentina (Fig. X.14). It was in former times an important emergency food for Indians in desert regions (Roig 1987).

MALPIGHIACEAE

## 19. *Malpighia punicifolia* L.

Synonym: *M. glabra* L., *M. uniflora* Tuss.
Barbados cherry, acerola, cerisier des Antilles

Its supposed origin is in Central America and the West Indian Islands. Acerola is a spreading shrub (2–5 m high) with dense ramification. The leaves are dark green, shiny, elliptic-ovate with an opposite insertion on the braches; they are short-stalked, and 2–6 cm long.
The flowers (1–2 cm diam.) are pink and sit in axillary cymes. The fruits are bright red drupes, slightly three-lobed, with a waxy

surface and a very tender skin, which is easily bruised. The fruits have a pleasant aroma and the size of a European cherry. The pulp contains 70% sweet-acid juice and three winged seed kernels. Acerola is considered as the fruit with the highest vitamin C percentage of all known species, and exceeds, for example, the orange by some 100 times. Besides a high content of ascorbic acid (1700–4000 mg) per 100 g pulp) it contains also a considerable quantitiy of malic acid. Fruit analysis gave the following result: 94% water, 4.7% carbohydrates, 0.4% protein, 0.2% minerals.

Thanks to its excellent vitamin content, the Barbados cherry has attracted the attention of international food processors who have promoted its cultivation in different continents. In Florida (USA) it was introduced as early as in 1880. The problem is that according to the method of processing, the ascorbic acid percentage varies considerably. Therefore commercial plantations should start with selected plant material, tested for its yield in fruits and vitamin C in advance (Fig. X.15).

Table 15 is a resumé of the ascorbic acid content of several fruit species, which are known as good vitamin sources. The values must be taken with certain care, because they originate from different sources:

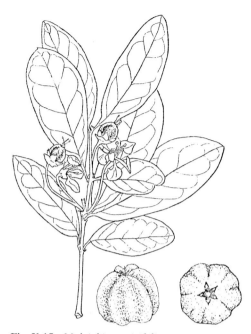

**Fig. X.15.** *Malpighia punicifolia*

shrubs and trees. Only *B. crassifolia* has aquired a certain importance as a semi-domesticated edible fruit, because the species has been dispersed by man from Mexico through all Central American countries, to Venezuela, Brazil and Paraguay. Medium-

**Table 15.** Ascorbic acid content in different fruit species

| Species | Ascorbic acid (mg) | Origin |
|---|---|---|
| *Malpighia punicifolia* (Malpighiaceae) | 1000–4600 | America |
| *Myrciaria jabotica* (Myrtaceae) | 700–2400 | America |
| *Phyllantus acidus* (Euphorbiaceae) | 625–1800 | Asia |
| *Anacardium occidentale* (Anacardiaceae) | 147– 348 | America |
| *Psidium guajava* (Myrtaceae) | 300– 486 | America |
| *Byrsonima crassifolia* (Malpighiaceae) | 90– 200 | America |
| *Citrus sinensis* (Rutaceae) | 37– 80 | Asia |
| *Rosa rugosa* —hips (Rosaceae) | 250–2900 | Europa |
| *Rosa rugosa*, Germany wild (Rosaceae) | 800–2900 | Europa |

## 20. *Byrsonima crassifolia* (L.) HBK.

Nance, murici, moureiller, chaparro

Several *Byrsonima* species have been described from South and Central America. Natives eat the pulp from wild-growing

sized trees (3–10 m) with fissured bark, or 3–5-m-high shrubs. Leaves are ovoid, opposite, simple, 8–15 cm long on short petioles.

Inflorescences are 15–20 cm long racemes with many yellow-red flowers. The fruits are small (10–15 mm diam.) drupes which be-

**Fig. X.16.** *Byrsonima crassifolia*

existence of this curious tree, which produces a latex juice which Indians consume like milk, it sounded so fantastic that only the authority of this famous naturalist could disperse the doubts. Humboldt and Bonpland observed the species in prov. Carabobo of Venezuela. The trees, called by natives Sande, Guaymaro, Palo de Leche, are 15–30 m high and form an enormous extended canopy. The leaves are simple, oval-oblong (10–30 cm long) and sit on short petioles. The inflorescenses are spherical receptacles with many male flowers, and only one pistilate flower without perianth. The fruits are small berries with one seed inside, and are edible. But much more important for the nourishment of Indians who cross the forests is the abundant latex, which exudes from the trunk when the bark is incised. This milky juice does not coagulate. Natives have used this creamy juice since time immemorial. Also Europeans have tried it during their tropical expeditions

come yellow when ripe. Natives appreciate them greatly for their sweet pulp and peculiar aroma. Therefore the fruits are commonly sold at all native markets of Central America (Fig. X.16).

---
MORACEAE
---

## 21. *Brosimum* spp.

The pantropical "mulberry" family has given to mankind several important fruit trees, to mention only the Asiatic genus *Artocarpus* with the distinct breadfruits and jackfruits which in some tropical countries now represent the daily standard diet. The worldwide-dispersed *Ficus* species belong also to the Moraceae. In the neotropics we have two genera which are still underexploited: *Brosimum* and *Pourouma*, both of paramount importance as food for the Indians: of the genus *Brosimum* have been described several neotropical species. We mention only the following:

### B. galactodendron HBK
Avicuri, sande, ạrbol de leche

This is the famous "milk-tree" of Alexander v. Humboldt. When he described in 1825 the

**Fig. X.17.** *Brosimum galactodendron* "the milk tree" of the Venezuelan rainforests

without any side-effects. The "milktree" has a rather wide areal, from Ecuador, Colombia to Venezuela, and recently in Asia (Java, Sri-Lanka), where the kernels are eaten and the latex consumed like milk (Fig. X.17).

### B. alicastrum Swartz

Synonym: *B. bernadetteae* Woods., *B. latifolium* Standl., *B. terrabanum* Pitt.
Breadnut, ramon, ujushte, ojoche, apompo
2 n = 26

Tall trees (24–40 m high) in tropical forests distributed from Ecuador, Colombia, Panama to Guatemala, Mexico (as far as Veracruz), Yucatan, also reported from Jamaica. The trees are dioecious.
The bark, when cut, also exudes latex (similarly to the above-mentioned species, but not as abundantly).
The leaves are oblong-ovate (10–25 cm long) with prominent veins (16–18 pairs) sitting on short stout petioles, with amplexical, 5–mm-long stipules.
Inflorescenses are subglobose heads (3–6 mm diam.). The anthers are circulate and centrally peltate, about 1 mm in diameter.

**Fig. X.18.** *Brosimum alicastrum,* the "breadfruit" tree of Central America

Fruits, when ripe, are yellow (30 mm diam.) globose, with edible pulp. The seeds are covered with a papery testa. The seed kernels are very nutritious, they fall down and cover the ground under the trees with thousands of small seeds. In their food value they compare favourably with maize, because their percentage of essential amino acids is considerably higher, especially triptophane. The calculated 14% of amino acids is four times higher than in corn. Additionally the leaves are good cattle forage. (The Spanish name "ramon" means: browse for forage) (Fig. X.18).

*B. alicastrum* trees are abundant near Maya ruins. Puleston (1972,1982) studied this association near the ruins of Tikal in Guatemala and came to interesting conclusions. He revealed that the ramon tree had been deliberately cultivated by Maya Indians, as an important alternative food, when, for example, maize failed. As is known, the Maya civilization achieved its greatest development rather late, i.e. between the years 300–900 of our epoch. Admirable constructions were built in dense tropical forest regions, under living conditions very difficult for human occupation, in the Northern Petén of Guatemala, but Puleston (1982) demonstrated that when such a marginal environment was skilfully utilized, especially its plant resources, it could have sustained such a high culture. He concluded that the Mayas created a state society with one of the highest regional population densities in the pre-industrial world. The key to this success was the utilization of the seed crop produced by *B. alicastrum*. Traditionally it has been thought that *Zea mays* was the sustenance of the ancient Maya culture. But Puleston doubted this, and stated that the nutrition of the Maya people could not have depended on the maize – milpa – cultivation alone. It needed completion with *Brosimum,*, the "breadfruit" of the Amerindians. Whilst maize could have produced there an average of 324 kg/ha per year, the ramon tree can produce five times more per year and unit of land. Thanks to the enthusiastic research work of Puleston, this underexploited useful food plant recently won the interest of Mexican centres of biological research (Peters and Tejeda 1982). We may hope that *B. ali-*

*castrum* will receive genetical improvement to become a modern food plant for the tropics in general.

## 22. *Morus rubra* L.

Red mulberry

This fruit tree deserves a short mention, because it is said to be of American origin, whilst other *Morus* species are Old World species. It extends from the Southern states of the USA to Central America.

## 23. *Pourouma* spp.

Uvilla, imauba do vinho, guarumo (Kuna: cuábar), caimaron

Several edible *Pourouma* species are found in the rainforests of Peru, Colombia, Guayana and East Brazil. When forests are cut down, the natives let the *Pourouma* trees grow, because they appreciate their fruits. Sometimes one even finds them planted in their house garden. The following taxa have been described by botanists.: *P. cecropiaefolia* Mart., *P. guianensis* Aubl., *P. mollis* Trec., *P. cecropiaefolia* Mart., *P. aspera* Trec, *P. acumninata* Mart. and *P. sapida* Karst.

Perhaps some of them are identical species. Only one (*P. aspera*) is of Central American origin; the others belong to the humid hot region of South America (Patino 1964). The trees are dioecious; according to the species, some are small, others reach 25 m in height, e.g. *P. guianensis.*

They produce quantities of large (20–45 cm) leaves, which are cordate or trilobate and have white, scurfy pubescence beneath. The fruits are borne on long racemes. They resemble grapes (3 cm diam.), have a purple colour and a sweetish yellow pulp and are used for fruit drinks and the preparation of a tasty wine.

It seems that *P. cecropiaefolia* Mart is best indicated for future cultivation because it has a prolific fruit-set and produces heavily for several months. Besides, it begins early (after 3 years) with fructification and hy-

bridizes easily. We observed that Kuna Indians plant *Pourouma* for its grape-like fruits which they call cuabar.

## 24. *Eugenia* spp.

This very numerous pantropical genus with many useful species has also some authentic neotropic representatives. They produce small fruits, often in big quantities per tree and with high vitamin content; but in spite of such advantages they are still underexploited in the Americas. We mention some of the promising species:

### *E. pungens* Berg

Guayabo, mato, guabiju

The trees are 5–10 m high and live along the rivers in the humid Chaco of North Argentina, South Brazil and Paraguay. They produce hugh quantities of tasty fruits with the size and aspect of vine grapes. Their aromatic pulp can be used for conserves and fresh consumption and represents 26% of the fruit.

### *E. brasiliensis* Lam.

Cagaiteira, Pitomba

Small trees with white flowers and shiny black fruits of the size of cherries, which are appreciated by the natives of Brazil. This species could easily be selected and domesticated for industrial use in jam production (Fig. X.19).

### *E. myrcianthes* Nied.

Iba-jayi

The trees are 8 m high and grow in Paraguay and the Southern states of Brazil. They produce relatively big fruits of the size of peaches, with yellow skin and sweet-sour fruits.

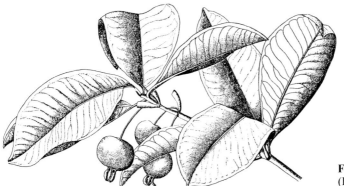

**Fig. X.19.** *Eugenia brasiliensis* (Hoehne)

### *E. tomentosa* **Camb.**

Cabeludinha

This small tree excels many other tropical fruits by its vitamin C content. According to analysis carried out at Campinas (Brazil), the whole ripe fruits had 931 mg ascorbic acid per 100 g and the pulp alone 1.730 mg per 100 g seed weight.

### *E. dombeyi* **Spreng.**

Grumichana

The trees are 5–8 m high, with dense foliage. Leaves are brilliant dark green and 6–12 cm long. The white flowers have large peduncles and produce dark red, sperical fruits (2–3 cm diam). The fruits are eaten fresh or may be used for different conserves and jellies.

### *E. uniflora* **L.**

Pitanga
2 n = 22

A bushy tree (2–8 m tall), often used as hedge due to its compact foliage. Leaves oval-elliptic, 2–4 cm long with acuminate apex. Flowers are solitary. The fruits (2–5 cm diam.) are spherical with a characteristic superficial ribbed segmentation. At the beginning yellow, they have a wine-red skin when ripe. The pulp is red and has a strong acid taste and agreeable aroma. The

fruits have a high vitamin C content and are also rich in calcium and phosphorus. Due to its high vitamin C percentage, *E. genia uniflora* has recently attained worldwide distribution.

## 25. *Psidium guajava*

Guava, guayaba, goyavier, Guave
2 n = 22

From the hundred American taxa, we select only this species as the most cultivated (Fouqué 1976).
Small shrubs and trees (3–5 m high), easy to recognize by the peeling brown bark on a green stem surface. In America numerous local races and selections exist. Wild-growing guavas are mostly of poor fruit quality and contain many seed kernels.
The side branches are quadrangular and have opposite, pinnate-nerved leaves. The lamina is broad elliptic (3–15 cm long and 2–6 m broad) with many prominent glandular veins and pubescence beneath.
Flowers are borne in the leaf axils as solitary, or in two- to three-flowered cymes with short petioles. The petals are reflexed (1–2 cm long), white and enclose many stamens. The fruits are berries, surmounted by persisting calyx lobes. They vary in form (pyriform, elongate, ovoid) and size (4–12 cm large but often much smaller). They have the peculiarity of containing in the mesocarp several "stone grains". These esclereidic cells produce a "sandy" consis-

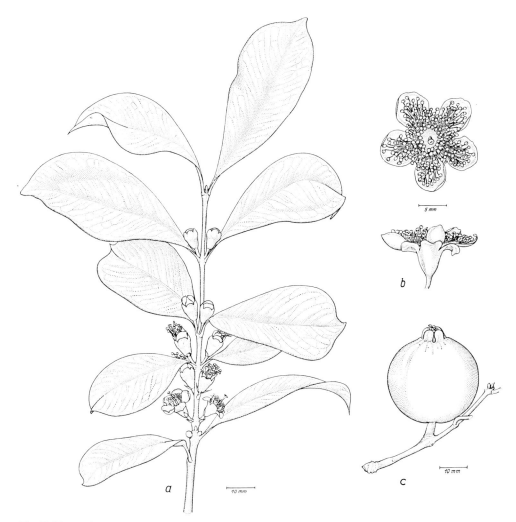

**Fig. X.20.** *Psidium guajava* tree in the neotropics (Mansfeld 1986)

tence of the pulp, a factor which has been eliminated in better cultivars. The numerous seeds are embedded in an aromatic pulp which is yellow, pink or red. The seeds are reniform and 3–5 mm long. "Seedless varieties" now exist mostly of triploid genome structure, which of course must be propagated vegetatively. Selected varieties have a high content of ascorbic acid (= vitamin C), on average 1000 mg per 100 g fleshy fruit. Guava has become a worldwide appreciated fruit. Cultivars which are free from seed kernels are an important material for fruit in-dustries, and are planted now in South Africa, California, Florida and India. They always suffer from the problem of attacks by the fruitflies *Anastrepha, Dacus dorsalis* and *Ceratitis*, which provoke rotting fermentation.

The yield in good varieties may be high, for example we observed in the Island of Trinidad 20 t/acre. Fruit analysis gave the following results:

78%–80% water, extract 12–13%, total sugar 5–11%, protein 0.7–1.3% minerals 0.4–0.9%, fruit acid 0.3–0.5%, ascorbic acid 200–300 mg %. Guava fruit is also a fair source of vitamin A and calcium, iron and phosphorus (Herrmann 1983).

### 26. *Acca sellowiana* (Berg) Burret

Synonym: *Feijoa sellowiana* Berg
Feijoa, guayabo, goibeira

Small trees from South Brazil, adapted to cooler climate in the subtropics, often growing in hill regions of South Brazil ( = "Planalto").
The leaves are obovate with short petioles. The flowers have large pink petals and many stamens. Fruits are rather large (3–8 cm

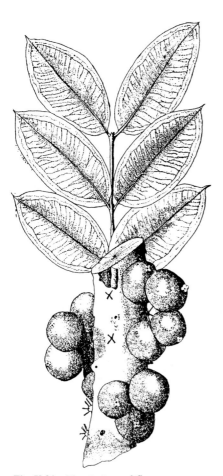

diam.), often with a wrinkled green skin. Natives collect the fruits and use them in home industry for conserves and beverages. The pulp presents 50% of the fruit, and is cream-coloured. There exist biotypes with hundreds of small seeds and others which have only one to three; these latter should be selected.

### 27. *Myrciaria cauliflora* (DC.) Berg

Yba-puru, camboim, jabuticaba

Trees only 2–4 m tall, growing on river banks in North Argentina, Paraguay and South Brazil. Leaves elliptical, dark green with greyish underside. Flowers pink with big petals, with very short peduncles directly on the trunk. Fruits are black-violet, aromatic (2–3 cm), similar to vine grapes. South American natives have collected and used the fruits since prehistoric times and even tried to domesticate the trees, which now sometimes grow in the local house gardens in Paraguay (Fig. X.21).

---
OLACEAE

This pantropical family is represented in South America by only a few genera. Only one, *Ximenea*, has acquired any economic importance for its fruits and oil-rich seeds (Fig. X.22).

**Fig. X.21.** *Myrciaria cauliflora*

**Fig. X.22.** *Ximenia americana*

## 28. *Ximenia americana* L.

Mountain plum, manzanilla, cagalera, manzana del diabolo.

2 n = 26

Probably of Central American origin, exists e.g. in Panama in wild state, but at present spread through all countries of the neotropics.

Shrubs of 2–4 m height or small trees with hanging branches, spiny, with high tannin content.

Leaves elliptical, entire, alternate (5–10 cm long).

Flowers in axillar racemes, with four hairy petals.

The fruits are yellow-red drupes, ovoid (3 cm diam.), edible, with one oil-rich seed kernel. The pulp is sweet-sour. Natives of tropical regions of America like the fruits for its acid-sweet taste. The fat of the seeds is extracted for fabrication of candles.

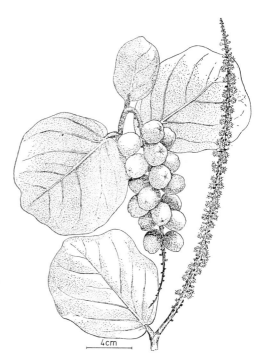

**Fig. X.23.** *Coccoloba uvifera* (Aristeguieta 1962)

POLYGONACEAE

## 29. *Coccoloba uvifera* (L.) Jacq.

Uvero de playa, uvero

Often abundant on the sea beaches of the Atlantic and Pacific Ocean in Central America and the Antilles. Trees 6–8 m high with a smooth, hairless appearance. Leaves large, simple, cordate (15–30 cm), with visible strong nerves. Flowers greenish, small on long inflorescences. Berries numerous (30–40) on the racemes. Fruits spherical (2 cm diam.), violet when ripe; with an acid-sweet pulp which encloses one seed kernel. Still collected only from the wild, but the species deserves domestication, especially for colonization of large sea beaches (Fig. X.23).

The genus *Coccoloba* includes other tree-forming species, like *C. venenosa;* but as the Latin epithet indicates, this is a poisonous plant. Natives of Puerto Rico call it "calambrena", due to its nerve-paralyzing effect.

RUBIACEAE

This cosmopolitan botanical family is represented in the neotropics with about 80 genera. They have certain common morphological characters, such as a four-lobed calyx and corolla, and opposite leaves with stipules. Many stimulating and pharmaceutically useful plants (*Cinchona, Ipecacuanha, Coffea* from Africa) belong to this family, but the quota of edible species is surprisingly low. We mention here the following:

## 30. *Aliberta edulis* Rich.

Trompillo, guabillo, madroño

Very extended from Mexico to the Amazonas river system, where natives domesticated this and three related species as fruit trees in their house gardens, or still collect the fruits from wild stands.

### 31. *Genipa americana* L.

Genipap, nandipá, jagua, caruto

Medium-sized (4–10 m) tree with dense foliage and lemon-sized fruits. Leaves opposite, simple with strong nervature. Inflorescences corymbose (3–12 flowers together) corollas yellow, five lobes (14–18 mm long). Fruits when ripe with a sweet aromatic pulp (5–8 cm diam.). Natives preferentially use the unripe fruits to obtain a sap which colours the skin dark. Genipa is widely used in Indo-America for body painting. During their initiation period, young women of the Kuna or Choco tribes paint their whole body black.

### 32. *Psychotria* spp.

Several species in the rainforests of Panama, Colombia, Venezuela, extended to tropical regions of Peru-Bolovia. The small fruits are eaten by natives.

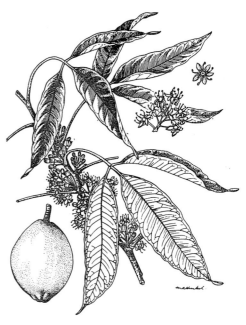

**Fig. X.24.** *Casimiroa edulis* (Photo Fouqué 1974) (Kunkel 1984)

---

RUTACEAE

### 33. *Casimiroa edulis* Llave & Lex.

White sapote

Medium-sized trees, similar to orange-trees. Botanically related to genus *Citrus*. Habitat in Central America. Leaves shiny dark green, elliptic-obovate. Fruits with green skin and sweet-tasting yellow pulp, with a slight turpentine flavour. Selections of trees free of turpentine should be undertaken for improvement of the fruits (Fig. X.24). This family is spread over the whole world with more than 800 species, mostly herbs and small shrubs. Only a few are represented in the neotropics. Seeds contain a toxic substance which may induce sleep. The genus *Coccoloba* includes tree-forming species with edible fruits, but some are considered toxic, e.g. *C. venenosa* (= "calambrena") in Puerto Rico.

---

SAPINDACEAE

This large pantropical family is represented in the New World with 18 genera. Without doubt, *Paullinia* is one of the most important as a stimulant due to its high alkaloid content.

As fruit trees we mention the following species:

### 34. *Melicocca bijuga* L.

Genip tree, mamon

Distribution in many tropical countries from Central America, Caribbean Islands to Paraguay
Tall tree often loaded with quantities of green fruits. Leaves composed of several leaflets. Flowers small on long (panicles) inflorescences. Fruits have a sweet-sour aril, which is very appreciated by inhabitants of tropical countries. Fruits (2–3 cm) are drupes which enclose one seed in a fleshy pulp (Fig. X.25).

**Fig. X.25.** *Melicocca bijuga* (Aristeguieta 1962) (Photo Fouqué)

## 35. *Talisia olivaeformis* **Radlk.**

Yellow genip, cotopris

Small trees (6–12 m) with sporadic presence in Colombia, Venezuela, Trinidad. Fruits elipsoid (2–3 cm), with slightly wooden pericarp and a hard seed. When ripe, orange coloured with a sweet pulp of aromatic taste. Several other similar species exist in Guayana and Brazil.

### *Talisia esculenta* **Radlk.**

Area of diffusion from NW Argentina, Bolivia to Brazil and Mexico. Trees of small (4–6 m) size, with small (2–3 cm diam.) berries. They have a white, tasty pulp and are offered on native markets as refreshment.

---

SAPOTACEAE

---

The botanical family of Sapotaceae has a pan-tropical dispersion, with a strong representation in Asia and Africa, and some genera with neotropical origin. Several species have long been cultivated on a worldwide basis, which has complicated its systematical grouping and provoked many synonyms. Baehni (1965) tried to put order into the taxonomy of the *Sapotaceae*, and we follow his proposals, for example, discarding the further use of the epitheta *Lucuma* and *Achras*.

## 36. *Calocarpum* spp.

Several species are used for their latex, e.g. *C. viride* Pitt. (= zapote verde) of real importance since prehistoric times are:

### *C. mammosum* **Pierre [= *C. sapota* (Jaq.) Merr.]**

Many synonyms, like *Achras mammosa* L., *Lucuma mammosa* Gaertn., *Pouteria mammosa* L., *Sapota mammosa* Mill
Mamey sapote, mamey

Its origin is the tropical forests of Guatemala, where it is an important source of

**Fig. X.26.** *Calocarpum mammosum* (Kranz 1981)

"chicle". Large trees (25 m) with abundant flow of latex when incised. The branches are covered with a brown pubescence. The leaves are in alternate position (10–15 cm long) simple, obovate with well-marked nerves. Flowers are hermaphrodite, borne near the stem in leaf axils, small, with yellow five segmented corollas (Fig X.26).

Fruits are big (10–18 cm long), ovoid, with a brown lignified and hard epicarp, containing a delicious salmon red sweet pulp and one big seed kernel. The seed is 5–7 cm long, glossy brown, with a typical long "scar" which covers nearly the whole surface. The pulp is rich in vitamins A and C. This tree had been domesticated by Central American Indians in pre-Columbian times. Oviedo found mamey cultivated in Honduras when he visited the tribes of Nahua.

## 37. *Chrysophyllum* spp.

In the tropical regions of America several species exist, some of them cultivated for their pleasant-tasting fruits. We mention the following taxa, some of them with uncertain taxonomic position: *Ch. gonocarpum* Engl. native in North Argentina, *Ch. excelsum*

Hub. from Guajara in Amazonia, *Ch. oliviforme* L., *Ch. balata* Ducke, *Ch. guyanense* Eyma, *Ch. auratum* Miq., *Ch. glabrum* Jaq. Outstanding importance as fruit tree has:

### *Ch. cainito* L.

(with a dozen synonyms)
Caimito, caimo starappel, caimo morado, caimitier

Its origin is Central America and perhaps also the West Indian Islands, where it is of ancient use as a shade and fruit tree and for its latex. The tree is generally 10–15 m tall, but may reach 30 m in height.

The leaves are glossy green, elliptic-ovate (15–18 cm long), with a coppery-brown underside, which makes a striking contrast to the dark green surface. The numerous flowers in axilar clusters with yellow or purplish small corollas. Fruits subglobose (8–10 cm diam.) with an aromatic white pulp. When the fruits are cut in cross-section the seeds appear around a central axis in a star-like fashion, which gave it its popular name. The seeds are brown and flat.

## 38. *Manilkara* spp.

This genus exists in both tropical hemispheres; most are of Old World origin. In the neotropics are native (Fig. X.27):

**Fig. X.27.** *Manilkara achras* (Leon 1968)

### M. achras (Mill.) Fosberg

Synonym: Achras zapota, sapota achras

2 n = 26

Of Central American origin, cultivated by Indians. Trees 18–20 m high, producing a considerable quantity (1000–2400) of edible fruits. The latex is highly appreciated for fabrication of chicle and chewing gum. Early taken by Spaniards to the Philippines and then to Malaysia. Now planted in tropical regions of all continents.

### M. zapota (L.) Van Royen

Naseberry, sapodilla, nispero, nèfle d'Amérique, Brei-Apfel

Medium-high (10–25 m) trees with a symmetrical canopy, areal extended from Central America to Florida, where this species has been selected for better fruit quality.
Leaves, simple 5–14 cm long, assembled at the top of branches. The small flowers sit on short pedicels in leaf axils, whitish, tomentulose. Fruits need 4 months to mature, globose (5–10 cm diam.), rusty brown outside, with yellow green flesh and a pleasant taste, with 4–10 hard black seeds, which easily separate from the pulp. The trunk can be tapped every 2–3 years for obtaining a latex which contains 20–40% gum. "Chicle", extracted from *Manilkara*, is still in great demand as raw material for North American industry which produces the popular "chewing gum". The bad custom of spitting around with "chiclets" seems to be an old Indian habit.

### M. chicle (Pitt.) Giley

Chicozapote

Native to South Mexico where the wild-growing trees are tapped for the latex, which produces a soft gum for masticating.

### M. bidentata A.D.C.

Native to Amazonia and the Islands of Trinidad and Tobago. A large, wild evergreen tree, sometimes reaching 35–40 m height. The natives collect the sweet fruits

and use the rich-flowing latex sap from the trunks to prepare balata chicle. Before the discovery of "plastic" substances by the chemical industry, the South American "balata" (= gutta) was of paramount importance and was exported in many thousands of tons.

### M. inundata (Ducke) Monachino

Synonym: *M. surinamensis* Mig.

Also very important for extraction of balata. Unfortunately the "chicleros" destroy the whole tree to tap the latex. Growing wild in Guayana and Brazil. The "true balata" contains 80% gutta and the rest is recines.

## 39. Pouteria spp.

Synonym: Lucuma

According to Baehni (1965), the genus *Lucuma* belongs here as a synonym.
This difficult genus for taxonomists includes some of the most delicious fruits of the neotropics. We shall try to order them systematically in the following way:

### P. caimito (Ruiz & Pavon) Radlk

Synonym: *Achras caimito, Lucuma, caimito* Lucuma, abiu, caimo

This tree of 10–15 m height is indigenous to Colombia and Brazil but has been intro-

**Fig. X.28.** *Pouteria caimito* (*lucuma*) (Kranz 1981)

duced also to Chile. Obviously dispersed by migrating Indians from the rainforests of Amazonia northward, passing the Darién mountains to Panama. The fruits are very nutritious and much appreciated by the natives (Fig. X.28).

### P. campechiana (HBK) Baehni

Synonym *Lucuma campechiana* HBK
Zapotilla, canistel

This species most probably originated in Central America (tropical forests of Mexico). In Cuba it is called "zapote amarilla". The tree is of medium size (5–10 m) and always grows in the vicinity of water, riversides and gallery forests and resists alkaline soils.
Leaves are brilliant green, oblanceolate, 10–18 cm long. Flowers sit in leaf axils. Fruits (5–7 cm diam.) are orange-yellow; can be eaten fresh or in preserves. The pulp is considered to be highly nutritious and totally lacking in acid flavour.

### P. domingensis (Gaert.) Baehni

Mamey colorado

Under this name this fruit tree is widespread in the Antilles, but it seems (Meyer 1957) that it has been introduced also to Paraguay and NE-Argentina.

### P. sapote (Jacq.) H.E. Moore & Stearn

Mamey colorado, sapote mamey

This impressive fruit tree with dense leaf canopy produces abundant fruits, some 1–2 kg heavy. The exocarp is reddish-brown, the pulp is red and contains one big black glossy seed kernel. The flowers are very small and sit direct on the stem.
The species has been introduced into Florida as garden plant with several useful selections, like Copan, Tazumal and Pantin.

### P. ucuqui Pires & Schultes

This is a local species from the valley of Rio Vaupes and Caqueta in Colombia. The Indians use the fruit and the latex of the trunk.

### 40. Vitex triflora Vahl

Taruma, aceituno del monte

Amongst several South American *Vitex* species, (e.g. *V. cymosa, V. gigantea, V. divaricata* and *V. pseudoolea*), we consider *V. triflora* as the most attractive. Its areal is the subtropical part of Brazil and Paraguay, where it produces huge quantities of olive-like fruits. The stems (5–10 m high) have a smooth bark. The leaves are digitate, with lanceolate glossy leaflets.
The flowers sit on racemes close together and have blue corollas. The fruits are drupes of the size of an olive, or cherry, with an agreeable sweet taste (Fig. X.29). The trees produce an impressive yield of fruits and deserve the interest of the preserves industry. The pulp encloses one seed kernel.

**Fig. X.29.** *Vitex triflora,* a richly bearing fruit tree from Paraguay

# B. Fruits from Herbs and Bushes

## 41. *Monstera deliciosa* Liebm.

This liane, with its tasty fruits, is a typical climber in the dense neotropical forests. In general these woods are poor in edible fruits, therefore for explorers and naturalists the infructescences of *Monstera* may really appear as a "delicious rescue amidst the green hell". At least we enjoyed it. One must only be careful to avoid the sharp crystals of ox-alate in the green parts. Future breeding of this ornamental plant for floricultural purposes should pay due attention to this problem.

The plant has thick, climbing stems. The oval leaves are large (60 cm long and 40 cm broad) with long petioles. The lamina is interrupted along the lateral nerves. Flowers sit inside a white spathe and develop slow-growing fruits (Fig. X.30).

## *M. perusa* (L.) de Vries

Similarly to the above-mentioned species, this liane is abundant in the rainforests of

**Fig. X.30.** *Monstera deliciosa,* an aromatic fruit from tropical rainforests

Panama. Natives eat its sweet-acid fruits and use the dried aerial roots for weaving baskets.

To this vast pantropical family belong some epiphytic plants which are appreciated as ornamental and fruit-producing vines. We mention here the American genera: *Anthurium, Monstera, Philodendron* and 42. *Montrichardia*. Even if none of these Araceae lianes has acquired economic importance on a world scale, they represent an appreciable emergency food for the inhabitants of neotropical rainforests. Besides they are used in Central and South America as garden varieties and for making jelly and fruit conserves.

### 42. *Montrichardia arborescens* Schott.

Natives of Venezuela and Amazonia call it "moco-moco" or "boro-boro". This liane has succulent, 4-5-m-long stems and prefers as habitat the estuaries of the Orinoco and Amazonas. Its leaves are big and shiny and always clustered at the upper part of the stem. The inflorescences have white spathes and develop edible fruits.

### 43. *Philodendron bipinnatifidum* Schott.

This species is the best known of the perhaps 200 *Philodendron* taxa described from America, which are mostly used as garden ornamental. Its ripe fruits have a pleasant aroma, when eaten fresh or cooked. The plants extend from North Brazil to Paraguay, where the natives use also the dried aerial roots for weaving baskets.

Climbing lianes in the humid forests, producing large fruits, which are used in Brazil for making jelly and are sometimes also eaten as fresh fruits.

None of these Araceae fruits has aquired economic importance; they are used rather as an emergence food.

### 44. *Ananas comosus* (L.) Merr.

*Synonym: Ananas sativus* (Lindl.) Schult.
Pineapple, ananas, piña, abacashi, Ananas
$2n = 50$

NAME, ORIGIN AND DISPERSAL

The English term is rather misleading, because the fruit is neither an apple nor a pine. Therefore we prefer the name ananas, which is of Guarani origin. These skilful Indians selected wild-growing ananas, and most probably took them during their long wandering northward, until they reached the arawak regions. In their idiom the fruit is called "taino", which means "noble fruit" (Fig. X.31).

Its South American origin is assured, but ananas had reached the Caribbean region, and most probably the eastern coast of Mexico, before the Spanish conquest. It was Columbus himself who reported on the 4th of November 1493 that natives of the Island Guadelupe offered him this delicious fruit,

**Fig. X.31.** *Ananas sativus.* Mixed cultigens offered on market place in Venezuela and selected by local peasants

on his second voyage to America. Acosta (1590) and Oviedo (1535) described ananas plantations from the east coast of Panama and some Caribbean islands. Some English pirates and slavetraders of the sixteenth century also found ananas. Hawkins (1565) sailed into the harbour Sta. Fe (100 km westward from Cumana, in Venezuela) with his ship *Jesus of Lübeck* and bartered from the natives: ... "pines of the bigness of two fists, the outside whereof is of the making of a pine-apple, but it is soft like the rind of concomber and the inside eaten like an apple, but it is more delicious." Raleigh in 1595, entered the Orinoko river 300 km upstream, where he observed that Indians cultivated ananas. Pigafetta (1519) reported it from Brazil.

France was the first European country to cultivate this exotic fruit. According to Gibaut (1912), the court of King Louis XV received two ananas plants in 1730, which in Versailles glasshouses in 1733 produced several fruits of pleasant aroma. Offspring of these were presented some years later as royal gifts from France to other European dynasties. A hundred years later the number of these clones numbered thousands. We may thus declare the European glasshouses as the "centre of origin" of worldwide ananas cultivation. Edible ananas cultivars are of South American origin and were introduced rather late to Central America. In addition to historical arguments, there exists also the fact that Middle American natives used and developed other edible Bromelaceae, like *Aechmea bracteata*, or *Bromelia pinguin* and *B. karatas*, but at this early stage they did not know *Ananas comosus*. The use of two species of *Aechmea* (*A. bracteata* and *A. magdalena*) is very ancient in Central America, where the natives call them "Izchu". They eat the acid-sweet fruits raw or cooked, and extract the long fibres of the leaves for making their hammocks. The fruits are 5 cm long and are used in Panama for the preparation of alcoholic beverages. The areal of *Bromelia pinguin* and *B. karatas* extends from Mexico through the arid regions of San Salvador, Nicaragua, Costa Rica to Panama. The plants are acaulescent, with long (150–200 cm) 5-cm-broad leaves with many sharp spines on the margins.

They contain numerous fibres which give excellent strings. Young leaf shoots are cooked as vegetable. The fruits are produced in long inflorescences, of bright yellow colour with red seeds inside the sweet pulp. Natives collect the fruits and prepare them fresh for beverages or cook them for preserves (Fig. X.32).

In South America, numerous wild-growing species of the genus *Ananas* exist, but the discussion as to which of them was the direct ancestor of *A. comosus* is not yet resolved (Collins 1966). A decade ago I had the occasion to study the two presumptive centres

**Fig. X.32.** *Bromelia pinguin*, another Bromeliaceae used by American natives

of origin, one in the Orinoco region of Venezuela and the other in Paraguay (Brücher 1977). Without entering here into details of the problems, we mention shortly the following taxa, which indicate the existence of a Southern centre of species diversity:

*A. bracteatus, A. fritzmuelleri, A. guaraniticus, A. microcephalus, A. macrodontus (= Pseudoananas sagitarius), A. paraguariensis.*

### A. bracteatus Rom. & Schult.

2 n = 50

The Paraguayans call this wild-growing ananas "karaguata-ruhá", which refers to its tall growth. The leaves are 60–120 cm long and 5 cm broad, always with strong dentate-spinose margins. The natives use the cooked leaves as an effective abortive medicine. The fibres are useful for making hammocks and yarn. One plant yields 3–4 kg green leaves, which after a special retting process give 150 g of fine fibre (Fig. X.33). The inflorescenses are produced on long thin stems, which have the tendency to turn against the light. The syncarps are small and bear at their top and sides several leaf bunches. The fruitlets contain hard-coated seeds. I was told, however, that some vari-

eties have only few seeds and are multiplied for this reason. It is obvious that in former times the Tupi and Guarani Indians tried to select better biotypes. According to Bertoni, who dedicated many agrobotanical studies to ananas in Paraguay, 100 years ago, *A. bracteatus* was widely cultivated there by the natives, who were not acquainted at that time with *Ananas comosus* (*A. sativus* cultivars). The fruits had excelled in durability and tolerance to dry climate, but the fruit flesh was fibrous. Some botanists believe that *A. bracteatus* may be the ancestor or at least an important link in the evolution of *A. comosus*,but Bertoni is inclined to consider *A. guaraniticus* as the real precursor.

### A. guaraniticus Bert.

After having studied several wild-growing *Ananas* species, Bertoni (1919) declared with conviction "We conclude with certainty that *A. guaraniticus* is the wild ancestral form of *A. sativus*, which has long been sought ..." As this opinion has not been refuted until now, it deserves serious experimental study, based on original collections from Paraguay. The plant has narrow (12–18 mm) leaves, 80–120 cm long, with fewer spines (distance 1 cm on the leaf margin) than other *Ananas* species. The fruit is cylindrical

**Fig. X.33.** *Ananas bracteatus* from the Parana river region of Paraguay

(13–15 cm long and 4–6 cm broad) and sits on long (55–90 cm) stalks. The syncarpium has a fibrous, sweet pulp and produces seeds which are occasionally aborted. The bracts which cover the fruitlets are 18–22 mm long and pink-coloured.

### A. microcephalus (Baker) Bert.

Well known in Paraguay by the guarani name: Y-vira. I found the species growing wild in the Paraguayan "Chaco Humedo". The native families use the plant predominantly for its fibres and not for consumption of the fruits. In my observation, *A. microcephalus* flowers and fruits twice in the year. After flowering, the syncarpium neither increases its size nor produces "suckers" at its basis, nor a terminal leaf tuft (= crown). The bracts are whitish, denticulate and rather long (30–32 mm). The fruit has no aroma and produces many hardy black seeds. Several botanical varieties have been described since Baker declared this species to be the ancestor of *Ananas comosus*. Some of these Paraguayan biotypes (e.g. var. *robustus*) may be useful for future breeding work, due to their cold-resistance, claimed at minus 3°C, and fair fruit quality. We saw such varieties cultivated at the Parana in small native gardens.

### A. ananasoides (Baker) Smith

The distribution of this species reaches from Central Brazil (Matto Grosso, Minas Gerais) to the northeast (Paraiba, Pernambuco) and crossing the Amazonas river to the Gayanas. It prefers open grassland, xerphytic "serrados" and sunburned rocky environments, thanks to its high tolerance of dry climate. The plants form dense rosettes with 60-cm-long leaves. In the centre is a 40–50 cm high stalk with a syncarpus. When it becomes ripe, the fruit breaks away easily from the stalk. The syncarpus is small (8–10 cm long and 4 cm thick) and contains many polyeders with short bracts acuminate. Each fruit produces hundreds of black seeds of the size of a grain of rice.

### A. erectifolius L.B. Smith

The southern limit of this wild ananas is the Equator. It grows in the hot-humid climate of the confluents of the rivers Amazonas and Orinoco. I found it abundant at the Upper Orinoco, where the Waika and Makiritare Indians use it much more for extracting the fibres than for fruit consumption. The natives told us that there exist spontaneously growing plants with spineless leaves. With the help of a guide we found such examples at the confluent of the Rio Ocamo with Rio Orinoco (Fig. X.34).

The long (80–100 cm) leaves were really "spineless" with exception of some irregularly distributed small spiny protuberances on the margins. As the Indians have used this ananas since time immemorial as fibre supplier, we believe that our find was a semi-domesticated mutation, selected by native people somewhere on the Orinoco. We were

**Fig. X.34.** A delicious tasting *Ananas clon* with smooth leaves selected by Makarikare Indians in the Upper Orinoco region

strengthened in this belief when we observed that most of the troublesome seeds in the ripe fruits were not well developed and had aborted before becoming hard and ripe. Most probably wandering Indians spread such semi-domesticated which they call "Guragua" or in Spanish "Piña montañera".

## *A. parguazensis* Camargo & Smith

This species was discovered by some Venezuelan and Brazilian botanists in the frontier region between Venezuela and Guyana in the nearly inaccessible mountains of Pakaraima (highest elevation: Roreima with 3000 m alt.). According to Steyermark, wild ananas grow up to 1200 m a.s.l. on the so-called Tepuys. Southward, in the Amazonas teritory, on the borders of Rio Siapa (200 m alt.) the same species appeared. Finally Camargo and Smith (1968) described this interesting wild-growing ananas as a new species for science from collections made at the confluent of the Rio Parguaza with the Orinoco river in the State of Bolivar (Venezuela). The authors do not discard the idea that this find may be the ancestor of cultivated ananas. The habitat is humid forests and most places in the shadow of dense trees, where Indians collect them.

The leaves are long, 140–160 cm, with many spines. In contrast to other *Ananas* species, these spines have a retrorse position in the basal part of the leaves. The fruits are small (10 cm diam.) and develop an excessive quantity of seeds. In view of these primitive characters, *A. parguazensis* merits the status of a true wild species (Velez 1946).

## *Pseudoananas sagitarius* Smith

Synonym: *Ananas macrodontus* Morren, *A. sylvestris* Muell.
2 n = 100

This species is native in the Southern States of Brazil and in the humid Chaco of Paraguay. It develops very long (3 m) leaves with extremely sharp spines. The inflorescence-bearing stems are very tall (2 m) and produce only small syncarps, which are only

**Fig. X.35.** *Pseudoananas sagitarius* from the Paraguayan Chaco

seldom consumed by natives. For cytological reasons, this species must be excluded from further discussions about the origin of *A. comosus* (Fig. X.35).

Several thousand kilometers North of Paraguay there exists another "hearth" of genetical diversity in *Ananas* . (We avoid the obsolete expression "gene centre"). This is situated between Amazonas and Orinoca, not in the jungle part, but in the wide river border region with rocky and sandy soils and often xerophytic vegetation, or on the edges of shrub forests.

This is the habitat of the following species: *A. ananassoides, A. lucidus, A. erectifolius, A. microstachys* and *A. parguazensis.*

### MORPHOLOGY

*Ananas* is a perennial monocotyledonous herbaceous plant. It has a short stem (25–35 cm), which produces a single terminal inflorescence, which develops after 12–14 months in a syncarp fruit, (Smith 1955) which represents the edible part of the cultivated plant.

The thick stem has very short internodes, from which 60–80 leaves emanate, forming a dense rosette. The long leaves (60–100 cm) are arranged in a typical spiral with a 5/13 phylotaxy. This means that every 15th leaf in the whorl is inserted exactly over the initial leaf with five turns around the stem. This leaf arrangement gives to the ananas plant an ornamental aspect. The lancette-like leaves are not flat, but form a channel which collects and conducts rainwater to the basis of the plant. Wild forms have spiny leaf margins (also some primitive cultivars), whilst the variety Smooth Cayenne is "spineless". The *Ananas* leaves (as all members of the Bromeliaceae) possess trichomes, which act in the absorption of water and nutritional solutions, together with the useful spiral arrangement of the leaves. All these characteristics indicate that the physiology of the *Ananas* plant has been developed under semi-dry conditions. It belongs to the few cultivated plant species in the world which possess such a high grade of adaptability to an arid climate,

The pineapple leaf contains very strong fibres, which extend lengthwise and are connected to the vascular bundles. The possession of fibres was an important characteristic of the *Ananas* plant for the Indians, when they selected and domesticated this species. Thanks to the high quality of *Ananas* fibres, natives of the Philippine Islands still produce a special textile for the weaving of very fine cloth , the valuable "mantillas".

The inflorescence, at the early state of the syncarpus, contains 100–200 trimerous flowers with liguli-shaped small (5–15 mm) petals.They are white or purple in colour and after anthesis remain on the fruitlets. In wild ananas each fruitlet produces seeds which are covered with a hard testa, whilst the edible varieties are seedless. Such mutants of fleshy syncarps without seeds were discovered at a very early stage by Indians (Maipures, Piaroas) in the Orinoco region and propagated as clones, which gave rise to the cultivar Cayenne.

ECONOMY, VARIETIES

The present worldwide-distributed, high-yielding ananas variety Smooth Cayenne has a complicated history, spiced by events that are hardly known outside of South America. First of all, Smooth Cayenne does not have its origin in Cayenne, the coastal town of French Guayana. The French botanist Perrotet arrived there in 1819 and took five ananas plants from the harbour of Cayenne to France. They grew well in a greenhouse of the Royal Palace at Versailles and multiplied rapidly by side-shoots. They represented at this time extremely valuable gifts to other European royalties, who cultivated the – seedless – plants carefully as a special attraction in their orangeries. Within a few decades, the five plants from Paris had multiplied into thousands and were sent to many tropical countries for commercial propagation. In 1858 there were reports of "Cayenne" plantations in Australia, by 1860 they had arrived in Florida, in 1863 they were already growing on the Azores, and in 1870 on Jamaica. The great impact, however, occurred in 1886 when Ananas Cayenne arrived in Hawaii (Py and Tisseau 1965), transforming the living conditions in these islands.

Under optimal growing conditions, ananas plantations give high yields, i.e. the clone Singapore Spanish 30 t/ha in the first year. The var. Cayenne yields even more: 35–40 t/ha.

World production has doubled during the last decade. Thailand leads with 1.8 mio t and the Philippines with 1.2 mio t, mostly elaborated as conserves. Brazil and Mexico contribute 0.6 mio t, mostly with the var. Abacaxi.

---

CACTACEAE

The "marvellous world of Cactaceae", as floriculturists and gardeners of all continents call this botanical family, offers several species useful for mankind. First of all, the natives of marginal zones use them for food and drugs (Hoffmann 1982).

Cacti have succulent stems, often many-branched, giving them the appearance of "leaves". The succulent tissues are correlated with the need for water storage and reserve food. These perennial plants are nearly

**Fig. X.36.** *Opuntia ficus-indica,* a spineless mutation from California appreciated for its sweet fruits and useful fleshy stems (Photo Lyman Benson)

always protected by an elaborate system of spines, which are in reality transformed specialized leaves. Flowers are composed of showy perianths with colourful petals and hundreds of stamens. On maturity, they develop fleshy fruits with many seeds in the chambers.

Cactaceae are an exclusive neotropical family comprising many hundreds of different species which are represented from low coastal regions and tropical woods up to maximum elevations in the Andes at 4000 m altitude. Many of them produce edible fruits, some have even entered the fruit trade, like certain species of the genera *Cereus, Nopalea* and *Opuntia.* Interesting for special reasons are also: *Cereus hexagoneus* (Queen of the Night) and different species of *Echinocereus, Hylocereus* and *Pereskia.*

### 45. *Nopalea cochenillifera* (L.) Salm.

This cactus species, which is nearly spineless, centuries ago aquired historic importance in Mexico and Guatemala as an industrial plant. It is the host plant for cochineal insects, which produce a red dye. Even in the last century – before synthetic dyes existed – trade with dried insects amounted to millions of kg. Today, *Nopalea* is used as animal forage in dry zones and also quick-growing hedges; the fruits are eaten.

### 46. *Opuntia ficus-indica* **Mill.**

Prickly pear, tuna, Kaktusfeige

The genus *Opuntia* is represented with more than 300 species from the southern states of the USA, the Antilles, Mexico and the whole of Central and South America down to Patagonia (Fig. X.36).

The early use of its fruits is documented by findings of seeds in coproliths from the Ocampo culture in Mexico. Even today the fruits of various species are collected. They form a polyploid series from $2n = 22$ (*O. robusta*) to $2n = 88$ chromosomes (*O. megacantha, O. ficus-indica*) (Benson 1969). This latter has attracted the attention of agronomists and plant breeders. This "prickly pear" has existed for two centuries also as a prickleless mutant. Discovered in Mexico, such spineless examples were introduced to California in 1769 by Franciscan missionaries, and are still today called "mission cacti". This *Opuntia* has a double use. Its egg-sized acid-sweet fruits ("Indian fig") are appreciated by natives of the arid zones in America, but have also entered the fruit markets of North America, the European Mediterranean regions and South Africa. There are estimations that more than 12,000 ha are planted with this cactus. Analysis of fruits indicates the following values: water 89%, pulp 10%, vitamin C 42/mg/ 100 g, acidity 0.12%. *O. ficus-indica* is so easily adapted to semi-arid conditions on other continents that it has even invaded the hills around the Acropolis in Greece.

Quite unexpected this "noxious weed" acquired recently in the Republic of South Africa considerable economic importance. The "prickly pear" native from Central America is used now as a hostplant for the mass rearing of the cochineal insect, *Dactylopius coccus*, for red dye extraction. The reason is that the use of synthetic red aniline dyes has been questioned for health reason. The potential gross income from 1 ha of "prickly pear" for red dye production is calculated in 15,000 US $, which may represent an appreciable income for handlabouring natives.

---

CUCURBITACEAE

---

This pantropical family includes 90 genera with more than 750 species; many of them in one way or another useful for mankind. Among them are such venerable cultivars as *Lagenaria, Citrullus, Cucumis* and last but not least *Cucurbita*, which has accompanied early man in the Tropics for several thousand years. The northern hemisphere is, however, very poor in Cucurbitaceae: only the genera *Ecballium* and *Bryonia* grow wild there.

## 47. *Cucurbita* spp.

This neotropical genus consists of 17 species distributed in both Americas. The majority of the wild species are indigenous to Central America whilst *C. andreana*, the wild form of *C. maxima* (winter squash) has its natural areal south of Capricorn, and is found even in the temperate "pampa humeda" of South Argentina.

Several of the cultivated cucurbits have been an important item of diet since pre-Columbian times for the natives of North, Central and South America. They used the fruits in different ways; the flesh was cut into strips and sun-dried; the ripe fruits could be easily stored, as several species remain many months without decaying. They were also used as cooked vegetable, and the fat-rich seeds were eaten fresh or roasted.

Cucurbits belong to the first cultivars of mankind. This is true for the palaeotropics (*Lagenaria* and *Cucumis* in Africa and India, as well as for the neotropics (several *Cucurbita* species, *Sechium, Cyclanthera, Sicana*). Modern archaeobotany has been successful in differentiating excavated cucurbit remains and determining the correct species. Fortunately the peduncles and seeds of *Cucurbita* species differ from each other and are therefore a most useful tool for determination of archaeological finds.

*C. ficifolia*: seeds almost round (18–20 mm), flattish and pebbled surface with visible margin; mostly black, sometimes cream-coloured.

*C. mixta*: some var. with large (34 mm) seeds, broad seed margin, white or silver-grey.

*C. moschata*: seeds small (14–22 mm), with shredded margins of different colour; testa white or brownish.

*C. maxima*: seed size 13–27 mm, sometimes fissure in their coat, colour white or tan with margins of different colour.

*C. pepo*: seed size 15–23 mm. Colour white or cream, margins of the same colour as coat. Sometimes seed coat very thin "naked" seed in oil-cucurbits.

Such anatomically differentiated features allowed, for example, to determine archaeological material from excavations in Huaca Prieta (North Coast of Peru) as a mixture of *C. ficifolia* together with *C. moschata*. Besides there were rests of Lima-bean and cotton from the same stratum, indicating that a sort of diversified primitive agriculture already existed there since preceramic time.

Excavations at Pampa de Ventanilla, on the central coast of Peru, dated as early period V (pre-ceramic), disclosed huge quantities of *Cucurbita* shells, found together with fish bones, mussels and shore birds. Lanning (1966) determined them as *C. ficifolia* and (very important) *C. andreana*, the wild ancestor of *C. maxima*. These different finds allow the conclusion that coastal Indians did primitive farming together with fishing and bird-hunting.

Funerary vases in the form of crook-neck squashes with warty surface have been found at Chimbote and Valle de Santa on the Peruvian coast, probably belonging to the species *C. moschata*.

On the other hand, excellent remains have been found in Mexico. These cucurbits (probably *C. pepo*) have been dated at 7000 years B.P. (Table 16).

sis effect. Less difficulties arose in crossings between the annual *C. moschata* and the perennial *C. foetidissima*. In spite of considerable physiological and morphological dif-

**Table 16.** Archaeological Cucurbita in the Americas

| Area or site | Species | Age (earliest) | Remarks |
|---|---|---|---|
| Southwestern USA Cutler and Whitaker 1961 | *C. pepo* | 300 B.C. at Tularosa Cave, NM | Rich, overlapping record |
| | *C. mixta* | 380–1340 A.D. | |
| | *C. moschata* | 900 A.D. | Late, scanty |
| Great Plains, west of Mississippi River, USA Cutler and Whitaker 1961 | *C. pepo* | 1400 A.D. | Abundant, many sites |
| | *C. moschata* | 1700 A.D. | Late, scanty |
| Ozark Highlands, MO, USA Kay et al. 1980 | *C. pepo* | From 4000 B.C. to time of contact | No *C. moschata* or *C. mixta* Many sites |
| Ocampon Caves, Tamaulipas, Mexico Whitaker, Cutler and MacNeish 1957 | *C. pepo* | 7000–5000 B.C. | Found at all levels |
| | *C. mixta* | 200–900 A.D. | Late, scarce |
| | *C. moschata* | 4900–3500 B.C. | Late, scanty |
| Tehuacan Valley, Puebla, Mexico Cutler and Whitaker 1967 | *C. pepo* | 2000–1000 B.C. | Late, scarce |
| | *C. mixta* | before 5200 B.C. | Numerous at most levels |
| | *C. moschata* | 4900–3500 B.C. | Earliest known *C. moschata* |
| Valley of Oaxaca, Mexico Whitaker and Cutler 1971 | *C. pepo* | 8750 B.C.– 700 A.D. | Earliest *C. pepo* to be reported |
| | *C. ficifolia* | 700 A.D. | Only one seed, if identification correct. |
| Rio Zape, Durango. Mexico Brooks et al. 1962 | *C. pepo* | 600–700 A.D. | Abundant |
| | *C. mixta* | 700 A.D. | Late, scanty |
| Huaca Prieta, Peru Whitaker and Bird 1949 | *C. moschata* | 3000 B.C. | Warty rinds, fringe-seeded form |
| | *C. ficifolia* | 3000 B.C. | First archaeological *C. ficifolia* to be recorded |
| Ica, Peru West and Whitaker 1979 | *C. maxima* | ca. 1800 B.C.– 100 A.D. | Earliest *C. maxima* reported |
| | *C. moschata* | 600–1100 A.D. | Relatively late, fringe-seeded form |

Artificial hybrids between different South American and Central American *Cucurbita* species are not easy to produce. Most difficult to cross are the large-fruited *C. maxima* × *C. pepo* . Thousands of pollinizations were necessary to obtain finally four living and flowering $F_1$–plants. Weiling (1960) used embryo culture and improved the results, especially when *C. maxima* was the female partner. Backcrossings with the female *C. maxima* sometimes gave the hetero-

ferences (root system, leaves, fruits, cucurbitacines), Weiling (1959) observed a high degree of bivalent formation during meiosis which indicates relationship. With respect to cytogenetical evolution, the different genera of the Cucurbitaceae display a wide spectrum. We suppose that the basic chromosome number of this botanical family is $x = 5$ or $x = 7$. The lowest chromosome number has been determined in the Asiatic cucumber (*Cucumis sativus*) with $n = 7$. The

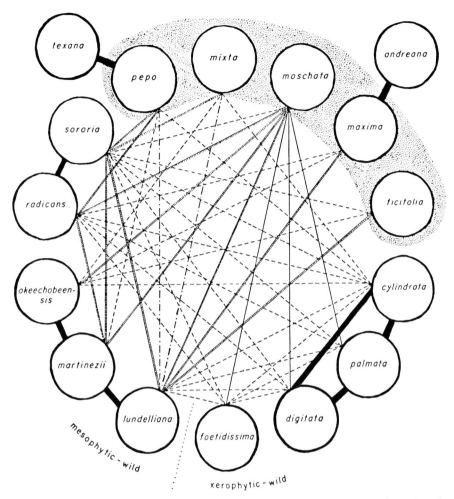

**Fig. X.37.** Explanatory diagram of the different cross-relationships in American *Cucurbita* species. *Dotted area* cultivated types; *thick black line* completely fertile hybrids; *medium black line* partly fertile hybrids; *thin black line* sterile hybrids; *dashed line* unsuccessful cross; *dot-dash line* fruits, but without viable seed

culmination is the genus *Cucurbita* with $2n = 40$, considered as allopolyploid. Between them are placed *Lagenaria, Citrullus* and *Momordica* with $n = 11$, followed by *Benincasia, Sechium, Cucumis melo* with $n = 12$ and *Luffa* with $n = 13$ chromosomes.

According to Weiling (1959) the genome formula for *C. maxima, C. mixta, C. moschata* and *C. pepo* is: AABB, different from *C. ficifolia* with AACC, and *C. lundelliana* with AAWW. The results of species crossings performed by Whitaker and collaborators suggest that *C. lundelliana* may have been involved in the evolution of several cucurbits (Whitaker and Bemis 1965). It has a wide spectrum of compatibility, as is shown in the following diagram (Fig. X.37)

### *C. ficifolia* Bouché

Malabar gourd, courge à feuilles de figuier, cayote, chilacayote, Feigenblatt-Kürbis

$2n = 40$

This perennial cucurbit was considered of Afro-Asiatic origin ("Malabar"), until Whitaker established its home definitively in Central America, where it is closely related to the wild species *C. lundelliana* Bailey. The negroes of the Basuto mountains (Lesotho) are still convinced that it originated there, due to its wide dispersion in this region, and its excellent survival under slight night frosts, which I could personally verify.

Natives of the highlands of Mexico call it "chilacayotl" and prefer this cucurbit also for its frost resistance. *C. ficifolia* must have been introduced very early – in pre-ceramic times – to South America, because its remains have been excavated at Huaca-Prieta (Coastal Peru) in a horizon dated at 2000–1000 B.P. Its southern limit today is North Argentina, where natives of remote valleys of prov. Salta and Jujuy esteem this "cayota" for the durability of its hard shell (when ripe) and its spaghetti-like, fibrous pulp. Its main use is in the preparation of preserves and jams, used like *C. mixta*.

The perennial plant develops vines, several metres long, densely covered with large (25 cm diam.) lobed leaves and many fruits (20 cm long). The fruits are shiny green with white blotches and a rather thin peduncle. Seeds are in general black, flat and broad eliptical. Due to its high resistance to soil diseases, the plants are commercially used for grafting. It is essentially a short-day plant.

## *C. lundelliana* Bailey

Peten gourd

$2n = 40$

This wild species is of considerable importance in the phylogeny of cultivated pumpkins and squashes. It is native to Guatemala, Honduras and the southern part of Mexico. It crosses with *C. moschata, C. mixta, C. maxima* and *C. pepo* and often gives fertile offspring. For this reason Whitaker concluded: "*C. lundelliana* may have had a significant role in the origin and domestication of the cultivated cucurbits of America."

## *C. mixta* Pang.

Silverseed Gourd, cushaw, ayote

$2n = 40$

Not known in the wild state and often mistaken for *C. moschata*. The two species had to be separated taxonomically due to the different peduncle, which is nearly round in *C. mixta*, rather corky and not enlarged at the connection with the fruit. This species developed sterility barriers to other Central American cucurbits during its rather recent domestication. No archaeological material exists dated earlier than 1000 years B.P.

Originally from Mexico and Guatemala, but spreading continuously southward, reaching its present meridional limit in cultivated fields of the province Mendoza (SW Argentina).

Monoecious annual plants with 2–4-m-long vines, with five-angled stems. Leaves cordate, slightly lobed, with white blotches at the surface. Flowers yellow with long sepals. Androecium columnar; stigmas enlarged green yellow. The hard-shelled fruits have various colours and are mostly greenish-white striped and have a crook-neck form. The cv. coreano is yellow-brown and has a crook-neck form. The pulp is yellow with an agreeable sweet aromatic taste. The seeds have an obtuse funicular attachment, are tan or silver-white with a sharply defined margin.

## *C. moschata* Duch. ex Poir.

Pumpkin, crookneck, courge musquée, ahuyama, anco, joko, Moschuskürbis

$2n = 40$

This cucurbit received its name because it smells slightly of musk. Due to its good performance in hot and moist climate it was distributed in pre-Columbian times throughout the whole of the neotropics. For example, in the tropical region of East Bolivia, Cardenas (1969) observed a great diversity and even found a bitter wild form near Beni, called "joko amargo".

In Panama it is the most common cucurbit, offered the year around at the vegetable markets under the name "uyama" or "aullamo".

**Fig. X.38.** *Cucurbita moschata.* Widely dispersed in Central and South America and preferred by natives for its compact pulp and high field resistance to diseases

Proof of prehistoric cultivation in Mexico are abundant remains in the Ocampo caves (2900–1500 B.P.) and from the Tehuacan valley.

Its early presence in South America is documented by ceramics from the Chimu epoque, depicting "crookneck" forms, and finds in Huaca-Prieta (3000 B.P.). For the Seminole and Mikosukee Indians who have inhabited the Everglades (Florida) since time immemorial, *C. moschata* is of great importance in daily life and nourishment, so we may conclude that some cultivars originated there. The plant can be easily recognized for its dense and soft hair indument. Leaves are round-shaped. Fruits have a soft shell and varying forms. Their pulp is dark yellow or orange. The peduncle is five-angled with a broad attachment on the fruit. The seeds are thick with pronounced margins. *C. moschata* crosses easily with *C. pepo, C. mixta, C. foetidissima* and even with the Southern species *C. maxima*. It is considered therefore as a basic species in the phylogeny of the American cucurbits (Fig. X.38). According to Whitaker and Davis (1962) "C. moschata is the indispensable cog through which the species of Cucurbita are related, in tracing back the origin of the cultivated cucurbita..."

### C. pepo L.

Summer pumpkin, summer squash, marrow
2n = 40

NAME AND ORIGIN

This pumpkin has the most northern range of all edible cucurbits. According to some reports from shortly after the Spanish Conquest, it was cultivated in the southwestern states of the USA and reached even as far north as Canada. In Mexico it was abundant and from this country several archaeological proofs of early domestication exist. The oldest find (dated at 14,000 years B.P. with $^{14}$C methods) is from Caño del Diabolo, Tamaulipas, and was discovered by Mac-Neish. Further excavations from Oaxaca (6700 B.P.) and from the Puebla II horizon in the valley of Tehuacan (3000–2000 B.P. revealed *C. pepo* remains. In North America the oldest *C. pepo* samples are from the Ozark highlands (Montana), dated at 5000 years B.P. Hundreds of other sites exist in the states of Arizona and New Mexico, the Tularosa caves and the Cordova cave, with pre-Columbian *C. pepo* remains (Fig. X.39).

**Fig. X.39.** *Cucurbita pepo.* Paraguayan native with a high productive landrace

The wild ancestor of the summer pumpkin is *C. texana* (Scheele) Gray, which was found as early as 1848 along the Guadelupe river in Texas, but whose importance for plant genetics was not recognized so that no seeds were available until Correll (1963) rediscovered the species at various places at several watercourses in south-central Texas. Recently its hybrids have been investigated to eludicate the genetic basis of the *C. texana-C. pepo* complex from the perspective of al-

lozyme variation (Decker 1985) (Fig. X.40). It was shown that the morphological differentiation between the ancestor (*C. texana*) and the actual cultivars of *C. pepo* is considerable and that it is clearly accompanied by heterogeneity at the molecular level (Decker and Wilson 1987). The existing polymorphism in the *C. pepo* group reflects its long pre-Columbian history and domestication by Indians at distant places. Several thousand years of separate cultivation in Middle America resulted in notable differences of horticultural characters (fruit size, colour, pulp consistency, skin surface etc.) and even in the creation of turban-shaped non-edible ornamental biotypes.

### MORPHOLOGY

Annual monoecious herbs; primitive forms several metres long-running vines, modern varieties bushy. Stem hard, angled and covered with many coarse bristles. Leaves cordate, often deeply lobed with serrate margins. Flowers emerge from leaf axils, bright yellow with short sepals. Masculine flowers numerous in comparison with female flowers, which have small yellow stigmas.

The fruits show great variation in size, form (crook-neck turban, snake form) colour and flesh ("spaghetti" type) and sugar content. Their pulp consists of 90% water, with a low percentage (1%) of protein and 8% carbohydrates. The seeds separate readily from the pulp and have a smooth margin. They are very nutritive, with a fat content of 40–50% and approximately 30% protein. The high fat content has stimulated various plant breeders to select "oil pumpkins". The well-known geneticist Tschermak-Seyssenegg (1871–1962) used the recessive factor "malakosperma" to create his famous "schalenlose Oelkürbis" with its valuable oil. His work was continued by Schuster (1977).

a

b

**Fig. X.40. a** *Cucurbita texana*, the wild form of *Cucurbita pepo*. **b** A fancy gardenform of *C. pepo*

### *C. maxima* Duch.

Winter squash, zapallo, courge, zipinka, Riesenkürbis

2n = 40

**Fig. X.41.** *Cucurbita maxima* landraces together with the wild-growing *C. andreana* (in centre) showing the 1000 × increase in fruit size as result of Indian domestication

NAME AND ORIGIN

To avoid the continuous deplorable name confusion in different cucurbit species all with the same English term "squash", one should call this species "zapallo". This is a kechua word, widely used in Latin America for this genuine South American crop. *C. maxima* has its origin outside the real tropical latitudes, because its wild form *C. andreana* grows in the temperate zones of Argentina and Uruguay (Fig. X.42). From there, different Indian tribes may in remote times have spread seeds to Bolivia/Peru. *C. maxima* never crossed the Isthmus of Dairén northward, and it is doubtful that it had even been cultivated in pre-Conquista times in Colombia or Ecuador.

Seeds and peduncles of *C. maxima* have been recovered from many places in the southern parts of America; there are, moreover, excellent reproductions in the ceramics of Chimu and Mochica, and examples from Viru-Valle (Peru), Ancon and Pachacamac. More than 100 seeds have been excavated at Ocucaje, Valle de Ica and 220 at Chulpaca (period "Medio Ica", dated between 1300 and 1400 A.D.).

All these findings are irrefutable proofs of a South American origin and early domestication in pre-Columbian times. It is noteworthy that *C. maxima* is isolated in nature by genetic barriers of incompatibility from all other cucurbits. Even artificial hybrids between *C. maxima* and other species are very difficult to obtain.

The genetic variability of primitive landraces of *C. maxima* in the inter-Andine valleys of North Argentina and in adjacent Bolivia is enormous and rather difficult to classify. Colours of flesh may vary from greyish white, yellow and orange to nearly red; still more differentiated, however, are the shape and colour of the fruit skin (round, turban-formed, long, with irregular protuberances etc.).

MORPHOLOGY

In view of the above statement, it is rather difficult to give a generally valid morphological description. Plants can have running vines several metres long or may be compact and bushy. Stems are round in cross-section and soft. Leaves are cordate, large (30 cm diam.) and only slightly lobed. Flowers are bright yellow. Fruits are large, even extremely large (150 cm diam.) with smooth or very wrinkled surface (Fig. X.41). The pulp is yellow with fine grained, tasty flesh. Seeds

**Fig. X.42.** *Cucurbita andreana,* the ancestral species of cultivated *C. maxima* (Photo taken in Argentina)

are difficult to separate from the dense flesh; they are obtuse with smooth margin, oil-rich and nutritive.

There was decade-long discussion about the ancestral form of *C. maxima.* Finally it has been accepted that *C. andreana* Naudin is the wild precursor. I have always opposed the idea that *C. andreana* may be "an escaped feral form" of *C. maxima.* Over many years I have traced the natural wild habitats of *C. andreana* in several provinces of Argentina. The wild zapallo grows there in undisturbed mesophytic and sometimes even xerophytic environments: on the border of the Calden forests (Prosopis caldén) in the prov. San Luis, amidst the grass steppes of the Pampa (prov. Buenos Aires, prov. La Pampa), on sandy river banks of prov. Cordoba (Rio Tercero) and on the boundary between prov. Mendoza (Rio Desaguadero) and San Luis. At present *C. andreana* has invaded also the disturbed and ploughed lands in big estancias in the Pampa, where it even became a noxious weed. Its dense leaf masses cover potato plantations and sometimes cross with cultivated *C. maxima,* which causes considerable problems for zapallo producers, for example in the prov. Cordoba, due to its bitter-tasting offspring. *C. andreana* has hard-shelled fruits, which survive in the soil for several months, per-

haps even years, continuously producing new seedlings. For this reason, this wild pumpkin is vastly extended in the South American plains. This region was inhabited before the Spanish conquest by nomadic Indians (Charruas, Puelches and Tehuelches), who collected the fruits for consumption. Their custom of roasting the fruits of this wild cucurbit and eating them still persisted in the times of the "gauchos". They used to collect the fruits and put them in the hot ash of their camp fires. This was the only "vegetable" in the monotony of pure meat feeding of these pampa cowboys. A British traveller in 1847 observed this strange use of wild cucurbits (William MacCann, cited by Millan 1968), and described how the cowboys roasted the fruits in the hot ash. Millan discovered also "non-bitter" mutants of *C. andreana* in samples from prov. Santa Fé. The same has been reported recently by Boelcke (1981) from spontaneous wild "zapallitos" in the prov. Buenos Aires. These findings are a clear indication that occasionally in free nature segregations of recessive biotypes occur which are free of bitter-tasting glucoside. I presume that in former times some Indians knew the places where such mixed populations with "sweet plants" (= exempt of cucurbitacine) of *C. andreana* grew. With a simple organoleptic test they

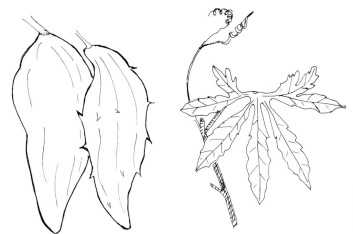

**Fig. X.43.** *Cyclanthera pedata*, a climbing cucurbit with small fruits used as vegetable in Bolivia and Peru

could separate them and use the non-bitter forms for future domestication in their own gardens. This was the beginning of a new cultivar: *C. maxima*. A still surviving antique land race is "zipinque", now rarely cultivated in the Parana region of Argentina.

## C. foetidissima **H.B.K.**

Chilicote, buffalo gourd

This perennial species can reach an age of 30–40 years, thanks to its enormous root system. It has a high content of glucosides in all its green parts. The seeds are very oil-rich and also contain 30% protein. In view of its perspectives as a future oil plant, this species is treated in more detail in Chapter IV *Oil Plants* of this volume.
Finally we have to mention some Cucurbitaceae of American origin, but of rather limited practical use. These are:

## 48. *Cyclanthera pedata* **(L.) Schrad.**

Pepino hueco, caihua, achokcha

A dozen taxa have been reported from Bolivia and Peru, so we suggest that the origin of the "var. *edulis*" of *Cyclanthera* may be in the Ceja de Montana, a moist, evergreen region at 2500 m altitude, famous for its

wealth of useful species. From there Indians may have taken seeds for further domestication in their own gardens. Even if we consider this cucurbit to be of slight nutritional value, we have always been astonished during our travels in Bolivia and Peru to find so many "achokcha" fruits for sale on the native markets. The fruits are curved, shiny green, with contrasting black seeds in a white pulp. The annual plants reach 4–5 m and climb on other herbs and bushes. Flowers are yellow, the leaves (6–15 cm) are digitate, with 5–7 leaflets elliptical and with dented margins; they sit on long petioles. The fruits (12 cm long) are half-hollow, oval oblong, sometimes flat and curved. The natives of the Anden countries appear to like the fruits as a vegetable, but we found them rather tasteless (Fig. X.43).

## 49. *Lagenaria siceraria* **(Mol.) Standl.**

Bottle gourd, tecomate, calabaza
2 n = 22

Name and Origin

Widely used in the neotropics for thousands of years. We include this Cucurbitaceae in this book, even if its original cradle may be the African continent. Biologists suppose that the fruits floated on the oceans and maintained the viability of their seeds for many months. Some ethnologists do not discount a very early migration of primitive people between the continents, which spread the bottle gourd.

Remains of *Lagenaria* have been found in Peru and Mexico. In the first case we have to mention Huaca Prieta, with 5000 years B.P. The archaeological finds of Mexico are even older (Ocampo Caves 9000 B.P., and Tehuacan 7500 B.P.). This is an astonishingly early time for a cultivated plant in America. Even if it is not very convincing that "boat-people" from Africa brought the bottle gourd to the western hemisphere, we must ask why it did not arrive before the Pleistocene, if the natural dispersal of floating fruits was so easy? If *Lagenaria* had in one or another way come from Africa, the puzzling question remains how it crossed the whole American continent and appeared so early on the Pacific side of this continent? Lathrap (1977) cited the caves of Pikimachay, in the

Ayacucho Bassin (13,000 years B.P.?) as the earliest finding place in his controversial book *Our father the cayman our mother the gourd* . Several archaeobotanists are still working on these enigmatic questions (Schwerin 1970, Pickersgill and Heiser 1977).

Before the invention of pottery in America, *Lagenaria* shells were of essential utility for Indian life; we may even call them the first household utensils. They served as water bottles, bowls and fish net floats. Its high estimation among the Indians has survived until our days, when native craftsmen make pyro-engraved utensils and other attractive objects, which are offered at the native markets in Ecuador and Peru.

The peculiar ritual importance of *Lagenaria* fruits as penis sheaths, selected for this purpose independently in Africa, South America and New Guinea (Heiser 1973) still represents a puzzle for ethnologists (Fig. X.44).

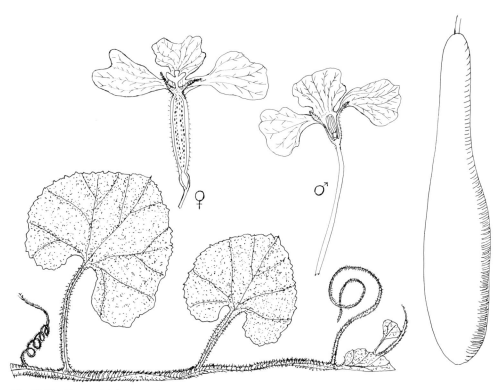

**Fig. X.44.** The world famous "bottle gourd" *(Lagenaria siceraria)*

**Fig. X.45.** Indian wood carving designs on a round variety of *Lagenaria siceraria*

MORPHOLOGY

Herbs with long running, or climbing branches. Stems are covered with gland-tipped hairs and many tendrils. Leaves 10–30 cm long with cordate base, lamina oval-cordate with dentate margin and hairy on the underside. Flowers emerge from the leaf axils, with five free petals of white colour (an exception in Cucurbitaceae) and 2–4 cm long. Flowers are insect-pollinated.

Male and female flowers are separated. Fruits have very diverse form and size. Small ones are suitable for children to play with as rattles or musical instruments (3–4 long), whilst others exceed 100 cm, often completely round. Fruits may be eaten when young, in spite of the somewhat bitter taste. When ripe, their very durable rind contains many (2-cm-long) seeds of different colour, mostly yellow-brown, but also white and black. Seeds have peculiar "ears" on the top and are often winged. With these characters *Lagenaria* seeds can be easily differentiated from other cucurbits (Fig. X.45).

## 50. *Luffa* spp.

Sponge gourd, luffa, torchon, Schwammgurke
$2 n = 26$

The genus includes six to ten taxa which are mostly of Old World origin, but one species, *L. operculata*, is native to the neotropics. Its main application is as sponges or as filter material and shock absorbers (Heiser 1979). Non-bitter selections also exist, however, which are eaten fresh or cooked as vegetable. Due to the habitual similarity of the different species which now have worldwide distribution, it is not easy to separate the different taxa.

### *L. operculata* (L.) Cogn.

Monoecious plants of neotropical origin with creeping or climbing stems, acutely 5–10-angled. Leaves suborbicular (12–15 cm long), at the base deeply cordate, 5–7 lobate with dentate margins. Pistillate flowers solitary, yellow. Staminate flowers in small racemes on 12-cm-long rachis. Fruits green, longitudinally ribbed. When ripe covered by thin exocarp inside with dense sclerenchym fibres. The seeds are brown, compressed and winged, and contain 46% oil and 40% protein. The Old World species are *L. acutangula* (L.) Roxb. and *L. cylindrica* (L.) Roem; the latter often designated as *L. aegyptiaca* Mill.

## 51. *Sechium edule* (Jacq.) Swartz

Chayote, christophin, chocho, cayotle, xuxu

2 n = 24

### NAME AND ORIGIN

The common designation "chayote" for this vegetable has been derived from the Nahuatel language. Most probably the region of domestication has to be sought in Central America, where also the highest grade of genetical variation exists. Several wild species of *Sechium* have been reported. I described an unnamed wild form from a Venezuelan forest (Brücher 1977) and Newstrom (1986) mentions *S. compositum* from South Mexico to Guatemala as probable wild ancestor.

Given the high esteem for this plant in Middle America, combined with a remarkable wide phenotypic fruit variation ,and its pre-Columbian use among the Azteks, we accept a Central American origin (see also Kunkel 1984).

*S. edule* has aquired worldwide distribution in all tropical countries as a favourite home-garden vegetable. One of the reasons is that chayote does not suffer – as other *Cucurbitaceae* – from fungus diseases or parasites; the production in roots and fruits is also high, one single plant producing 300–500 fruits in the course of one year (Fig. X.46).

### MORPHOLOGY

Perennial, climbing, monoecious herbs, with 6–12-m-long vines, emerging from tuberous

**Fig. X.46. a** A wild form of *Sechium,* growing in a forest of Venezuela. **b** Cultioars of *Sechium edule* extended on a read of wires

roots. Tendrils are large and branched. Leaves are ovate and lobed (10–25 cm diam.) inserted on long petioles and with a rough hairy surface. Flowers are small, with greenish colour. Female flowers are solitary, whilst the male flowers sit in peduncled clusters.

Fruits are round or pear-shaped, covered with spiny protuberances; but garden selections may be completely smooth. According to the different variety, the fruit size (mostly 8–16 cm), shape and colour vary extraordinarily. Mostly the fruits have deep longitudinal furrows with white, green or dark violet skin. The flesh is whitish and contains only one big seed. The seed kernel lacks a strong seed coat and begins germinating when still inside the fruit. This characteristic may be unique in the whole cucurbit family. Its hypocotyl emerges from the apex and may even produce some precocious rootlets inside the fleshy fruit. For this reason chayote seeds cannot be stored. We consider this "vivipary" as a positive selection factor for quick germination and growth in the shadowy thickness of forests, once the fruit has fallen. After 1 year of growth the perennnial rootstocks acquire considerable weight (up to 50 kg) and contain a fair quantity of starch. They are eaten raw as a source of water, or more often boiled as vegetable. Chayote fruits resemble squash and appear in America in many local dishes. Their nutritional value is not very high and consists of 1% protein, 7–9% carbohydrates, fibres and some vitamin A and C.

### 52. *Polakowskia tacacco* Pittier.

Tacaco

This dubious semi-cultivated taxon is so similar to *Sechium*, and has also the typical one-seeded fruit, that botanists should have eliminated this monotypic genus. In the meantime, it has received the superfluous denomination "Frantzia tacaco". Its reduced area as semi-cultivar is Costa Rica - Nicaragua.

### 53. *Sicana odorifera* Naudin.

Secana, pavi, melon de odor, cojombro, cajuba

We observed wild-growing plants in the cloudforest of Tovar, some 100 km distant from Caracas (Venezuela). These were 10-m-long lianes, climbing on trees, with large leaves (30 cm long) and showy, orange-red-coloured, cylindrical (60 cm) fruits. Natives collect them for their pleasant fragrance. Their use dates back to pre-Columbian times, as we conclude from a report of Bernabé Cobo, who in the 16th century called it "calabaza del Paraguay". *Sicana* plants can be found in Bolivia, Paraguay, Brazil, Venezuela and also in the tropical part of Central America (Bailey 1937).

---
PASSIFLORACEAE
---

### 54. *Passiflora* spp.

Of this neotropical genus more than 380 taxa have been described since Linnés time. This obliges us to select here only the most promising of the valid species (Killip 1938, 1960). It is said that 50 are edible. *Passiflora* are in general climbing herbaceous vines, sometimes with a slight woody stalk and beautiful flowers and decorative growth habit. Before approaching the commercially most important *Passiflora* species, we enumerate some interesting wild-growing taxa, which may be useful as bearers of genetic resistance or special aromatic factors. They have different chromosome numbers, from $2 n = 12$ (*P. lunata*) to $2 n = 14$ (*P. alba*) to $2 n = 18$ (*P. edulis*), to $2 n = 22$ (*P. foetida*).

#### *P. cincinnata* Masters

Originates from Paraguayan Chaco, where the Indios Chaguancos collect the shiny red fruits with white aromatic pulp. We found them especially tasty and refreshing, and consider that the species deserves future selection. The plant is a strong climber on trees and produces there many large fruits (Fig. X.47).

**Fig. X.47.** *Passiflora cincinata* a wild-growing Paraguayan passion fruit, regularly collected by natives

### P. caerulea L.

Maracajú

This is a climbing plant with egg-sized fruits and has very showy flowers of different colour combinations. The pulp is yellow with red arils around the seeds and has a very aromatic taste, when eaten fresh. This species is often used as an ornamental garden flower.

### P. ligularis Juss.

Sweet granadilla, curuba, granadilla dulce, parcha

With a very extended areal, from North Argentina to Mexico, and preference for temperate climate in hill regions of 1000 m altitude. Vigorous climbers with 10-m-long stems and dark green , cordiform leaves (15 cm long). Flowers white-yellow with a bluish corona. Fruits on rather long (12 cm) peduncles, with a hard pericarp. Inside, black seeds surrounded by fleshy arils of agreeable aroma. The possession of hard shells could be of great importance for future selection as a transportable market fruit, in comparison to other passion fruits with wrinkled and unattractive exocarps. This species is grown successfully in Hawaii, called "granadilla" and considered by fruit experts to be the finest of all passion fruits.

### P. laurifolia L.

Bell apple, yellow granadilla, pomme liane, maracuja

A wild-growing liane which extended from NE South America to several West Indian Islands. The stem is round in cross-section. The leaves are cordiform (12–14 cm long). The flowers are beautifully coloured: red, white and blue with an aromatic fragrance. The fruits are ovoid (7 cm long), yellow with a soft pericarp, and covered with three long bracts. The pulp is soft, with many seeds. It is eaten in Jamaica and called there honeysuckle. Since the 18th century it has spread in Malaysia and become wild there.

### P. mollissima (H.B.K.) Bailey

Banana-passionfruit, tacso, curuba

Originally from the lower mountain region of the Colombian Andes, now introduced under this name to Hawaii and New Zealand, where it is grown for its tasty edible fruits. The whole plant is covered with a dense pubescence. The stems are several metres long, climbing on trees. The leaves are trilobate (10–15 cm long) on 6–cm-long petioles which bear many glands. The fruits are elongated (7 cm long, 3 cm thick) covered with a velvet-like indument; when ripe they have a whitish coloured pericarp; the pulp is yellow and of exquisite taste.

### P. mooreana Hooker

A rather rare species in the Paraguyan Chaco, where the Maka Indios have the curious name "how-how" for it. They use the unripe fruits, baked on hot stones, but it can be also eaten as ripe fresh fruit.

### P. vitifolia H.B.K.

Granadilla del monte, guate, tumbo

This liane-like passiflora has a wide areal, from Peru to Venezuela, Panama to Nicaragua. The stem is densely pubescent and a strong climber on shrubs and trees. The flowers are rather large (10–12 cm wide), with sepals and petals of red magenta colour. The fruits are puberulent, (6–8 cm diam.), greenish white with darker stripes and an attractive fragrance.

### P. edulis Sims

Passion fruit, grenadille, maracuja, granadilla morada, Passionsblume
$2n = 18$

This is the most commercially interesting species. Its origin is the hill region of South Brazil, not the tropical lowlands. For this reason commercial plantations have failed when undertaken in the hot humid tropics. The most convenient ecological conditions are higher altitudes, above 1500 m altitude in cool tropical mountain climate (Fig. X.48).
*P. edulis* is a vigorous climber which lives 3–10 years, forming finally then thick (10–14 cm) woody stems. The heavy load of the passion fruits is supported by many spirally coiled long axillary tendrils.
The leaves are deeply three-lobed (10–20 cm in diam.) on 5-cm-long petiols. The lamina is somewhat leathery and glabrous.

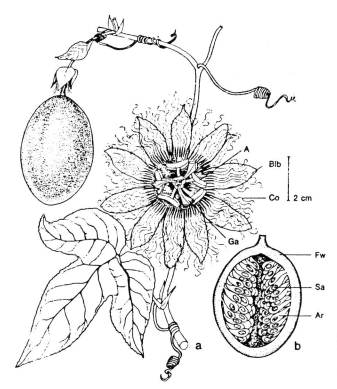

**Fig. X.48.** *Passiflora edulis* branches with flower and fruit (Rehm-Espig 1984, Ulmer, Stuttgart)

The flowers are anatomically rather complex. They are borne in leaf axils on 5-cm-long peduncles which carry three ovate bracts (1–2 cm long). Calyx tubular with five large fleshy sepals (2–3 cm long and 1–2 cm wide). Alternating with the sepals are five white elliptic petals. Then follows the so-called corona which is formed by two rows of waving (2–3 cm) sterile and thin filaments of different colour. At their basis are several rows of purple-tipped papillas and five fertile stamens which enclose a round gynophore. There are three large horizontal styles. To attract insects, some nectar is secreted at the base of the androgynophore which is self-incompatible. Besides honeybees and carpenter bees (*Xylocopa*), also humming birds provide pollination.

The fruits are globose (5–7 cm) berries, deep purple in the ordinary form of *P. edulis*, or bright yellow in the var. *flavicarpa*. The pericarp is thin (5 mm) with a white endocarp and greenish yellow mesocarp. The seeds are attached with short funiculi at the ovary wall, surrounded by thin-walled arils which contain an aromatic juice. This sweet-tart tasting juice is the commercial product of the passion fruit. It contains 8–10% sugar, 2–5% fruit acid, 0.2–0.3% pectin, 0.4–0.6 minerals, (phosphate and potassium) and also 1% protein and 1% starch. The vitamin C content is 30–50 mg per 100 g pulp (Herrmann 1983).

The first fruits can be expected 6–8 months after planting. An annual yield of 6–10 t/ha can be considered satisfactory, but there are also reports of 15 t/ha from South African plantations. The economic life span is 3–8 years. At present there is a considerable increase in the production of passion fruits in many tropical and subtropical regions of Africa, Australia, Oceania and America, but self-fertile high productive selections are still lacking.

### P. quadrangularis L.

Giant granadilla, badea, parcha, maracuja grande, barbadine

2n = 18

Of South American origin, this passiflora is now distributed in the whole neotropics and has been introduced for many decades in other continents. Besides its large soft fruits, the most distinctive character is the winged, sharp quadrangular stem. The plants are perennial vigorous climbers with enlarged fleshy roots. The leaves are big (20–25 cm long), oval with typical pale green colour. The flowers are large (10–12 cm diam.) with the characteristic complicated structure of different coloured sepals, fleshy sepals and red and white sepals, covered by the corona. The fruits are very large (20–30 cm × 10–15 cm), yellow-green with a thick pericarp (3 cm) and a fleshy insipid mesocarp. The seeds are surrounded by whitish acid-sweet arils. The large fruits are not as tasty as other passion fruits and are mainly used to prepare soft drinks (Fig. X.49).

**Fig. X.49.** The "giant passion fruit" = *Passiflora quadrangularis* cultivated by the author in Trinidad

*P. quadrangularis* is well adapted to hot humid tropical environments. Whilst living on the Island of Trinidad, we tried to produce hybrids with other $2n = 18$ chromosome *Passiflora* species to obtain offspring with better fruit taste. My successors at the German Seed Project Chaguaramas lost the breeding material.

The following pests and diseases limit the yields.

Several fruit flies of the genera *Dacus* and *Ceratitis* puncture the young fruits, which causes their premature abscission. In the mountain regions of Kenya the fungus *Alternaria* causes heavy losses in leaves and fruits ("brown spot"). *Fusarium* wilt causes the death of the vines, the roots are damaged by *Phytophthora cinnamomi*. Finally there is a severe virus disease, called woodiness, which is aphid-transmitted and possibly caused by cucumber virus 1. In consequence of the infection the fruits decrease in size.

ROSACEAE

## 55. *Fragaria chiloensis* (L.) Miller

Chilean strawberry, frutilla, fresa, frutilla salvaje, fraisier du Chile, Chile-Erdbeere

$2n = 56$

Strawberries have become a promising fruit crop on all continents since the last century, but only few people know that an essential part of the amphyploid genome of *Fragaria-ananassa* ($2n = 56$), the "garden strawberry" of international commerce has a South American origin (Fig. X.50).

The genus *Fragaria* consists of diploid, (*F. vesca* in Eurasia), tetraploid (*F. orientalis* in Asia), hexaploid (*F. moschata* in Europe) and finally octoploid (*F. virginiana, F. chiloensis*) species from America. This polyploid series in the genus *Fragaria* arose already in the Eocene period, when the differ-

**Fig. X.50.** *Fragaria chiloensis,* the South American wild strawberry growing on the coast of the Island Chiloé

ent wild species became separated by geographical and continental changes (Bringhurst and Gill 1970).

At present, an agglomeration of diploid and tetraploid species exists in the eastern part of Asia, whilst the only hexaploid (*F. moschata*) is restricted to Europe (Staudt 1968).

NAME AND HISTORY

The denomination *"chiloensis"* for the South American strawberry has been caused by an unfortunate misinterpretation. The species is not at all restricted to the Island of Chiloé, but rather has a geographical extension along the Pacific coast and cordillerean valleys of more than 20,000 km, interrrupted by several disjunctions, which gave rise to local races (forma "patagonica", "pacifica") and an interesting domestication centre in Ecuador. Indians selected there in early times shortday biotypes with different fruit colours: red, carmine, yellow, white and rather large fruit size, as we observed decades ago on the native market places of Ambato and Lacatunga.

The Indian aristocracy developed a high esteem for such strawberries, as we may conclude from an annotation by Garcilaso de la Vega (1557), an Inka descendant. It seems that he was astonished at the relatively small size of the European strawberries in comparison with those he remembered from Cuzco Peru. Another report was given by Ovalle (1646, a missionary who travelled in Central Chile and praised the local Chilean strawberries as superior to what he knew from Italy: "large like a pear and in red, yellow white varieties". Such reports stimulated Capt. Frezier (1716) to recover living plants from these supposed "giant strawberries from Chile". After a hazardous voyage of 6 months, sacrificing the last drinking water to water the five surviving *F. chiloensis* plants, Frezier arrived with his sailing boat at Cherbourg, and was protected there like a valuable treasure. However, to the utmost disappointment of French and Dutch planters, who multiplied the South American strawberry plants vegetatively, they did not produce any fruit, and Frezier was accused of fraud. The problem was only re-

solved when Duchesne discovered that the five original plants were females, lacking a pollinizator. After having intercalated some plants of *F. virginiana* in the strawberry rows, the pollination was restored and its fruits produced. From 1720 to our days a breath-taking evolution of the progenies of these artificially created intercontinental hybrids began, culminating in such outstanding creations as the cv. Senga-Sengana (1954 v. Sengbusch) with yields surpassing 15 t/ha under optimal growing conditions. .

In the whole world since the last century, more than a thousand selected varieties from the amphiploid hybrid *F. ananasa* have been created (Darrow 1966).

---

SOLANACEAE

## 56. *Cyphomandra betacea* (Cav.) Sendt.

Tree tomato, tamarillo, tomatillo, tomate d'arbre, Baumtomate

$2n = 24$

More than a dozen (some authors maintain 40) taxa of the genus *Cyphomandra* exist in South America. None of them has been recognized undisputedly as the ancestral form. Sauer, Bukasov and other authors even doubted the existence of a wild-growing ancestor. In Venezuela-Colombia various wild species exist, e.g. *C. diversifolia, C. glabra* and *C. meridiensis*, always in the cool-humid mountain region between 1400 and 2000 m alt., but their leaf anatomy and flower biology is rather distant from *C. betacea*. In Panama we saw an endemic species *C. allophyla*, with 2-cm-long greenish white striped berries, also morphologically too distant to be included in the ascendency. The most probable forerunner, however, is *C. bolivarensis*. The species was discovered by Steyermark (1966) on one of the many "tepuis" (2000 m alt.) in the frontier region between Venezuela (State Bolivar) and Guyana. We have already explained (Brücher 1977) the reasons which induce us to believe in closer phylogenetical relations between this wild species and the cultivated tree tomato. We mention here especially *C.*

a                           b

**Fig. X.51.** *Cyphomandra.* **a** Wild type from Venezuela. **b** cv. of *Cyphomandra betacea* from the mountains of Tovar, Venezuela

*hartwegii* (Miers) Dunal with a vast areal, extending from Central America to Ecuador-Bolivia. There the natives plant it in their gardens to extract the black sap as a good dye for ceramics (Fig. X.51). Recently *C. casana* from Ecuador entered the international fruit market, distinguished by yellow-green epidermis when ripe.

The habitus of tree tomatoes is not at all similar to the usual bush tomato. *C. betacea* is a decorative tree-like shrub (3–5 m high); the stem is really woody at its base. Plants are semi-perennial (2–6 years lifespan). Leaves large (30–40 cm) cordiform, simple, whilst wild species have imparipinnate, subdivided lamina.

Inflorescences are borne in leaf axils and produce many small (12–15 mm) pink flowers, most of which are shed early without giving fruits. The berries sit on 5-cm-long petioles and have the size of a hen's egg. The skin of the fruits is sometimes yellow, but in general carmine red, rather thin and with a slightly bitter taste. The pulp is sweet-sour and contains many seeds and unfortunately "stony" concretations which may hurt the gums when eaten raw. The vitamin C content is fairly high: 31 mg ascorbic acid in 100 g. The tree tomato has been distributed during the last decade to many countries outside South America with well-adapted precipitation in cool highland, without frost and is grown with success in Madeira, Teneriffa, Kenya, Java and especially in New Zealand, where advanced selections exist. Mutation breeding is recommended to eliminate the stony inclusions of the pulp.

## 57. *Physalis* spp.

$2n = 24$

The genus has only few representatives in the Old World; most species (probably 90 taxa) are American, with a principal centre of species concentration in Mexico (70) and a dozen in South America.

All species are characterized by a large calyx (= Greek: physalis), which covers or even completely includes the ripe fruit. The following neotropic species have acquired a certain economic importance *Physalis ixocarpa* Brot., *P. philadelphica* Lam., *P. peruviana* L., *P. edulis* Sims and *P. pubescens* L.

### *P. philadelphica* Lam.

Husk tomato, tomatillo, miltomatle (Aztec)

The "miltomatoes" have a millenarian tradition in Middle America as edible fruits. The Indians of the present Guatemala-Honduras used the wild-growing species *P. philadelphica* Lam., which has relatively large fruits, and transformed it into a garden crop. Similar importance was given to *P. pubescens* L. (called "tomatillo"), which after the Spanish invasion was spread over the whole pantropic belt. The fruits of these "miltomatoes" were the basis for many sauces and dishes of the Mayan and Aztec tribes. With the decay of these cultures, however, *Physalis* lost its importance and was replaced by the newly introduced tomato (*Lycopersicon esculentum*). Its fruits may be consumed raw, but "miltomatle" was (and still is in certain parts of Central America) a very important ingredient for meat dressings and sauces. In this character such *Physalis* species spread also to Asia and Africa.

The whole plant is covered with fine hair indument. The leaves are simple (3–5 cm large), with prominent nerves.

The inflorescences are borne in leaf axils. The yellow fruits are enclosed by enlarged calyx sepals.

### *P. peruviana* L.

Cape gooseberry, capuli, uchuva, aguaymanto (kechua)

The fruits can be eaten raw, in contrast to the above-mentioned species. Under the name "cape gooseberry" it has been planted for two centuries in South Africa, especially as an anti-scurvy fruit. From there it has spread to Java and Australia. The canning

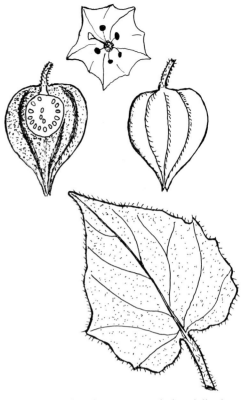

**Fig. X.52.** *Physalis peruviana*, industrialized as "cape gooseberry" used for vitamin-rich marmelades; fruits also fresh consumed

factories of these countries stimulate the cultivation of this plant, for its high vitamin C content and as basis for production of jams and jellies (Fig. X.52).

The "cape gooseberry" is in reality a cultivar on its own and rather different from the originally wild *P. peruviana*. The plant reaches 100–150 cm in height and needs support with a wire-pole system. The stems are weak, covered with hairs and glands. The leaves are cordiform (6–11 cm) and pubescent. The flowers are yellow, rotate (2 cm diam.). The calyx – small at the beginning – develops to a bladder-like organ, which completely encloses the ripening fruit; its size is 5 cm, whilst the fruit is only 2 cm large. The fruits are glossy yellow berries with many seeds. For this reason the new German denomination "Goldbeere" seems much more indicated than the objectionable old denomination "Kap-Stachelbeere", which is botanically incorrect. Recently its fruits have also entered the European fruit market, mostly importations from Kenya, South Africa, New Zealand and Hawaii. Selections of more frost-tolerant biotypes is indicated. In general the plants are robust, and easily produce adventive roots, which allows vegetative reproduction. The yield per plant has been calculated at 2 kg.

## 58. *Lycopersicon esculentum* **Miller**

Tomato, tomate, mexik: tomatle, Liebesapfel
$2 n = 24$

NAME, ORIGIN AND DISTRIBUTION

It seems strange that in the whole of South America no Indian word exists to designate this important neotropical fruit. From Middle America we know only "tomale", a Nahuatel word, which later spread over the whole world. To complicate matters further, *L. esculentum* was domesticated in pre-Columbian times in Central America, where no wild tomato species is native. On the contrary, a dozen wild species of the genus *Lycopersicon* exist in South America, several in Peru. On archaeological sites in Peru, however, remains of the tomato plant have never been found nor do any reproductions exist on ancient potteries in the Anden region, whilst many other cultivated plants have been depicted by Indians, or buried in their grave sites. We must conclude that neither the Aymara nor the Kechua planters and seedsmen accorded great significance to this Solanaceae, while so cleverly selecting so many other cultivars from this botanical family.

In conclusion: for lack of proof, we must discard the possibility that the rich Peruvian "gene pool" of wild-growing *Lycopersicon* is the cradle of the cultivated *L. esculentum*. On the other hand, domestication occurred precisely in a region, the so-called gene centre of Mexico/Guatemala, which is completely devoid of autochthonous wild-growing *Lycopersicon* species. I maintain the hypothesis that the tropical *L. humboldtii*,

collected by Humboldt in the year 1800 in North Venezuela and rediscovered at the same place at Lago Tacarigua 160 years later (see Brücher 1969 and 1977) may be the immediate ancestor, or at least one of the "missing links" in the phylogeny of *L. esculentum*. As this has not met with the agreement of North American tomato breeders, we can do no more than live with the opinion of Rick et al. ..."Thus although really critical evidence is still lacking, domestication in the Central America-Mexico region seems most probable" (in a personal letter, communicated March 24, 1983). Rick (1978), following the theory of Jenkins (1948) believes that the now cosmopolitan "weed tomato", *L. cerasiforme*, is the direct ancestor, in spite of some morphogenetic contradictions. But he did not exclude that also *L. pimpinelli folium* was involved.

There is agreement that *L. cerasiforme* is native in a narrow west coastal region of Peru, that it has very small fruits, no bigger than a cherry! and that this "cherry-tomato" spread as a weed throughout America. Jenkins and other botanists assume that this "weed" travelled more than 5000 km to Mexico and was domesticated there. Jenkins believes that Mexico (probably the region of Puebla-Veracruz with its present greatest varietal diversity) was the only region of domestication, that in the "cradle" of *L. esculentum* lived the immediate ancestor of cultivated tomatoes. We do not agree with this theory for several reasons, as we have explained earlier (Brücher 1977). We cannot believe that a "weed" whose small fruits never attracted the attention of South American Indian planters had to travel 5000 km northward before receiving due attention and domestication (Fig. X.53).

Further we must ask: if the "cherry tomato" really had been propagated over so many thousand kilometers as a "weed", what tutor crop accompanied it and what vectors (humans?) dispersed it? Weeds – as is known worldwide – depend on man and the disbalance he creates in closed natural habitats. We know neither any South American crop which migrated from the Pacific coast of Peru to Mexico which would have been suitable for such an event, nor which Indian tribe may have spread this weed.

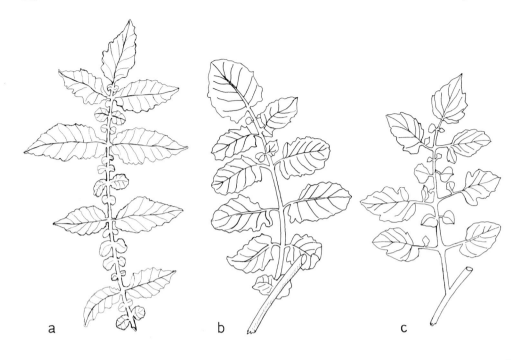

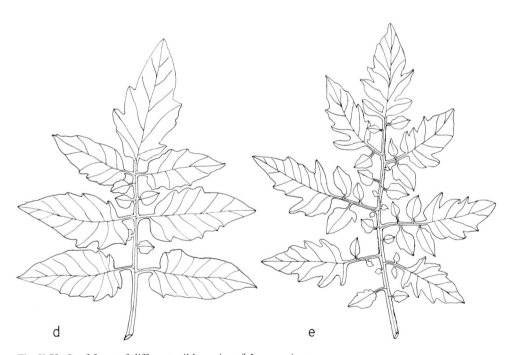

**Fig. X.53.** Leaf-form of different wild species of *Lycopersicon*
**a** *L. hirsutum* **b** *L. peruvianum;* **c** *L. pimpinellifolium,* **d** *L. humboldtii;* **e** *L. cerasiforme*

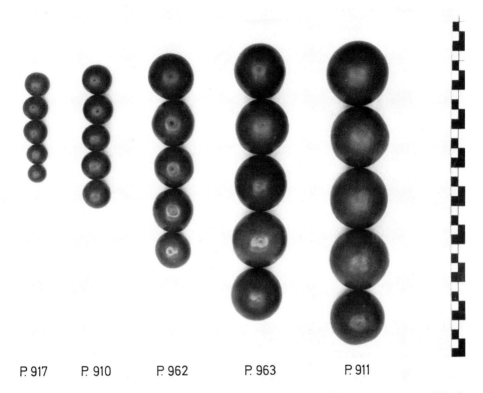

P. 917    P. 910    P. 962    P. 963    P. 911

**Fig. X.54.** "Laboratory-domestication" of the wildspecies *L. pimpinellifolium*. From very small fruit size (1 g), enlarging it to 17 g by artificial mutations in one decade and improving the whole plant habit (after Stubbe)

These phytogeographical implications can be avoided if we take as starting point a wild tomato, which grows naturally north of the Equator in a humid tropical environment. This is *L. humboldtii*, discovered in North Venezuela at 10° north of the Equator. Recently, we observed a similar biotype "on the other side of the Darién gap", in fields occupied by the Choco Indians, who have long inhabited both sides of the Isthmus of Darién and are known as avid wanderers. We were astonished, at the border of a rainforest with more than 6000 mm yearly precipitation, to find quite healthy primitive tomato plants free of the usual fungus and bacterial diseases which attack and even destroy *Lycopersicon* plants in such a climate (Fig. X.55). Perhaps these Choco tomatoes are remnants or descendants of a genuine wild tomato which originated north of the Equator. From there to Mexico it is only a distance of 10° more.

From the coastal region and low Andean valleys of Ecuador, Peru, Bolivia and North Chile the following wild species have been described: *L. glandulosum, L. chilense, L. hirsutum, L. minutum* and *L. peruvianum*, all with green fruits when ripe. They are separated genetically from the red-fruited *Eu-lycopersicon* group, with which they do not cross. They grow in the arid coastal region, in the so-called Lomas, where vegetation is limited to the misty months of August to November. Another wild tomato, discovered on the Galapagos Islands and called *L. cheesmanii*, is interesting for its high resistance to salty soils. Finally we mention the taxon *Solanum pennellii*, a "bridge species" between *Solanum* and *Lycopersicum*, which hybridizes with tomatoes and transmits its outstanding drought resistance, originating in the arid Garrua zone.

The *Eu-lycopersicon* group includes the taxa: *L. pimpinellifolium* ( = *racemigerum*),

**Fig. X.55.** Recent discovery of an Indian tomato (Choco tribe) growing in the adverse climate (8000 mm precipitation) in the rainforest of Darien (Panama), similar to *Lycopersicon humboldtii* (Brücher 1988)

*L. cerasiforme, L. humboldtii, L. parviflorum* and the cultivated *L. esculentum*.

Due to the perishable fruits of tomatoes, no pre-Columbian remnants of *Lycopersicon* exist in America and all archaeological and palaeontological studies have been unrewarding so far, but from the earliest European Herbals we are well acquainted with the aspect of the Indian tomatoes which reached the Mediterranean countries in the early 16th century. We consider of invaluable importance a handcoloured drawing by Oellinger (1550), which we found on p. 541 of the *Herbarium Oellinger* at the University of Nürnberg-Erlangen. It is quite obvious that the rather large tomato fruits and several flowers depicted have teratological deformations, similar to present-day Indian tomatoes, which have many carpels, an in-

creased quantity of flower segments and numerous sepals. Without doubt these were not simple derivates of *L. cerasiforme*, but advanced domesticates. A similar conclusion can be drawn from the pictures in old herbals, e.g. Dodoneus (1553), Matthiolus (1554) and Bauhin (1596), or from the instructive plant drawings of Gessner (1516–1565). Their tomato plants are neither wild nor undeveloped "cherry tomatoes" but large-fruited biotypes. Their irregular fruit shape with multi-valved, grooved berries, often fasciated, has survived in South and Middle America for centuries until today. An example is the Argentinian var. *platense*, a very old landrace with strong fibre bundles beween the carpels and tough skin. Such indigenous primitive varieties, by reason of their firm texture, in contrast to the tender skin of the "cherry tomato", survive the long transport journeys, which were done in former times on the backs of mules.

MORPHOLOGY

Annual herbs, with varying height; normally 60-cm-tall bushes, but also several metres (2–5 m) long stems, climbing on other plants in a tropical environment with undetermined growth. Selected cultivars have thick stems which become prostrate. Their branching system is sympodial with many axillary sprouts on the nodes. They have vigorous taproots with many adventitious roots. The whole plants covered with hairs and glandular trichomes which produce the typical tomato odour. Leaves imparipinnate, irregular (20 × 15 cm). Inflorescenses, frequently are borne in terminal clusters, often opposite to the leaves or between the nodes. The corolla is always yellow, stellate with five (or more) petals, recurved when ripening. The calyx possesses five narrow pointed sepals, which increase considerably during the ripening process of the fruits. The anthers are yellow and sit on short filaments. The tip of the pistil scarcely surpasses the level of the anthers, whose pollen fecundates the stigma very early, provoking a high degree of self-fertilization.

The fruits are bilocular in primitive tomatoes, but many-celled in advanced varieties. The berries are round, shiny red varying in size from 2–16 cm diam., often with irregular fleshy placentas and furrowed. The seeds are kidney-shaped and covered with short stiff hairs. They have a curved embryo, and contain 24% oil.

Unripe fruits contain the toxic alkaloid tomatine, which becomes converted during the ripening process. Perhaps for this reason *Lycopersicon* was considered a suspicious vegetable until the beginning of our century, more than because it belongs to the family of nightshade plants (*Belladonna, Hyoscyamus, Datura*) with very toxic compounds. As a personal remembrance of my childhood, I cannot forget that my father, a learned man with wide botanical knowledge, refused his whole life long – as many others at the time – to eat tomatoes. At that time the whole production in Germany did not exceed 25 ha. At present, the tomato is a worldwide accepted fruit with a world areal of more than 2,300,000 ha and a yearly output of approximately 60,000,000 t (not counting the innumerable house gardens with tomato plants).

ECONOMY

One of the first North Americans to sponsor tomato cultivation was none other than Thomas Jefferson, who lectured in 1782 on the advantages of this plant for Virginia. Nobody would have imagined at this time that Jefferson's beloved tomato would within only two centuries show annual yields of 8,000,000 t in the USA, at present by far the largest producer in the world. Until 1865, the USA did not dispose of its own varieties, but used seeds from European introductions, called large red-ribbed. Then there appeared such remarkable North American creations as Stone, Trophy and Ponderosa, which dominated the market in the USA around the beginning of this century. The cv. Marglobe was the most famous; selected in 1925, it still exists as a valuable breeding stock and for decades represented the most important tomato cultivar throughout the world.

The tomato fruit has aquired many uses; besides being eaten raw and fresh, it is consumed in juices, sauces and ketchup. A considerable part of the yield is industrialized as concentrate and even powder, stimulated by the increasing canning industry of Europe and North America and the high popularity of tomatoes. Strangely enough, however, this fruit is not particularly nutritious, consisting of 90% water and very low carbohydrate content. Only the increasing quantity of daily consumption in the households makes it a good source of minerals (phosphorus, potassium) and vitamins A and C.

*L. esculentum* has been grown outside America for the past 300–400 years, nevertheless its consumption and production have increased only at the end of this period. Whilst in the last century it was still a fancy exotic fruit (or vegetable) which did not even figure in the consumption statistics, its use in the USA was 8 kg per capita in 1920. Now its yearly consumption raw is four times higher. Also the yields per hectare have increased incessantly in the USA: from 13 t/ha in the year 1920, in California they exceed at present 50 t/ha in irrigated fields.

Absolute maximal yields are obtained from tomato hybrids grown in greenhouses. For example, from the Netherlands crops of 180 t/ha have been reported in glass houses. The USA are by far the largest producer in the world. Whilst in the year 1962 production was 4.5 mio t, 10 years later it was 6 mio t and in 1977 already 7.9 mio t. World production has developed as follows (in mio t):

| 1952 | 1962 | 1972 | 1980 |
|---|---|---|---|
| 9,876 | 16,358 | 32,200 | 48,000 |
| 1982 | 1983 | 1984 | 1985 |
| 51,742 | 55,803 | 60,273 | 60,825 |

DISEASES AND PESTS

In accordance with the leading theme of this book, we consider here mainly tomato diseases in the tropical environment. First of all there are several fungi which may be destructive when planting occurs during the rainy season and protective chemicals are easily washed off (Villarlal 1980).

*Phytophthora infestans* can – as in potato fields – within a short time completely destroy tomato plantations. As the fungus spreads very rapidly during wet weather, even prophylactic spraying with fungicides does not help very much. Unfortunately there is no genetic immunity in wild tomatoes for a successful cross.

*Fusarium bulbigenum* causes the widespread fusarium wilt. This is a soil-borne fungus which exists in several physiological races, a fact which complicates resistance breeding. We consider this latter, however, as the only efficient manner of combat, because the fungus survives chemical treatment, hidden in the soil and plant residues.

Working with *L. humboldtii* crossings on the Island of Trinidad, we found resistance to *Septoria lycopersici.* This typical tropical disease renders tomato production impossible during the rainy season. The leaves drop and the stems rot. Due to its origin in hot humid climate, *L. humboldtii* disposes of natural selection tolerance against this severe tropical disease.

*Sclerotium rolfsii* is also dangerous in a humid climate. It causes a quick wilt on the stem basis, which is encircled by a white mycelium which produces black sclerotia.

It seems advisable to collect natural immunity to such tropical tomato diseases in the tropical latitudes themselves. Recently, we were astonished to observe healthy primitive tomatoes in small fields of the rainforest of Darien under heavy cryptogamical pressure and temperatures constantly around 30°, with more than 6000 mm precipitation.

Early blight (*Alternaria solani*) and leaf mould (*Cladosporium fulvum*) proved very destructive in the Caribbean region, where we observed in extended variety tests good field resistance only in the cv. Floralu and Indian River. Genetic tolerance to wilt and blight has been introduced by crossings with the wild species *L. pimpinellifolium, L. peruvianum* and *L. hirsutum.*

BACTERIA

The situation is rather difficult with the bacterium *Pseudomonas solanacearum*, which is omnipresent in wet tropical soils, living on a quantity of different host plants. Partial resistance has been reported in some new tomato selections, like Saturn and Venus.

Viruses

From among the numerous viruses which attack *Lycopersicon* in the tropical environment, we mention yellow top virus, a dangerous disease in Venezuela, transmitted by *Bemisia.*

Tobacco mosaic virus is worldwide distributed. *L. hirsutum* resists infections and has been widely used as a source of immunity. Curly top virus is a widespread tropical disease. Genetical resistance has been observed in various wild tomatoes, like *L. hirsutum, L. peruvianum* and *L. pimpinellifolium*, but it is not easy to obtain resistant hybrids.

Breeding and Genetics

Working in South America with tomato selection, I tried on repeated occasions to use artificially produced polyploidy to increase plant vigour or fruit size. Reviewing the world literature on tomato breeding, I realize now that nobody has achieved practical results increasing the $2 n = 24$ genome. On the other hand, we observe an enormous gigas effect in *L. esculentum*, which is not related to genome increase. For me it is still an enigma how the Amerindians achieved

**Fig. X.56.** 1000 × increase in fruit size during the domestication process of the cultivated tomato (*Lycopersicon cerasiforme* to *L. esculentum*)

**Fig. X.57.** Two wild-growing tomatoes. *Lycopersicon cerasiforme* from Colombia. *Lycopersicon humboldtii* from Lake Tagarigua in Venezuela with considerably difference in fruit size

the considerable enlargement of the tomato fruit starting from a berry originally only half an inch long, in the course of their domestication.

I would guess that they made crossings. This was not easy, because as we know today (Butler, in Canada), there are 20–30 different genes at work to influence tomato fruit size. The gigas effect is equally surprising in other Indian-selected crops, like *Phaseolus* beans, *Capsicum* peppers, several root crops, *Cucurbita* fruits etc., to which amazing achievement only modern genetics has found the clue: the phenomenon of "disproportionate growth": If we take the case of tomato: the wild fruit has 1 g; highly selected cvs. have 1000 g = 1:1000! In my opinion we can find no modern breeding result in major crops with such a high efficiency quotient (Fig. X.56).

The enlargment of tomato fruits must have been based on a fortuitous combination of genes which acted favourably in a diploid genome, directed by planters with good observation of the offspring, always selecting the positive yielders. That this is possible has been demonstrated by an outstanding genetist of our century: Stubbe (1971), with *L. pimpinellifolium*. By using artificially induced gene mutations, Stubbe and his team at Gatersleben raised the 1 g fruit of the wild species in few decades to 17 g; increasing the size, and at the same time changing the morphology of the whole plant. He obtained, among other results, bushy plants instead of the 2–3-m-long creepers, reducing the great quantity of sympodial branches to only three to five. This "laboratory domestication" represents an instructive experiment of how under the guidance of man a wild plant

can be transformed to a sort of cultivar. What Stubbe achieved in few decades, the Amerindians did in centuries.

This trend continues. The areal of an originally neotropic plants has recently undergone impressive eco-physiological changes. Its geographic amplitude has been extended from Middle America to 54° lat. N. in Canada and 52° lat. S. at Rio Gallegos in Patagonia (Fig. X.57).

We still consider it impossible to obtain frost tolerance in *L. esculentum*, a goal which was followed in the Soviet Union for some time. The natural "gene fund" of the whole genus *Lycopersicon* does not contain hereditary factors for frost resistance. All wild species suffer from chilling temperatures.

## 59. *Solanum muricatum* Ait.

Melon pear, pepino dulce, kachun, poire-melon

2 n = 24

A bushy, many-branched herb, 60–100 cm high, annual, or semi-perennial according to climatical conditions. Leaves 6–13 cm long, simple, elliptical-oblong (not imparipinnate like *S. caripense*), sometimes small lobes at the base. Flowers violet, small, stellate-rotate. Fruits of different shape, mostly egg-shaped, or elongated (10–18 cm long) with violet stripes on yellow exocarp, and often a pointed apex (Fig. X.58).

Bitter described different "forms", but due to the extraordinary heterozygoty (*S. muricatum* needs pollination with other clones because it is almost self-incompatible), these local varieties do not deserve taxonomical rank (Fig. X.59).

**Fig. X.58.** *Solanum muricatum,* fruit motive used
for an antique Peruvian vessel

▲
◀ **Fig. X.59.** *Solanum muricatum* plant and ripe fruits

Contrary to an often-repeated statement that the "pepino fruit" does not produce viable seeds, we must declare that most cultivars contain some fertile seeds. Due to certain sterility factors, however, seed setting is always low. We multiplied *S. muricatum* plants by seed and recommend hybridization and genetical improvement of this pleasant neotropical fruit plant.

*S. muricatum* can be found in cultivation from Colombia, in the northern part of the temperate Andes, to Ecuador, Peru, Chile and Bolivia. It has been domesticated for thousands of years by Indians, as is documented by numerous excavated ceramics from the Chimu epoque and other horizons. It seems that farmers in New Zealand recently undertook *S. muricatum* plantation on a commercial scale and now export the fruits to Europe. Kranz (1981) described at-

tractive recipes for use of fresh "pepino" fruits. The wild ancestor has not been found so far. Assumptions that *S. caripense* may possibly be the wild ancestor must be discarded for morphological reasons, as has been stated already by Brücher (1966).

### *S. quitoense* Lam.

Lulo, naranjilla, morelle de Quito, Quito-Orange

$2n = 24$

Ecuador and Colombia are the principal regions of its cultivation. There the "lulo-fruit" has a certain local importance. It seems a rather young domesticate, and no archaeological finds are available. Historically "Lulo" was mentioned in the 17th century by Spanish explorers. Wild species from

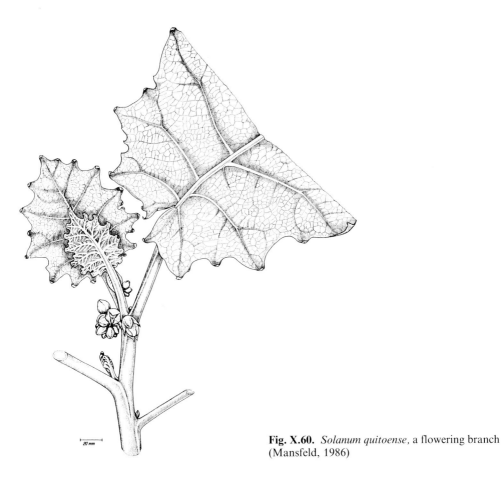

**Fig. X.60.** *Solanum quitoense,* a flowering branch (Mansfeld, 1986)

**Fig. X.61.** *Solanum quitoense,* ripe fruits on a Colombian market

which the Indians developed this cultivar are still under discussion. They should belong to the sect. *Lasiocarpa* of *Solanum*. In the opin-

before eating. The pulp is greenish and contains many small seeds.

*S. quitoense* cannot be cultivated in humid tropic lowlands. It needs a rather cool mountain climate in 100–1700 m altitude for optimal yields, which can give 30 kg per year (Fig. X.61).

Unfortunately, *S. quitoense* cultivars of Ecuador are very susceptible to soil nematodes, especially *Meloydogyne*. Even the fruit delivery to the local markets of Ecuador has diminished drastically for this reason. Therefore the selection of nematode-resistant cvs. is of paramount importance. Crossings with immune wild species should resolve this problem. Furthermore such new "lulo" selections should have reduced quantity of seeds and less hairs on the skin.

Analysis of the lulo-fruits compares favorably with other tropical fruits, e.g. oranges, as shown in Table 17.

**Table 17.** Comparative analysis: lulo and orange fruits (per 100 g fruit weight)

|       | Calories | Protein | Carbo-hydrate | Calcium | Phos-phorus | Niacine | Ascorbic acid |
|-------|----------|---------|---------------|---------|-------------|---------|---------------|
| Lulo   | 45 | 1.1 g | 11 g | 11 mg | 41 mg | 0.2 mg | 48 mg |
| Orange | 36 | 0.9 g | 12 g | 36 mg | 23 mg | 1.5 mg | 57 mg |

ion of Whalen (1981), the ancestral form is *S. candidum*, in spite of the fact that hybrids between *S. candidum* and *S. quitoense* cvs. are difficult to obtain. Crossings with other species of the sect. *Lasiocarpa*, e.g. *S. hirsutum, S. hyperhodium* and *S. straminofolium*, gave better results (Fig. X.60).

Bushy herbs 150–200 cm tall, stems ligneous at the basis, often many-branched. The whole plant is covered with a velvet-like pubescense (the hairs are typically star-shaped). The leaves are large (40 cm diam.), round-oval with very prominent nerves, which sometimes bear short spines. Flower buds are densely covered with short violet hairs. The corollas are stellate, whitish, outside very hairy, pentamerous (4 cm diam.). The fruits have the size of a small orange (= naranjilla) with a diameter of 4–6 cm, and are bright yellow-coloured when ripe. Their exocarp is densely covered with short hairs which must be removed by rubbing

**S. topiro Humb. & Bonpl.**

Synonym: *S. hyporhodium*

Orinoco apple, cocona, topiro, tupiru, Orinoko Apfel

2n = 24

Early European travellers observed this heavy-bearing fruit plant several hundred years ago in the Upper Orinoco valley (Diez de la Fuente, 1760), Humboldt and Bonpland (1800). During our stay in Upper Orinoco in 1967 we had occasion to study this species in more detail. It was neither wild, nor cultivated by the nomadic Indian tribes of this region (Waika, Carmarat, Makarikari), but isolated plants often grew at the edges of abandoned fields or near the river shore. Probably the seeds had originally been spread unintentionally by natives with their excrements, but they had clearly later protected the plants as useful. We took

**Fig. X.62.** *Solanum topiro* collected by the author in the Orinoco valley and selected on his experimental field in Trinidad

seeds from different places, and from the offspring selected biotypes with high yield and good fruit performance at our Depto.

de Genetica at Caracas and multiplied them later at our Experimental Station in Trinidad (Brücher 1977) (Fig. X.62). It seems that in the meantime further experiments have been undertaken in Peru and in Costa Rica. From Turrialba high plant yields have been reported (20–30 kg per individual). All efforts to improve this promising neotropical fruit plant should receive due support. While living in Venezuela, we undertook research for the wild ancestor of *Solanum topiro*.

The following wild taxa should be kept in mind for phylogenetical consideration in the *S. topiro* group: *S. alibile* Shultes, *S. asarifolium* Kunth & Bouché, *S. sessiliflorum* Dun. (= cubii in Brazil) and *S. violifolium* (= chipe-chipe in Venezuela).

As has been described by Pahlen (1978), topiro plants have a robust, many-branched growth (1–2 m tall). The bushes are semiperennial, with a ligneous stem basis and a strong root system. The whole plant is covered with a hairy indument. Leaves are large (50–60 cm), with strong nervature, simple, cordiform, with slightly lobed margins. The leaves are borne on short petioles and in a dense position on the main stem, covering flowers and fruits (Fig. X.62). The fruits have a diameter between 4 and 7 cm, and are enclosed in a strong skin, which according to the biotype is yellow or orange-red and only slightly pubescent. The pulp is creamy yellow and has a pleasant, sweet-acid taste (Herrmann 1983), somewhat recalling peaches, but quite different from *S. quitoense*. Analysis gave the following values: pulp-extract 11%, invert sugar 1.5%, fruit acid 1.6%, protein 0.5%, minerals 0.7%, (kalium 290 mg natrium 7 mg, calcium 40 mg, phosphate 80 mg), vitamin C 13 mg per 100 g, g, pectin (pectate) 0.6 mg. The pulp may be consumed raw, but also in many forms of refreshing conserves.

## References

Badillo VM (1971) Monografia de la familia Caricaceae. UCV, Maracay, Venezuela, p 205

Baehni C (1965) Mémoires sur les Sapotacées. III. Inventaire des genres Boissiera 11:1–262, Genève

Bailey LH (1937) The garden of gourds. MacMillan, New York

Balick MJ (1985) Useful plants of Amazonia, a resource of global importance. Chapter 19. In: Amazonia, Pergamon Press

Bemis WP et al (1978) The feral Buffalo gourd *Cucurbita foetidissima*. Econ Bot 32:87–95

Benson L (1969) The native cacti of California. Stanford University Press, p 243

Bertoni M (1919) Essai d'une monographie du genre ananas. Ann Cient, Paraguay, Ser II, 4:250–322

Boelcke O (1981) Plantas vasculares de Argentina. FECIC, Buenos Aires, p 400

Bringhurst RS, Gill T (1970) Origin of *Fragaria* polyploids II. Unreduced and double-unreduced gametes. Am Bot 57:969–976

Brücher H (1966) *Solanum caripense* HBK (subsect *Basarthrum*) in Venezuela. Feddes Repert 73:216–221

Brücher H (1969) Ueber *Lycopersicon humboldtii* WILLD. Ber deut Bot Ges 82:621–626

Brücher H (1977) Tropische Nutzpflanzen. Ursprung, Evolution und Domestikation. Springer, Berlin Heidelberg New York, p 529

Burkart A (1976) Monograph of *Prosopis*. Arnold Arb Harvard Univ Cambridge, p 113

Camargo F, Smith L (1968) A new species of ananas from Venezuela. Phytologia 16:464–465

Cardenas M (1969) Manual de plantas economicas de Bolivia. Ithus, Cocha-bamba

Cavalcante P (1976) Frutas comestiveis da Amazonia, 3rd edn. Belén, INPA

Collins J (1960) The pineapple. Leon Hills Books, London, p 290

Correll D (1963) Seed of native Texas gourd, *Cucurbita texana*, now available to geneticists. Wrightia 2:243–244

Crisci J (1971) Araceae (Flora argentina). Rev Mus La Plata 11:193–284

Croat T B (1978) Flora of Barro Colorado Island. Stanford University Press, p 943

Cutler HC, Whitaker TW (1961) History and distribution of the cultivated cucurbits in the Americas. Antiq 26:469–485

Darrow GM (1966) The strawberry. Holt Rinehart & Winston, New York

Decker DS (1985) Numerical analysis of allozyme variation. Econ Bot39:300–309

Decker DS, Wilson HD (1987) Allozyme variation in the *Cucurbita pepo* complex. Syst Bot 12:263–272

De Veaux J, Shultz EB (1985) Development of Buffalo gourd as a semi-arid land starch and oil-crop. Econ Bot 39:554–472

Engels JM (1983) Variation in *Sechium edule* in Central America. J Am Soc Hortic Sci 108:706–710

Fouqué A (1976) Especes fruitières d'Amérique tropicale. IFAC, Paris

Fouqué A (1982) Some passion flowers of French Guiana. Fruits 37:599–608

Gibault G (1912) Histoire des légumes. Paris

Heiser CB (1964) Origin and variability of the pepino (*S. muricatum*) a preliminary report. Baileya 12:151–158

Heiser CB (1973) The penis gourd of New Guinea. Assoc Am geogr 63:312–318

Heiser CB (1979) The gourd book. Norman, Univ Oklahoma Press

Heiser CB (1985) Etnobotany of the Anaranjilla *(Solanum quitoense)* and its relatives. Econ Bot 39:4–11

Herrmann K (1983) Exotische Lebensmittel, 2. Aufl. Springer, Berlin Heidelberg New York, p 175

Hoffmann W (1982) Nutzpflanzen aus der Familie der Cactaceae. Kakteen und andere Sukkulenten 33:58–61

Hunziker AT (1979) South American Solancaceae, a synoptic survey. In: The biology and taxonomy of the Solanaceae. Linn Soc Symp 7

Hunziker AT, Subils R (1981) Nuevos datos sobre los nectarios foliares de Cucurbita y su importancia taxonomica. Kurtziana 14:137–139

Jeffrey C (1980) A review of the Cucurbitaceae. Bot J Lynn Soc Lond 81:233–24

Jenkins H (1948) The origin of the cultivated tomato. Econ Bot 2:378–392

Killip EP (1938) The american species of Passifloraceae. Field Mus Nat Hist Bot 19:1–613

Killip EP (1960) Supplemental notes on the American species of Passiflorae with descriptions of new species. Contrib US Herb 35:1–24

Kirkpatrick K, Welson H (1988) Interspecific gene flow in *Cucurbita texana* vs. *C. pepo*. Amer J Bot 75:519–527

Kranz B (1981) Das grosse Buch der Früchte. Südwest, München, p 460

Kunkel E (1984) Plants for human consumption. Koeltz, Königstein. pp 393

Lanning E (1966) American aboriginal high cultures in Peru. 36 Congr Interameric Sevilla I:187–191

Lathrap D (1977) Our father the caiman, our mother the gourd. Emergence of agriculture in the New World. In: Reed (ed) Origins of agriculture. Mouton, Den Hague

Leon J (1966) Central American and West Indian species of *Inga*. Ann Mo Bot Gard 53

Mansfeld R (1986) Verzeichnis landwirtschaftlicher und gärtnerischer Kulturpflanzen. Springer Berlin Heidelberg New York, p 1204

Martin FW, Nakasone H (1970) The edible species of *Passiflora*. Econ Bot 24:333–343

Martin FW (1979) Vegetables for the hot humid tropics. Sponge and bottle gourds. SEA, US-DA, New Orleans

Millan R (1968) Observaciones sobre cinco Cucurbitaceas cultivadas o indigenas en la Argentina. Darwiniana 14:654–661

Morawetz W (1986) Remarks on karyological differentiation patterns in tropical woody plants. Plant system and Evol 152:49–100

Mori SA, Kallunki J (1976) Phenology and floral biology of *Gustavia superba* in Central Panama. Biotropica 8:184–192

Morton JF (1981) The chayote, a perennial climbing subtropical vegetable. Proc Florida St Hort Soc 94:240–245, Miami Univ USA

Newstrom L (1986) Collection of chayote and its wild relatives. FAO Plant Genet Resourc Newslett 64:14–20

Ohler JG (1979) Cashew. Koninkl Inst v Tropen, Amsterdam, p 260

Pahlen v d A (1977) Cubia (*Solanum topiro*) uma fruteira da Amazonia. Acta Amazonica 7:301–307

Patino VM (1964) Plantas cultivadas y animales domesticos en America Equinoccial. Imprenta Departamental. Cali-Colombia

Peters C, Pardo-Tejeda E (1982) *Brosimum alicastrum*, uses and potential in Mexico. Econ Bot 36:166–175

Peters RE, Lee TH (1977) Composition and physiology of *Monstera deliciosa* fruit and juice. J Food Sci 42:1132–1133

Pickersgill B, Heiser CB (1977) Origins and distribution of plants domesticated in the New World tropics. In: Reed (ed) Origin of agriculture. Mouton, Paris

Prance GT (1976) Flora of Panama. Part VI (Caryocaraceae). Ann Mo Bot Gard 63:541–546

Prance GT, Mori SA (1979) Lecythidaceae I. Actionomorphic flowered. Flora Neotropica, New York

Puleston DE (1982) Maya subsistence. In: Flannery (ed). Academic Press, New York

Py C, Tisseau M (1965) L'ananas. Maisoneuve & Larose, Paris, p 300

Rick CM (1976) Genetic and biosystematic studies on two new sibling species of Lycopersicon

from interandean Peru. Theor Appl Genet 47:55–68

Rick CM (1978) The tomato. Sci Am 239:76–78

Roig FA (1987) Los arboles indigenas de las prov de Mendoza y San Juan. Serie Cientifica 33:18, Mendoza

Roosmalen GM (1985) Fruits of the Guianan Flora. Inst of Syst Botany, Utrecht, Holland

Ruehle G (1963) The Florida avocado industry. Univ Florida Agr Exp St Bull 602

Schwerin K (1970) Apuntes sobre la yuca y sus origines. Bol Inf Anthropol 7:23–27

Schuster W (1977) Der Oelkürbis. Beih z Acker-Pflanzenb, p 53

Smith l (1955) The Bromeliaceae of Brazil. Smithson Misc Collect 126:1–290

Staudt G (1968) Die Genetik und Evolution der Heterozygotie in der Gattung *Fragaria*. III Untersuchungen an hexa-und oktoploiden Arten. Pflzücht 59:83–102

Storey W (1967) Theory of the derivations of the unisexual flowers of Caricaceae. Agron Trop Venez 17:273–322

Stubbe H (1971) Weitere evolutionsgenetische Untersuchungen in der Gattung *Lycopersicon*. Biol Zentralbl 90:545–599

Villareal RL (1980) Tomatoes in the tropics. Westview, Boulder, Colorado

Weiling F (1959) Übertragung des Merkmals "Weichschaligkeit" vom Oelkürbis *C. pepo* in fertile Artbastarde aus der Kreuzung *C. maxima* × *C. pepo*. Züchter 226: 22–25

Whitaker TW (1981) Archeological cucurbits. Econ Botany 35:460–466

Whitaker TW, Bemis WP (1975) Origin and evolution of the cultivated *Cucurbita*. Evolution 18:553–559

Whitaker T, Davis G (1962) Cucurbits. Leonard Hill Books, p 250

Whitaker TW, Knight J (1980) Collecting cucurbits. Econ Bot 34:316

Williams LO (1976) The avocados, a synopsis of the genus *Persea*, subg. *Persea* Econ Bot 31:315–320

Williams LO (1981) The useful plants of Central America. Esc Agric Panam, Tegucigalpa, Honduras

# Subject Index